MONARCHS

To the ejidos, Indigenous communities, and local landowners who safeguard the monarch butterfly's forests in Mexico, and to all those who dedicate their lives to protecting this extraordinary species

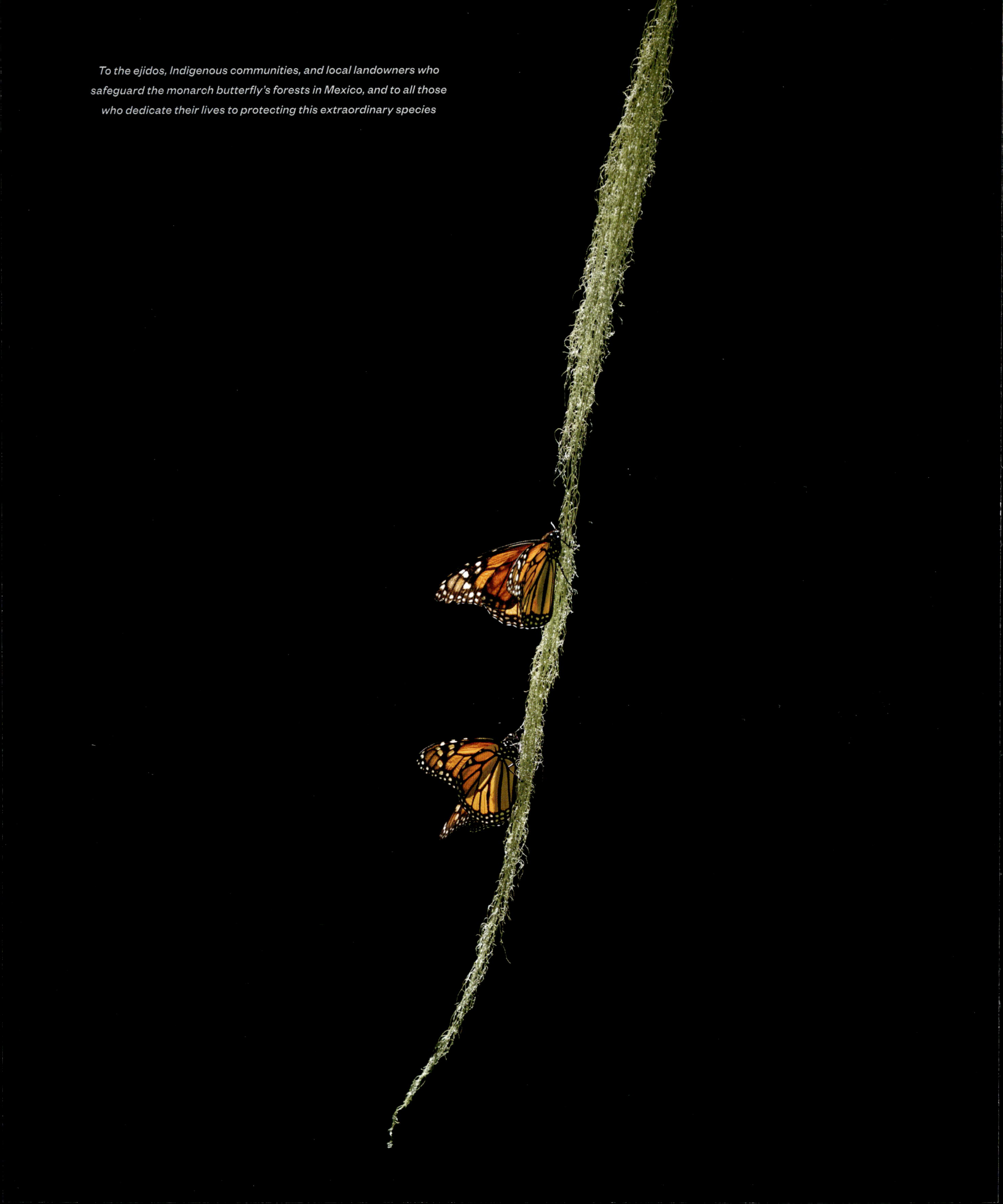

MONARCHS

JAIME ROJO

FOREWORD BY ELIZABETH KOLBERT

NATIONAL GEOGRAPHIC
WASHINGTON, D.C.

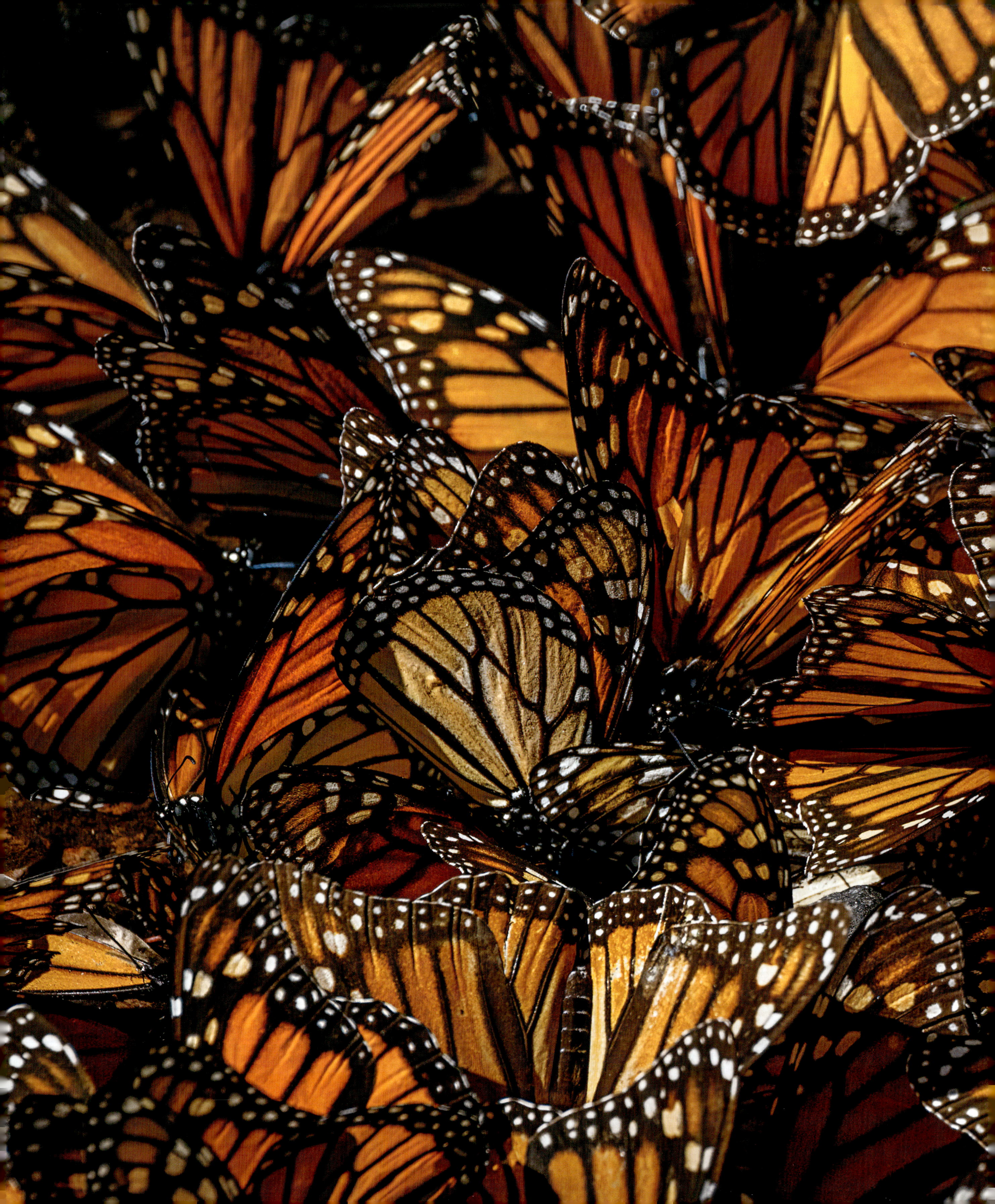

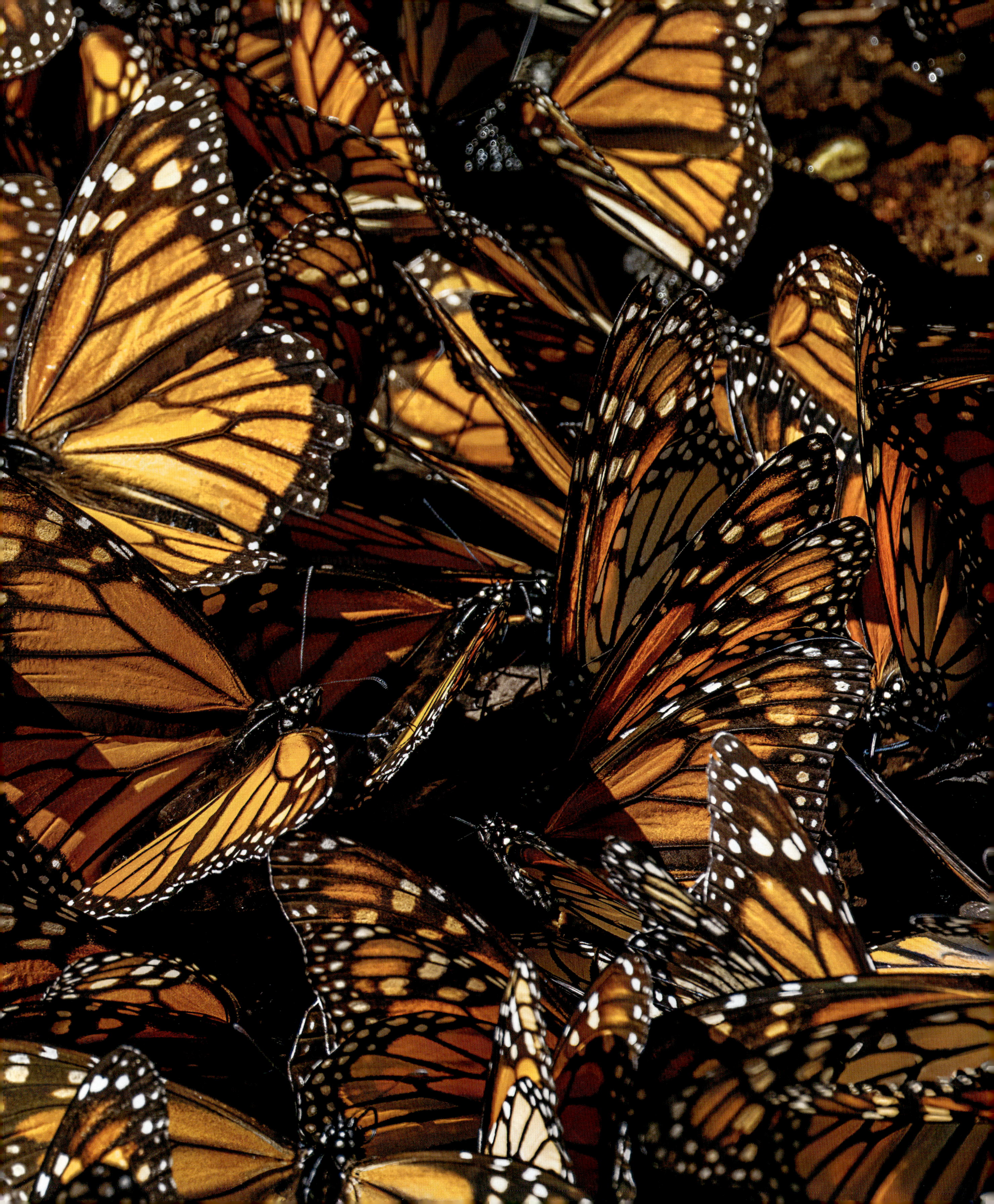

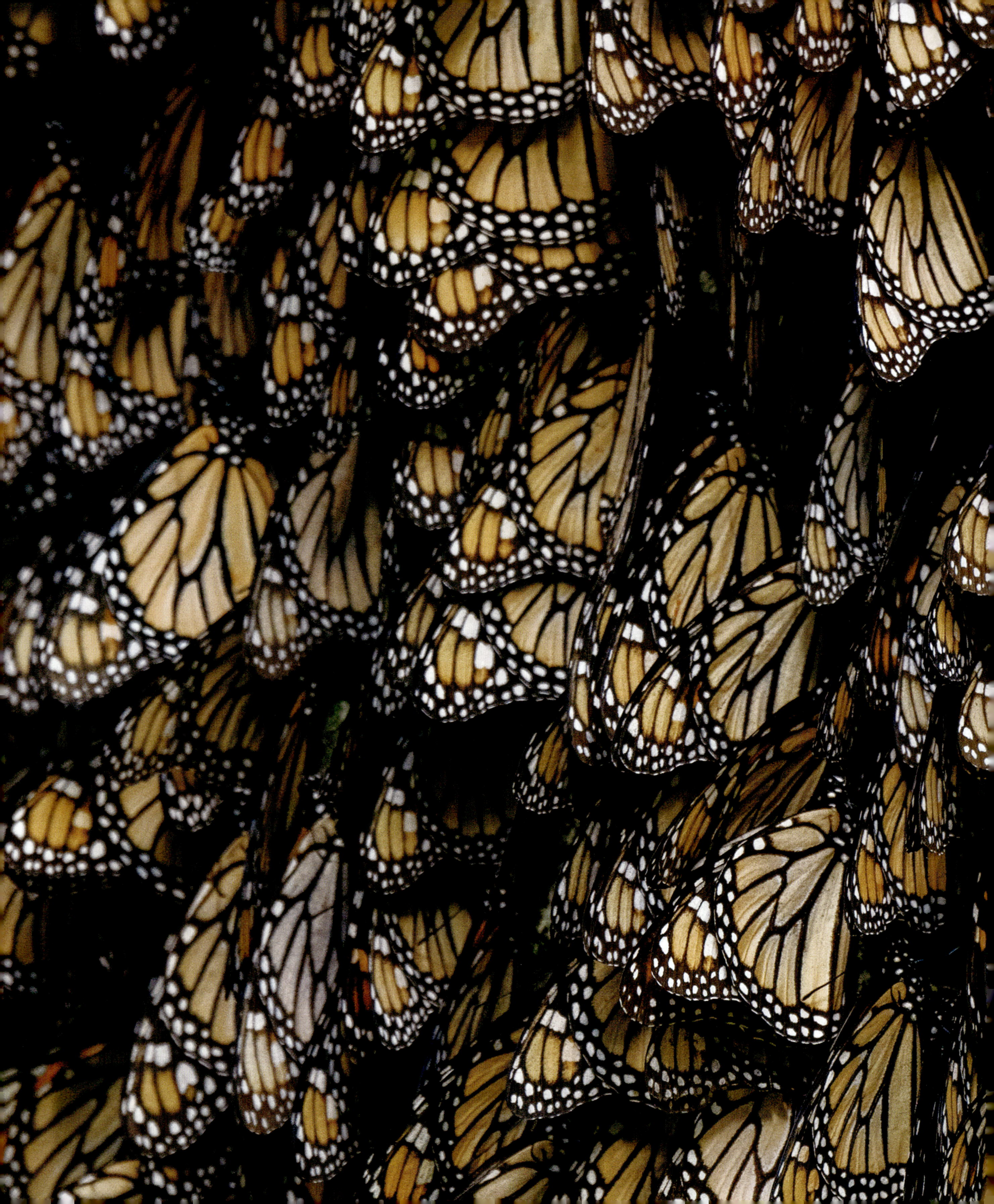

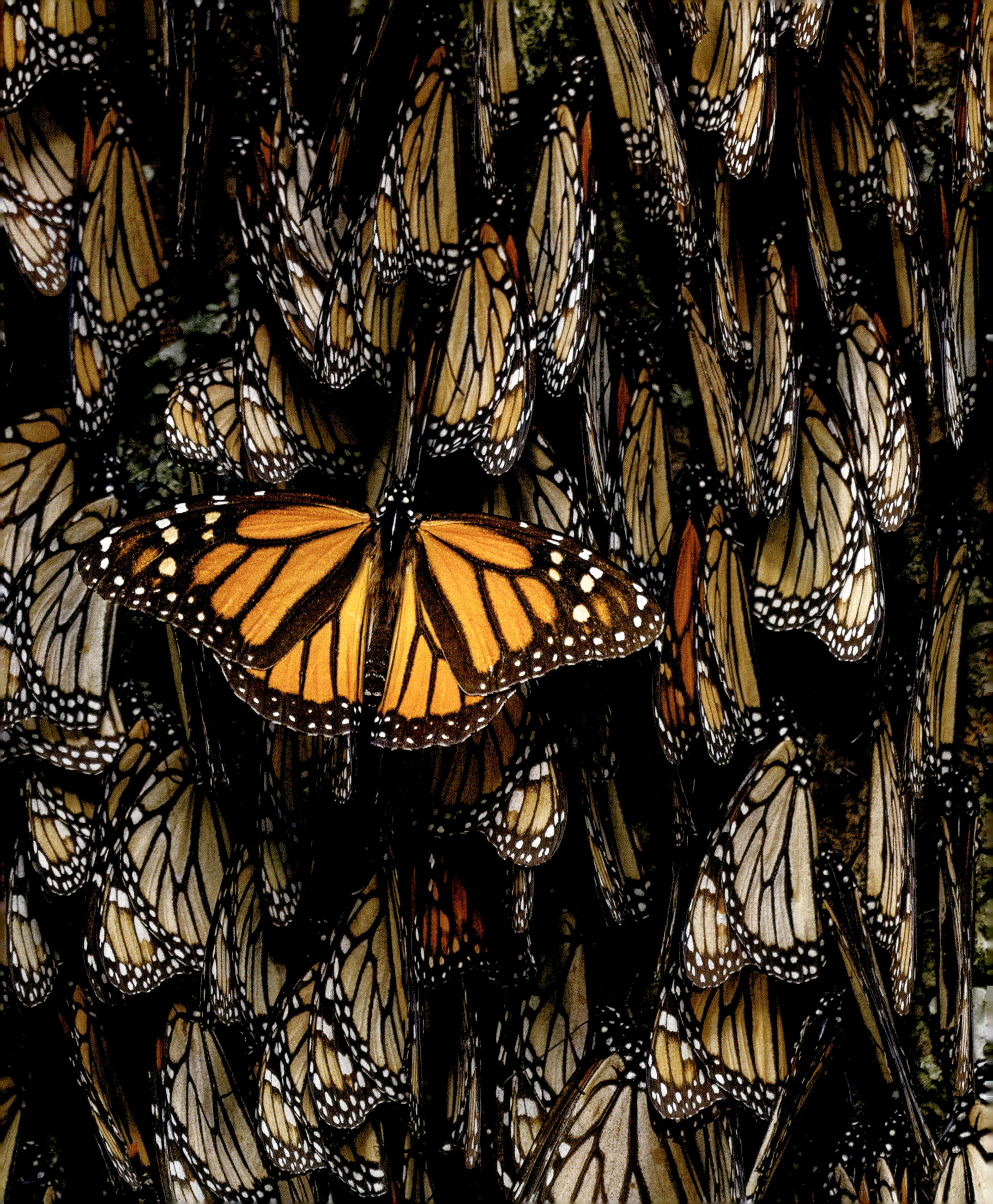

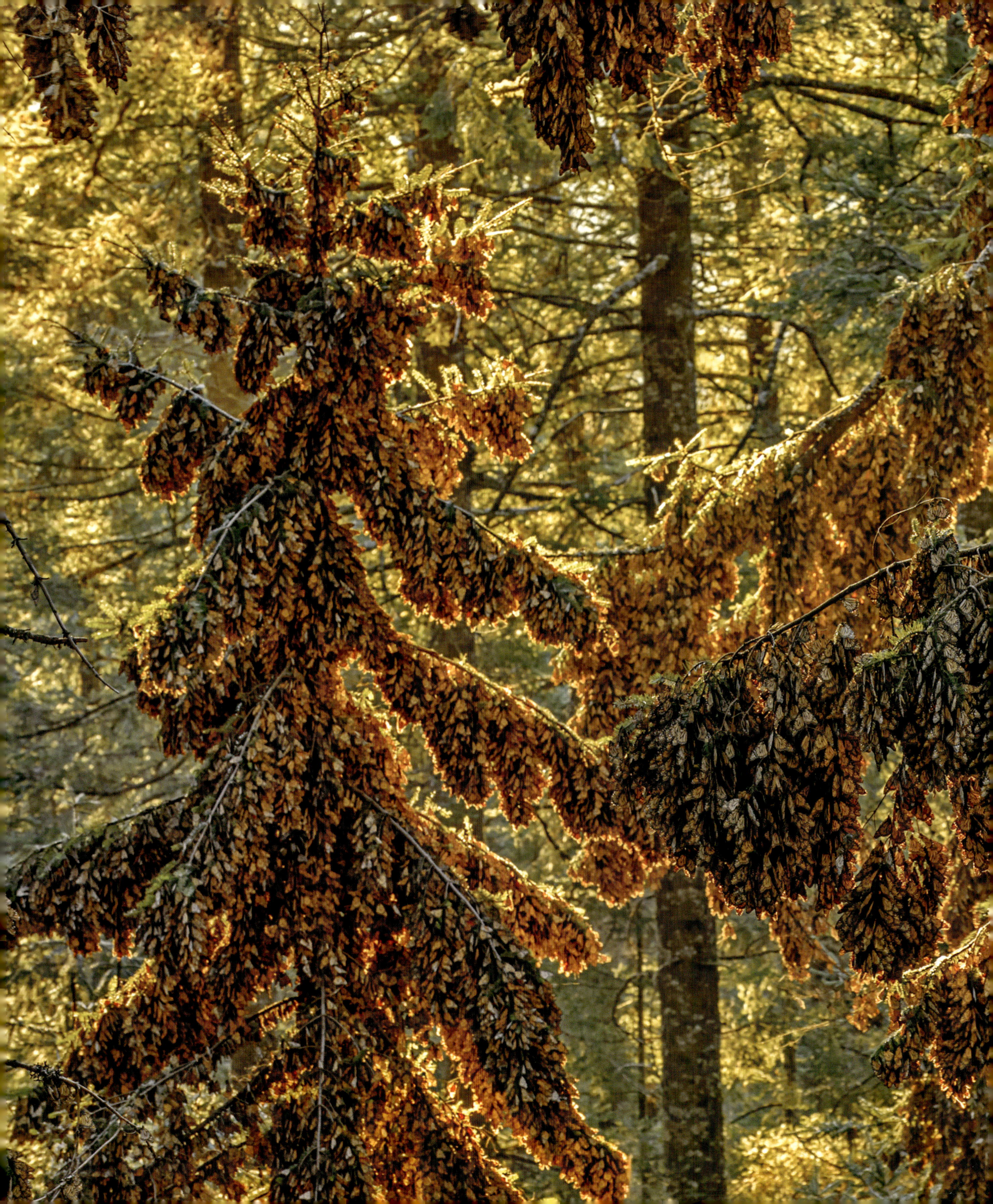

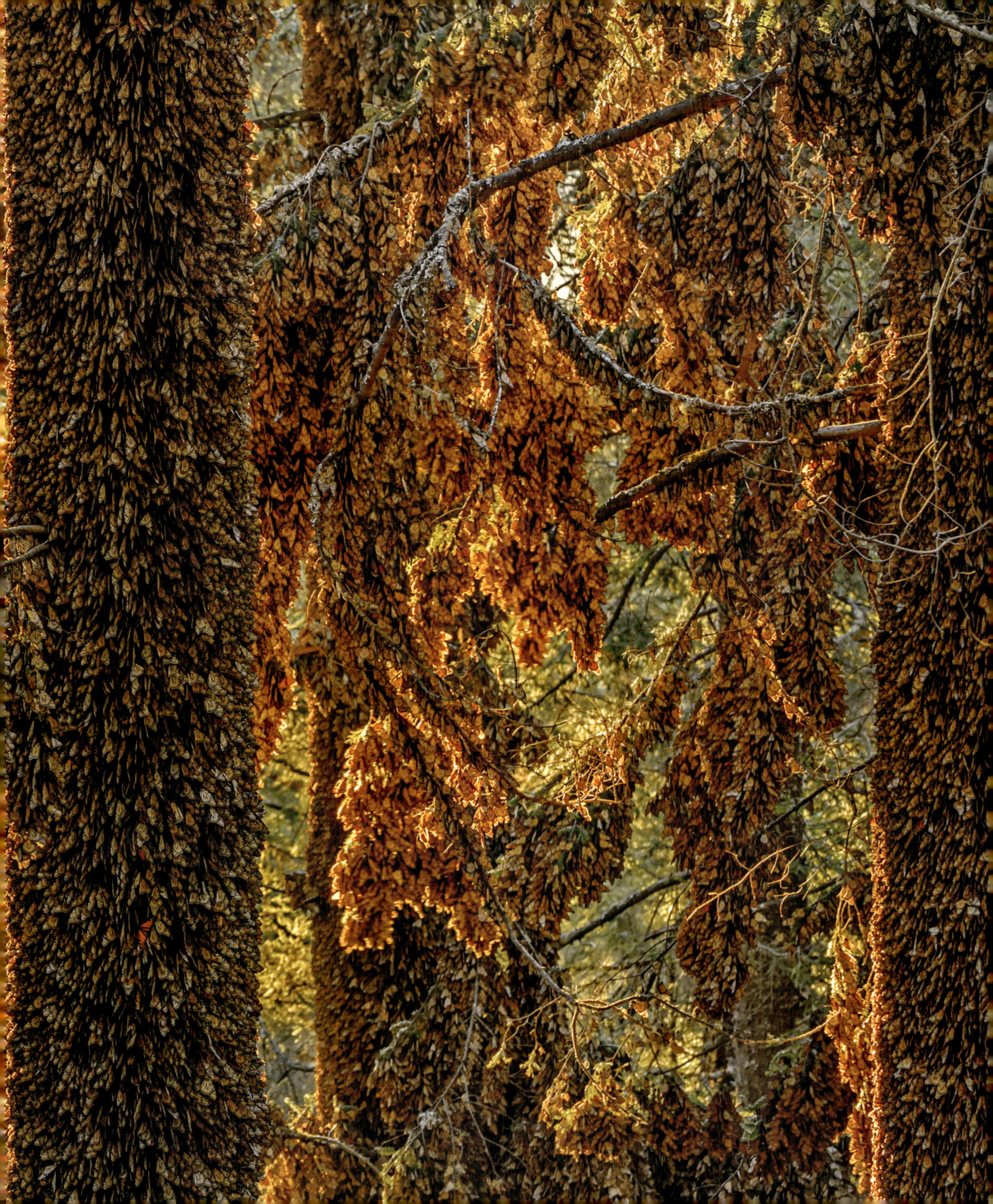

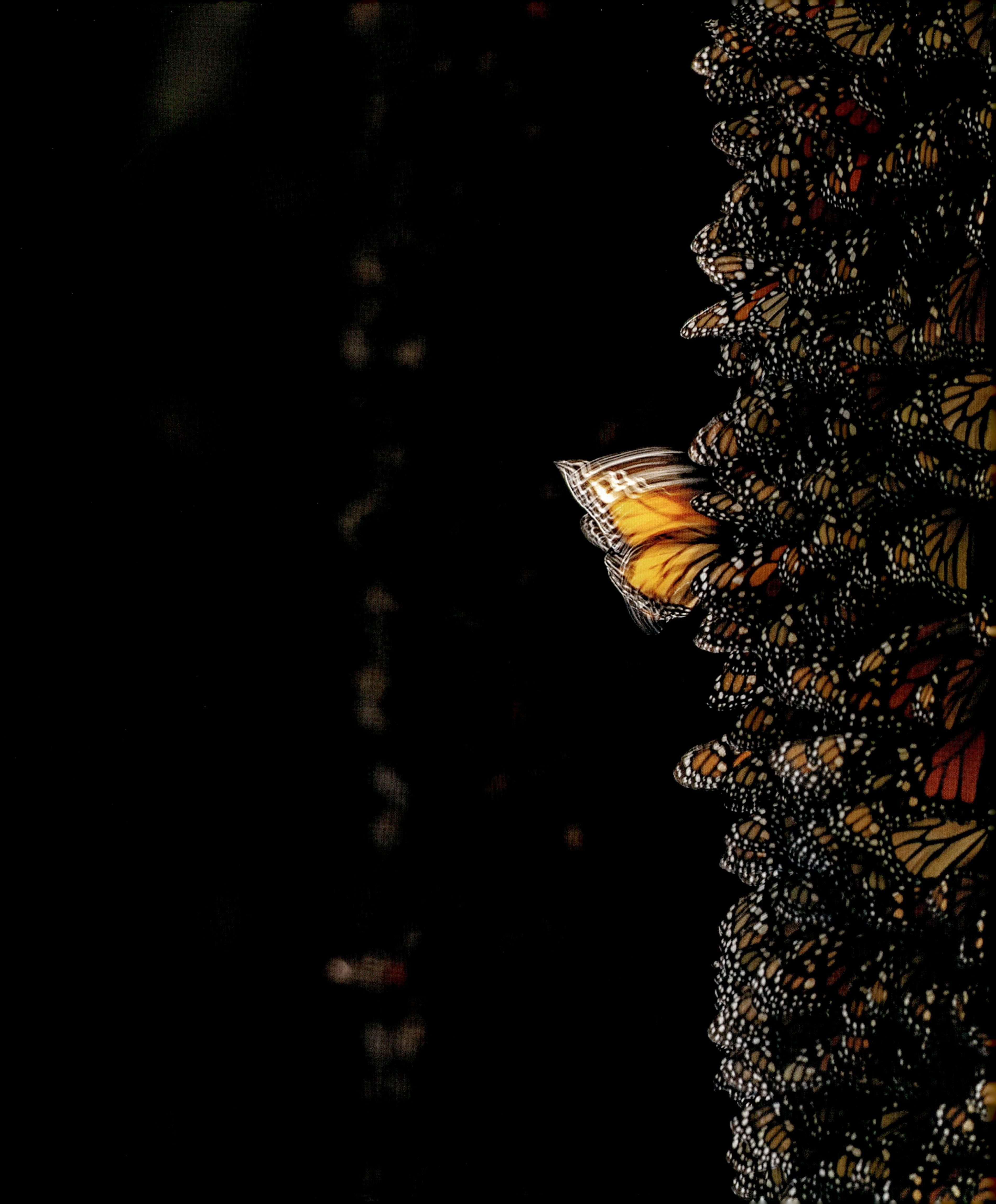

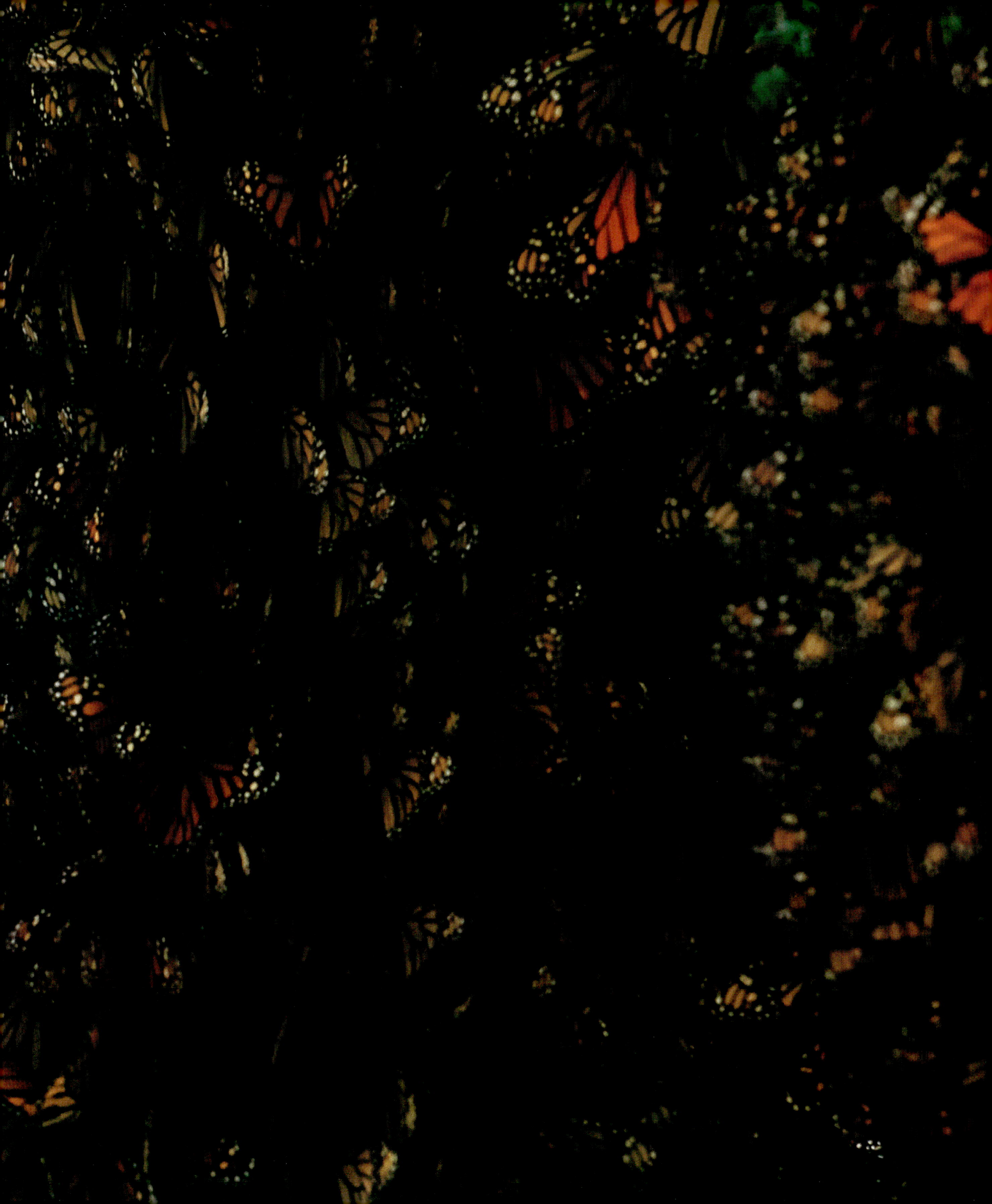

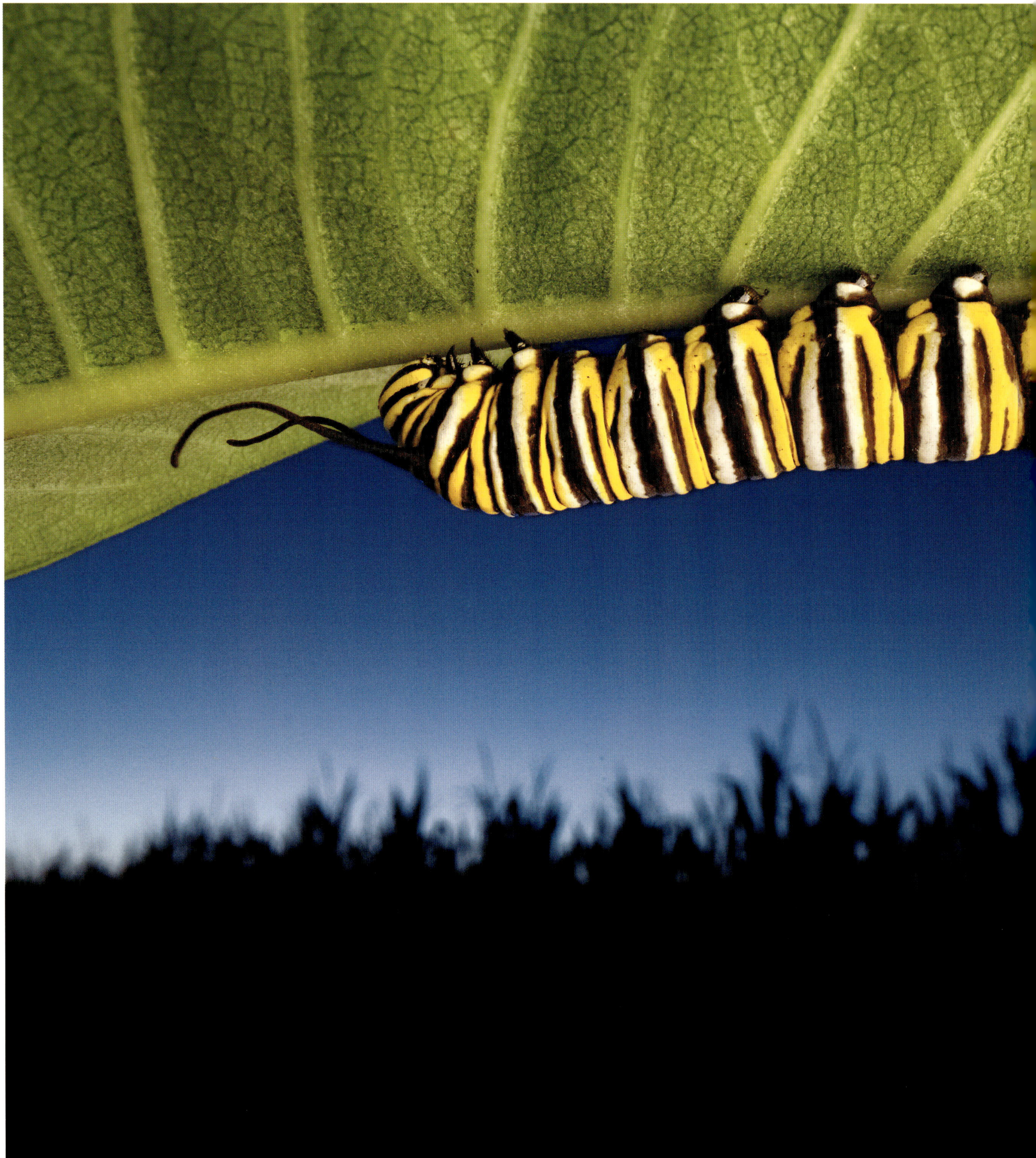

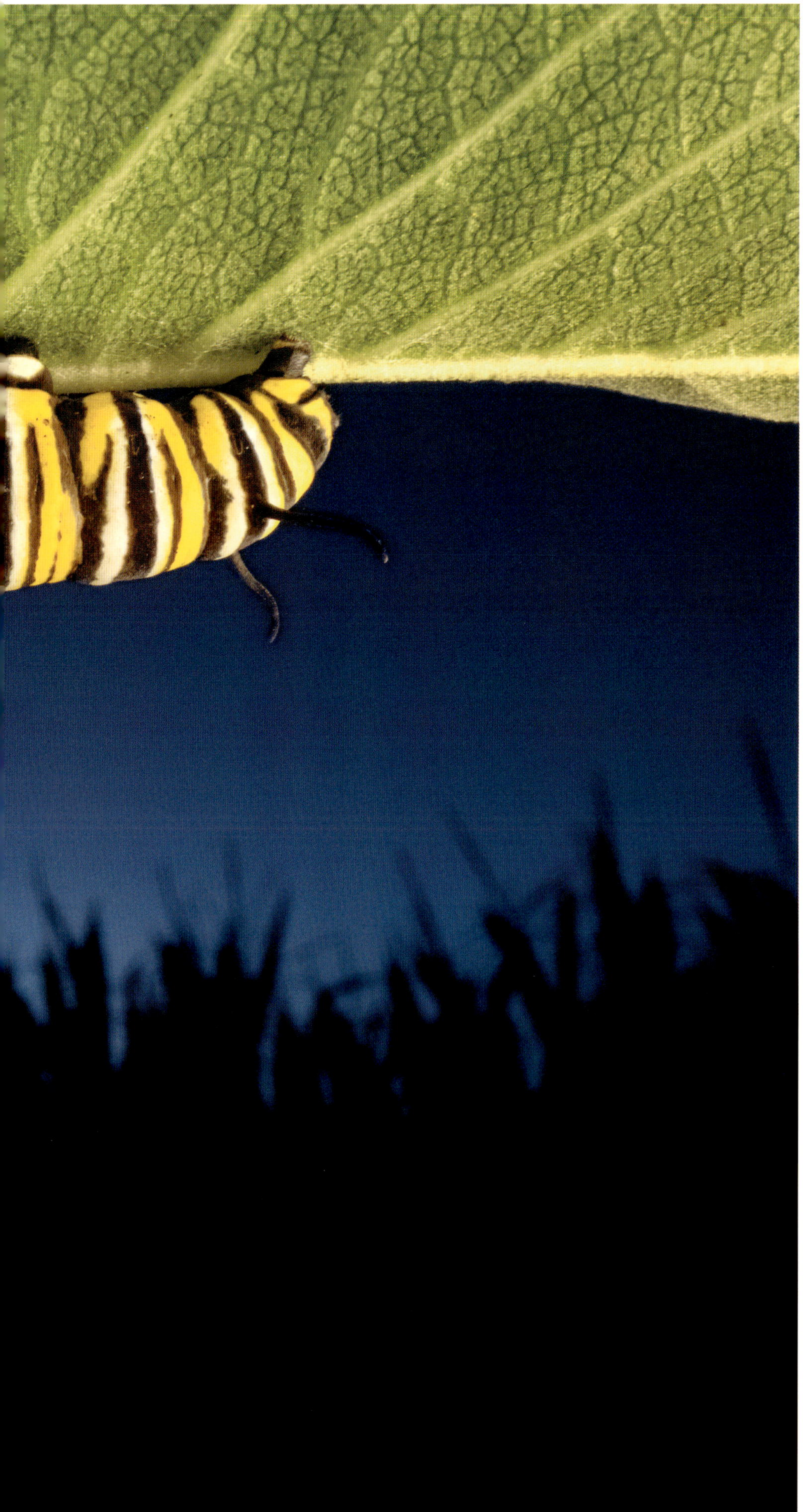

CONTENTS

A monarch caterpillar spends the night beneath a milkweed leaf near a cornfield in Henderson, Minnesota.

PAGE 2: Monarch butterflies perch on moss at El Rosario Sanctuary, Monarch Butterfly Biosphere Reserve, Michoacán, Mexico.

PAGES 4–9: The world's largest concentration of monarch butterflies can be found in Mexico's Monarch Butterfly Biosphere Reserve.

PAGES 10–11: A single monarch beats its wings in the last light of day at Piedra Herrada Sanctuary, state of México, Mexico.

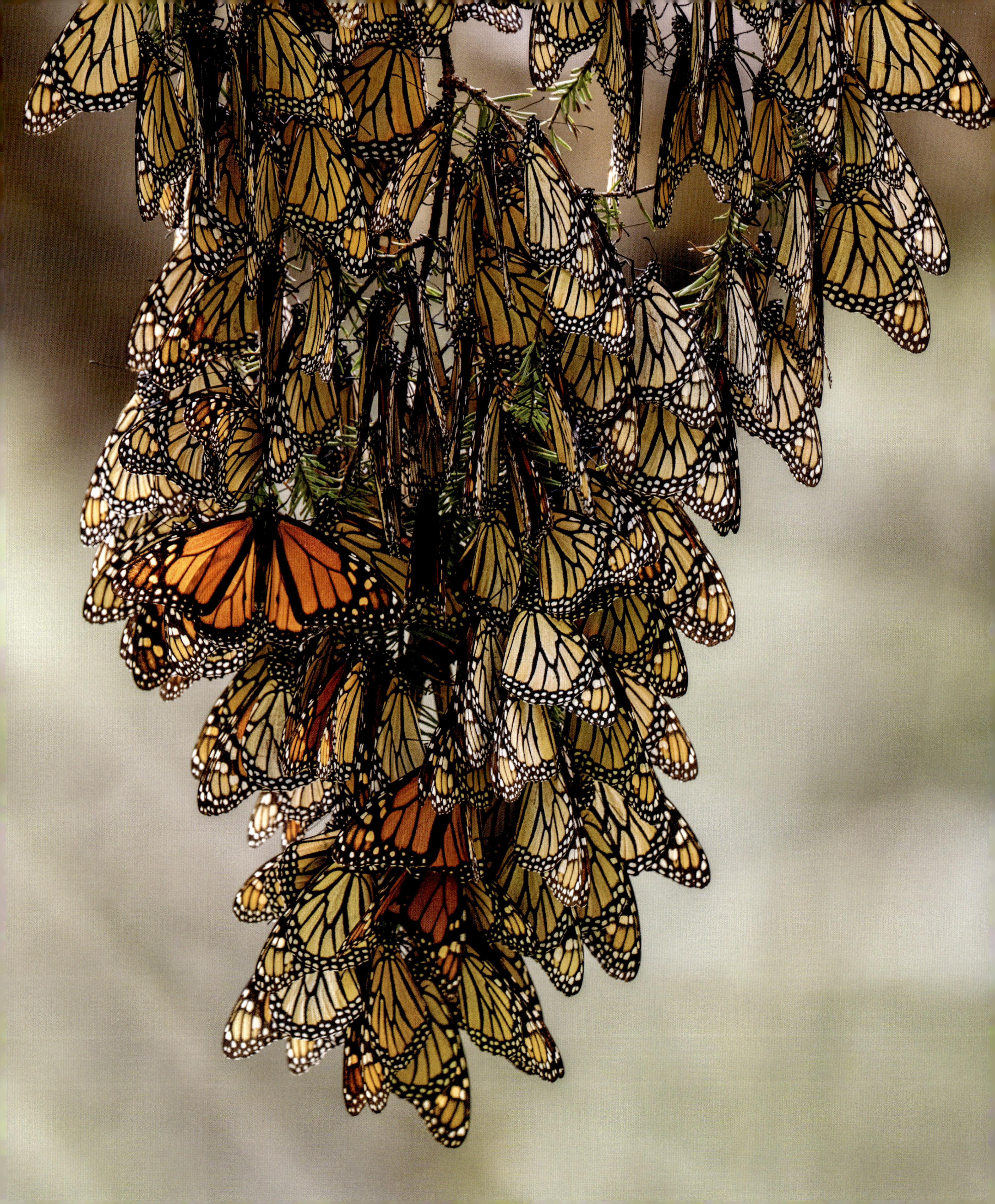

FOREWORD

BY ELIZABETH KOLBERT

I LIVE ON AN OLD FARM that long ago became unprofitable to plow. One of the few plants that thrives in its exhausted soil is milkweed, meaning the loss to agriculture has provided a leg up—or six of them—to monarchs. Every summer, I wade out into the milkweed in search of one or two caterpillars to place in an old goldfish bowl and raise to adulthood. In this way, the monarch's annual rhythm has also become my own.

Following a monarch's path to maturity is, perhaps, as close as a person can reasonably expect to get these days to witnessing a miracle. A newly hatched caterpillar—a so-called first instar—is a tiny, barely visible being. It soon sheds its skin—again and again (and again and again) until, as a fifth instar, it is more than an inch long and about as thick as a pencil. It then hangs upside down for a while in the shape of the letter J. Wriggling out of its last, stripy skin, it forms around itself a gleaming green chrysalis. Over the next 10 days or so, another transformation takes place: Inside the chrysalis, a butterfly takes form. The process ends, finally, with an escape. The butterfly breaks free of its shell and flies back into the milkweed.

The monarchs I like to watch, in western Massachusetts, are all migrants. As Jaime Rojo's photos so brilliantly document, they are participants in a remarkable long-distance journey that spans at least two and often three countries. As best as scientists have been able to determine, monarchs have been making this journey every summer and fall for a million years, though it is only in the last 10,000 years or so—since the end of the last ice age—that they were able to expand as far north as my field. As recently as the 1990s, the eastern population of migratory monarchs numbered close to 700 million. According to current counts, it has since declined by around 80 percent. On the basis of these trends, in 2022 the International Union for Conservation of Nature placed migratory monarchs on the so-called Red List as an endangered species. In 2024, the U.S. Fish and Wildlife Service recommended that monarchs be designated as threatened, which would afford them special protections. But this listing has not yet been made, and it seems unlikely that it will be anytime soon.

Monarchs are in trouble for a variety of reasons, and no one is entirely sure which is most significant. As adults, they lay their eggs only on milkweed, and as caterpillars, they eat only milkweed. This relationship is key to the species' survival. Milkweed contains toxic compounds known as cardenolides. Monarch caterpillars have evolved the ability not only to tolerate cardenolides but also to store them, and this stockpile scares off potential predators. But their reliance on these plants for survival also means anything that reduces the milkweed supply puts monarchs at risk. Where farmers in the United States used to leave hedgerows, or patches of scrub, between their fields, now they tend to plow from one edge of their property to the other. This more intensive form of agriculture has significantly cut back the space available to milkweed. Milkweed has also been hit hard by the development, in the 1990s, of genetically modified crops that can tolerate herbicides. These GMO crops can withstand the application of herbicides containing glyphosate, but milkweed cannot.

New classes of pesticides may also be contributing to the monarchs' decline. Climate change, too, is probably part of the problem: Warmer temperatures prompt the butterflies to fly north too early, before there is adequate food for them, and to migrate south too late, reducing their chances of reaching their overwintering grounds in Mexico.

These overwintering grounds, in the mountains of the Trans-Mexican Volcanic Belt, became known to the scientific community only in the mid-1970s, after a pair of Canadian scientists, Fred and Norah Urquhart, spent nearly four decades searching for them. (People who lived in the region, of course, already knew where the monarchs spent the season and what a spectacular show they put on.) The Monarch Butterfly Biosphere Reserve, in Michoacán, is supposed to protect the monarchs' seasonal home, but land clearing around the reserve poses an ever increasing danger. A recent study by two Mexican researchers found that avocado farming was encroaching on the monarchs' favored spots and concluded that it represents a "new threat" to the monarch butterflies' annual migration.

Monarchs are charismatic animals. Among Americans, they are, hands down, the most beloved insect. To spend time with Rojo's images is to understand why—alone or together, feeding or in flight, monarchs are just crazy beautiful. People want to help them, and there are all sorts of ways that just about all of us can do so. Planting milkweed in your backyard or in a local preserve can help make up for some of the habitat that's being lost. (There are many species of milkweed; if you decide to plant some, make sure to get a species that is native to your region.) Participating in summer butterfly counts can help scientists tease apart the trends that are driving monarchs' decline. Supporting groups in central Mexico that are working to protect local forests can help keep the butterfly's winter habitat intact.

In the end, though, the monarch's fate probably lies beyond the sort of intervention that any group of individuals, however well-meaning, can mount. Great forces are reshaping life on planet Earth. The use of pesticides and herbicides, the rise of monoculture farming, the spread of nonnative species, the progress of climate change: Each would on its own pose a considerable challenge. Monarchs are facing all of them all at once. With their marvelous, shape-shifting life cycle and fantastic, multistage migration, they remind us that life is resourceful and, at the same time, fragile.

A cluster of monarch butterflies rests on an oyamel fir branch at El Rosario Sanctuary.

PREFACE

BY JILL TIEFENTHALER

TINY PIECES OF STAINED GLASS fluttered across the midwestern sky, sunlight catching their orange-and-black-speckled wings. As the heat of summer gave way to fall, I distinctly remember watching the monarchs drift down to the purple milkweed blossoms that edged the fields of my family's farm. When I was growing up in Iowa, along a vital corridor for North America's monarch migrations, these delicate dancers marked the rhythm of my earliest seasons.

Life on the farm moved to nature's pulse, with every family member pitching in. My least favorite chore was walking the bean fields and pulling the milkweed that invaded our carefully planted rows. The fibrous stems stubbornly clung to the soil and to my hands. Yet I still recognized the irony in it all: This obstinate plant was the very reason those butterflies I loved returned each fall.

Today, I value those memories with new clarity, especially in my role as CEO of the National Geographic Society, where we dedicate ourselves to illuminating and protecting the wonder of our world. Is there a wonder more magical than the monarch?

This mesmerizing creature drew me to the work of National Geographic Explorer Jaime Rojo. We bonded over our shared love for these bright orange pollinators, and one of my most treasured possessions is a signed portrait of monarchs made by Jaime himself. It hangs in my home as a quiet but insistent call to action. Because, as Jaime often says, nature's smallest species reveal the greatest truths about our planet—and monarchs reveal a sobering one: The color of our world is fading. Nowhere is that more evident than in their story.

Each year, monarchs travel more than 5,000 miles across the continent, from the forests of Mexico through the United States to Canada—and back again. Along the way, milkweed provides vital nectar for the butterflies and a safe space for them to lay their eggs. It's also the only food their larvae can eat. While milkweed is resilient, it's not invincible. Herbicides, pesticides, and development have caused catastrophic losses. In recent decades, 165 million acres of monarch breeding habitat have vanished, contributing to a more than 90 percent decline in all monarch populations.

This is why Jaime's work is so important.

For more than 20 years, he's followed the monarchs along their full continental arc, returning season after season to prairies, pastures, forests, and mountaintops. For Jaime, standing among millions of butterflies in the cool oyamel fir trees, camera in hand, turned his hobby into a calling. It's an experience he says defies language.

And so, instead of telling, Jaime shows us.

Through his lens, we see both the spectacle and the fragility of migration: cobalt skies brushed with orange wings, sunlit prairies alive with motion, milkweed leaves trembling under the weight of eggs. His work illuminates this wonder of our world and calls us to help protect it.

It's a mission that perfectly aligns with the National Geographic Society, which has been studying and sharing the monarchs' story for more than six decades. Jaime is part of a long line of Explorers—storytellers, biologists, ecologists, entomologists, and technologists—who have devoted their work to understanding where monarchs go, what they need, and how their survival depends on us all. These Explorers have mapped migration routes, advanced genetic research, launched citizen science tagging programs, and captured some of the most iconic images of these butterflies.

To truly move the needle for nature and people, we need it all—science and storytelling. As a former academic, I understand the power of research to reveal truths. But what I return to again and again is awe. Science illuminates the path; awe inspires us to walk it. Photographs like Jaime's stay with us, sparking empathy and turning it into action. This is storytelling at its best.

Jaime's images have become catalysts for change. Conservation organizations across North America are using them to raise awareness, reduce herbicide and pesticide use, restore native grasslands, and advocate for the monarchs' protection. And not long after Jaime's photographs appeared on the cover of the January 2024 issue of *National Geographic* magazine, the U.S. Fish and Wildlife Service proposed listing the monarch butterfly as a threatened species—a pivotal moment shaped in part by Jaime's decades of tireless work, amplified in the most powerful way we know: through the iconic yellow border.

Even amid the urgency his photos bring, Jaime never leaves us without hope. He highlights everyday conservation heroes: volunteers recording wingbeats, classrooms discovering migration, and farmers planting milkweed. Every butterfly documented, imagination sparked, and seed sown reminds us that we are part of a vital thread in a growing tapestry of action and hope.

It's that hope that's trained my eyes to look for flashes of orange against bright blue skies whenever I visit my family's farm in late summer. Even though the sight has become rare, I know people like Jaime are tirelessly working to inspire change.

This book is the realization of that work. Through his vivid photographs, you'll feel the awe of the monarchs' epic migration, see the quiet beauty of milkweed-dotted prairies, and experience the hope of everyday people whose care makes conservation possible. Every page draws you into the fragility and wonder of the monarchs' journey—through North America and a world that's just waking up to all that's at stake.

As you turn the page, I hope you feel the same quiet urgency I do. Hold on to it, and let it compel you to join the tapestry of those protecting the life, color, and magic of our world.

Butterflies stream from the trees at El Rosario Sanctuary.

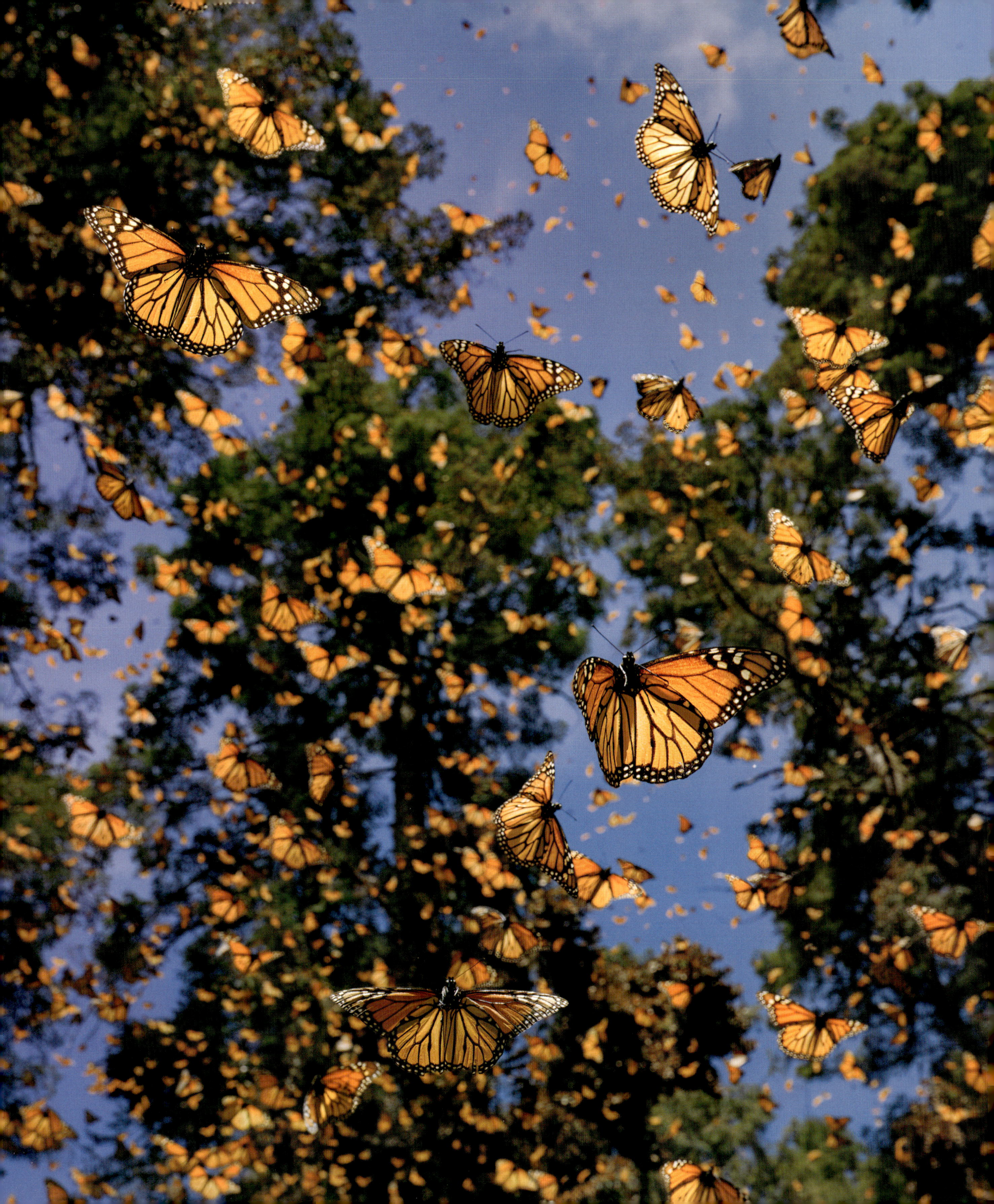

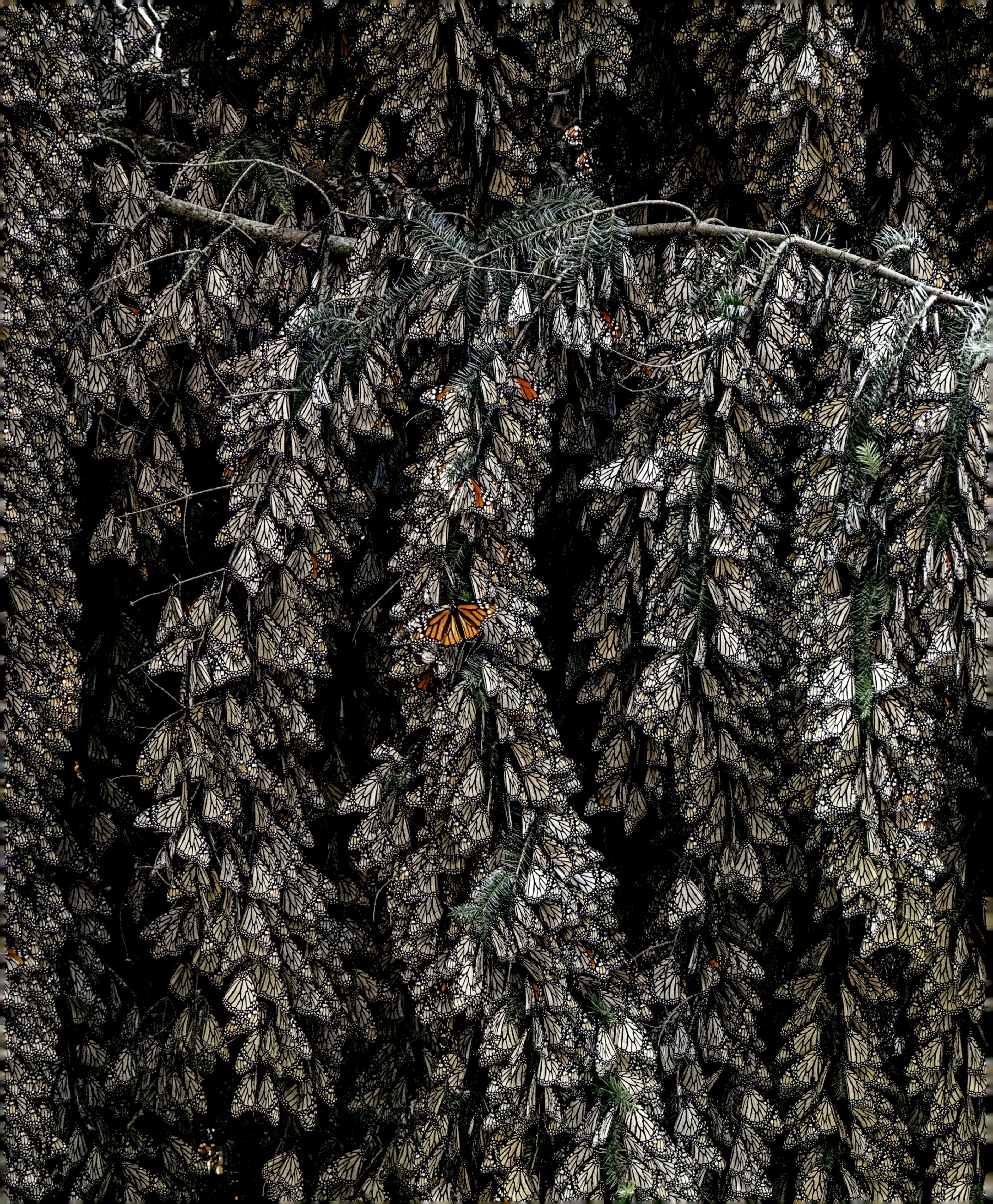

WHY MONARCHS?

MANY PEOPLE ASK ME, "WHY MONARCHS?" It is a question I have often asked myself, because I never intended to work with the same subject for half of my life. I am certainly not a butterfly expert, nor an insect or macro photographer. But monarchs are not typical butterflies, or typical insects for that matter. They gather by the millions during the winter months in Mexico, one of the great wildlife concentrations on our planet and a scene I was immediately drawn to. Each year, over multiple generations, they travel more than 5,000 miles across North America, from north to south and back again. And their decline has become a warning sign for our changing planet.

But for me, at the beginning it was simple: They were close to home, they carried international importance, and, above all, they were stunning—especially in great numbers. That allowed me to photograph them as a living landscape, a scale that felt more natural for me than macro. In the past two decades, I have worked on many other projects all over the world—from documenting scientists tagging Amazon river dolphins in Colombia to exploring the wild animals and landscapes of Andorra, in the Pyrenees. Still, the monarchs have been a constant theme in my career, a mirror reflecting my own transformations—from youth to middle age, from amateur to professional, from beauty seeker to conservation storyteller.

I first visited Mexico's monarch sanctuaries in 2004. The air was cool as I climbed the trail in Sierra Chincua, panting slightly because of the high elevation, surrounded by towering oyamel fir trees. All was new, and I was absorbing every detail: the sound of the birds, the smell of moisture in the forest. It was all a bit dreamy, maybe because I was still jet-lagged—just three days earlier, I had landed from Spain, my home country, with a college scholarship to finish my environmental sciences studies in Mexico. After hiking for a couple of hours, I finally saw my first monarch—alone, trembling on the forest floor. It must have fallen from a nearby tree, too cold to fly on that early February morning. I knelt to take some photos, and I remember being surprised by how large the butterfly was.

The sun was rising, and as the light crept through the trees, a faint murmuring sound, like a whisper, started to fill the forest. Then I looked up, and I couldn't believe my eyes. Millions of monarchs were flying around me. Many more hung in clusters from the trees, their grayish undersides so perfectly camouflaged against the forest backdrop that I had not even seen them. For a moment I rested there breathless, watching in awe how the whole forest became alive. What an unforgettable sight.

Monarchs soon became a regular subject in my work, and I started to learn more about their amazing life cycle: a multigenerational relay race and a natural history epic that starts with a tiny egg on a milkweed leaf; continues with a voracious caterpillar that increases its mass by up to 2,000 times, acquiring toxic defenses along the way; and ends with the extraordinary migration of the adult butterfly, its navigational skills yet to be understood by science. The more I discovered, the more I knew I needed to tell that story. I likewise discovered, with time, that my monarch images mirrored my own metamorphosis from a nature photographer only interested in making beautiful pictures to a conservation storyteller trying to make a difference. Monarchs have become my visual leitmotif, a safe harbor to which I return over and over. It wasn't a carefully planned operation. It just happened, and I went with the flow. And I am glad I did.

This is not a science book about monarchs, nor a photography manual on how I got my pictures. It is a visual journey of discovery and fascination—the story of how I came to create images to tell the complex, multilayered story of the monarch butterfly: its migration, its environmental threats, and the people working to protect it. The chapters follow the three major stages of the monarch's life cycle—spring and summer migration and breeding, fall migration, wintering season—pausing along the way to portray the people who work tirelessly to ensure the survival of this beautiful species and its habitat.

This book is also an invitation to everyone to fall in love with the monarchs in the same way I did more than two decades ago. To tell that story, I include a fourth chapter that briefly reviews my growing fascination with these butterflies. In all these pages, the images take the lead, and my words are only a complement—a peek into my creative process, my struggles as a photographer chasing impossible images, and the meaning monarchs have offered me along the way.

As a photographer with a background in conservation, I aspire to make a meaningful contribution with my images to the protection of our planet, and I wish nothing more than to be able to say that my images have somehow helped protect the monarchs. I may never know if they have. In the meantime, join me in these pages as we celebrate the beauty and awe of these beloved butterflies. This is my personal tribute to a species that has given me so much. I hope that, for a few minutes, you can get lost in the mystery of the monarchs—as I have been doing for the past 20 years.

Amid densely packed clusters of monarchs, a single butterfly opens its wings before settling in for the night.

PAGES 22–23: The last rays of sunlight shine on grounded butterflies after a snowstorm in Michoacán.

PAGES 24–27: Monarch butterflies on oyamel fir branches at the Monarch Butterfly Biosphere Reserve

MONARCHS ON THE MOVE

Monarchs are native across the Americas—from southern Canada to South America—and have accidentally dispersed to other parts of the world, forming small resident populations. But only North American monarchs undertake the famous long-distance migration, which can span more than 5,000 miles over multiple generations each year. In early spring, monarchs that overwintered in Mexico mate and fly north into Texas and the southern United States, laying eggs on milkweed from which the first generation emerges. Generations one and two each live only two to six weeks during spring and summer, breeding and dying as their offspring continue northward across the United States and southern Canada. The third and fourth generations—and in some years the fifth—are known as the super generation, which is genetically identical to but physiologically different from prior generations, capable of living up to nine months. In the fall, triggered by shorter days and cooler temperatures, these monarchs enter reproductive diapause, delaying sexual maturity to conserve energy for migration. This super generation journeys thousands of miles to spend the winter in the same forests of central Mexico their great-grandparents left in the spring—a place they have never seen. Monarchs are the only butterflies known to rely on this suspended state of reproduction to complete such an extraordinary two-way journey. A smaller, more localized migration also occurs in western North America, where monarchs overwinter along California's coast.

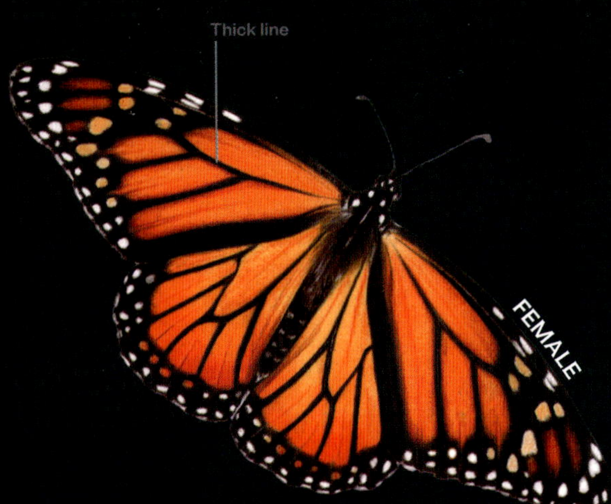

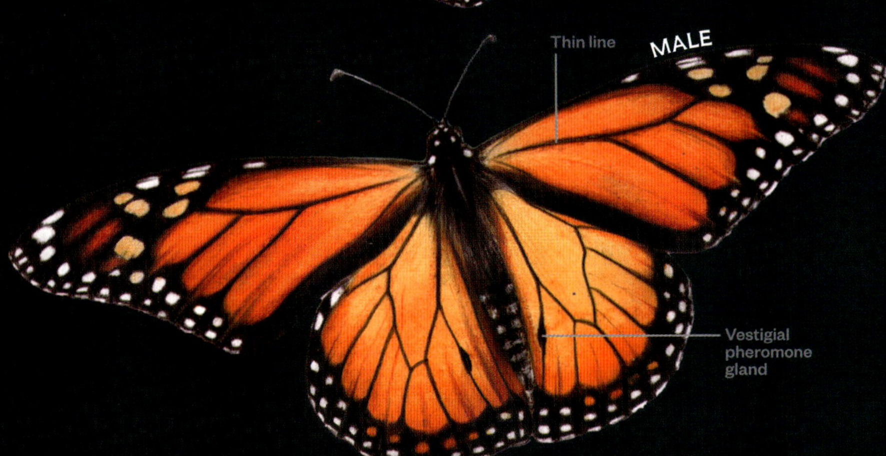

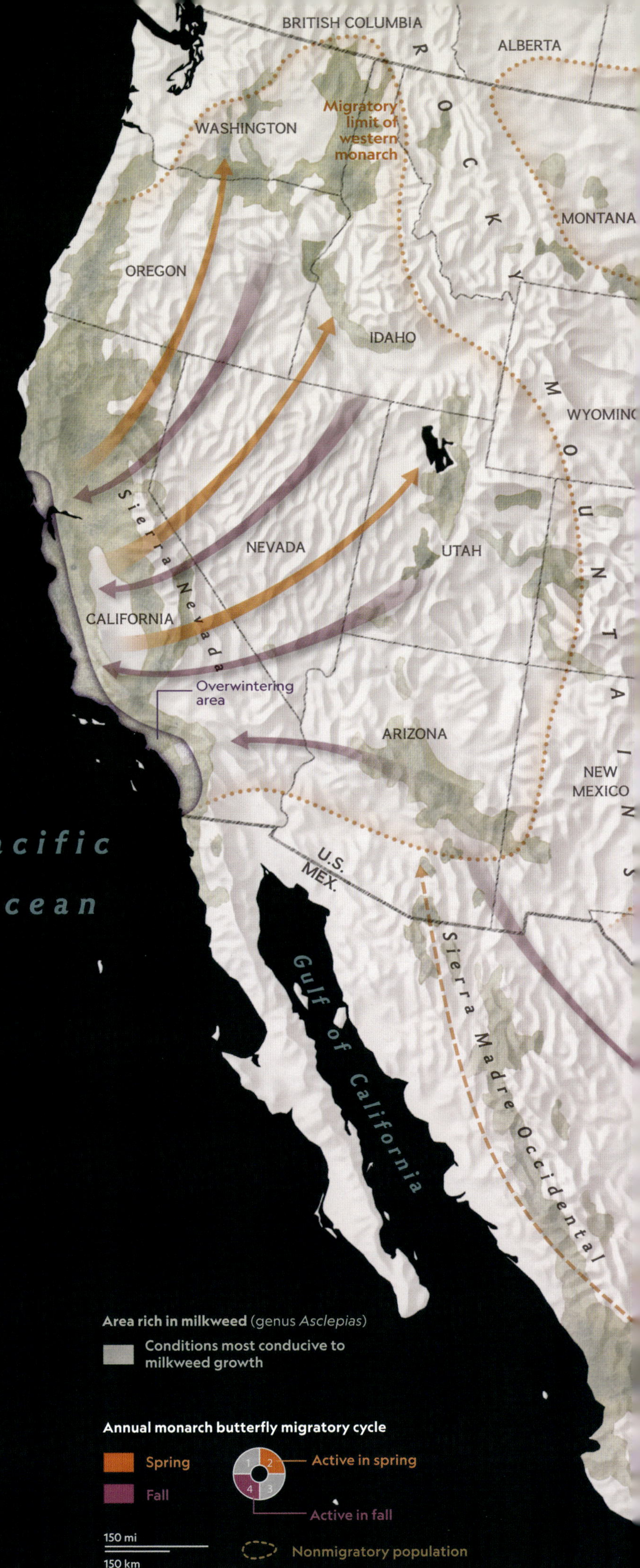

In 2025, the U.S. named the area between its coastline and maritime boundaries with Mexico and Cuba the Gulf of America.

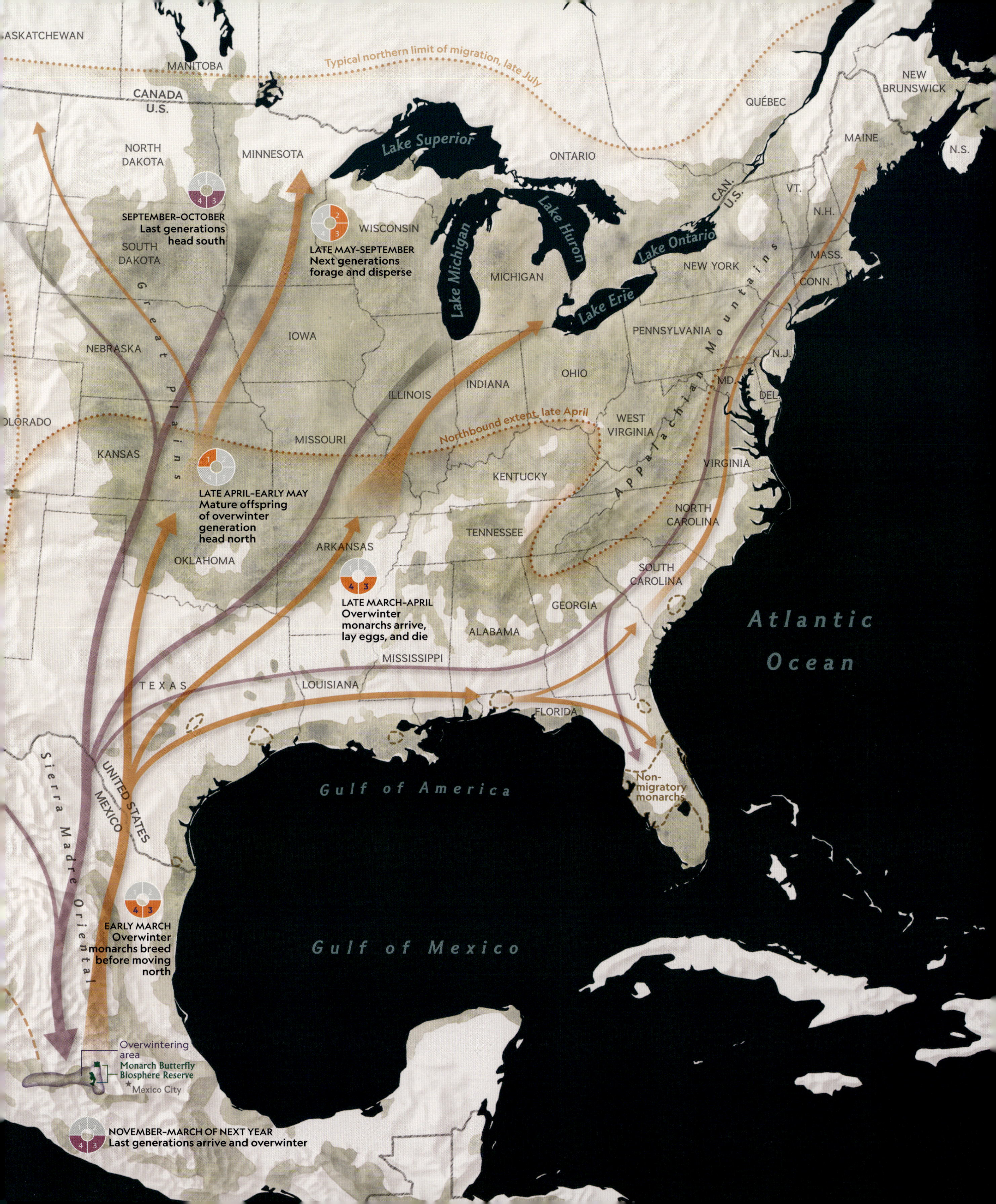

Typical northern limit of migration, late July
SEPTEMBER–OCTOBER
Last generations head south
LATE MAY–SEPTEMBER
Next generations forage and disperse
Northbound extent, late April
LATE APRIL–EARLY MAY
Mature offspring of overwinter generation head north
LATE MARCH–APRIL
Overwinter monarchs arrive, lay eggs, and die
EARLY MARCH
Overwinter monarchs breed before moving north
NOVEMBER–MARCH OF NEXT YEAR
Last generations arrive and overwinter
Overwintering area
Monarch Butterfly Biosphere Reserve
Mexico City
Non-migratory monarchs
Great Plains
Appalachian Mountains
Sierra Madre Oriental
Atlantic Ocean
Gulf of America
Gulf of Mexico
Lake Superior
Lake Michigan
Lake Huron
Lake Ontario
Lake Erie
CANADA
U.S.
CAN.
U.S.
UNITED STATES
MEXICO
SASKATCHEWAN
MANITOBA
ONTARIO
QUÉBEC
NEW BRUNSWICK
N.S.
NORTH DAKOTA
SOUTH DAKOTA
MINNESOTA
WISCONSIN
MICHIGAN
NEBRASKA
IOWA
ILLINOIS
INDIANA
OHIO
PENNSYLVANIA
NEW YORK
MAINE
VT.
N.H.
MASS.
CONN.
N.J.
MD.
DEL.
COLORADO
KANSAS
MISSOURI
WEST VIRGINIA
VIRGINIA
KENTUCKY
TENNESSEE
NORTH CAROLINA
SOUTH CAROLINA
OKLAHOMA
ARKANSAS
GEORGIA
ALABAMA
MISSISSIPPI
LOUISIANA
TEXAS
FLORIDA

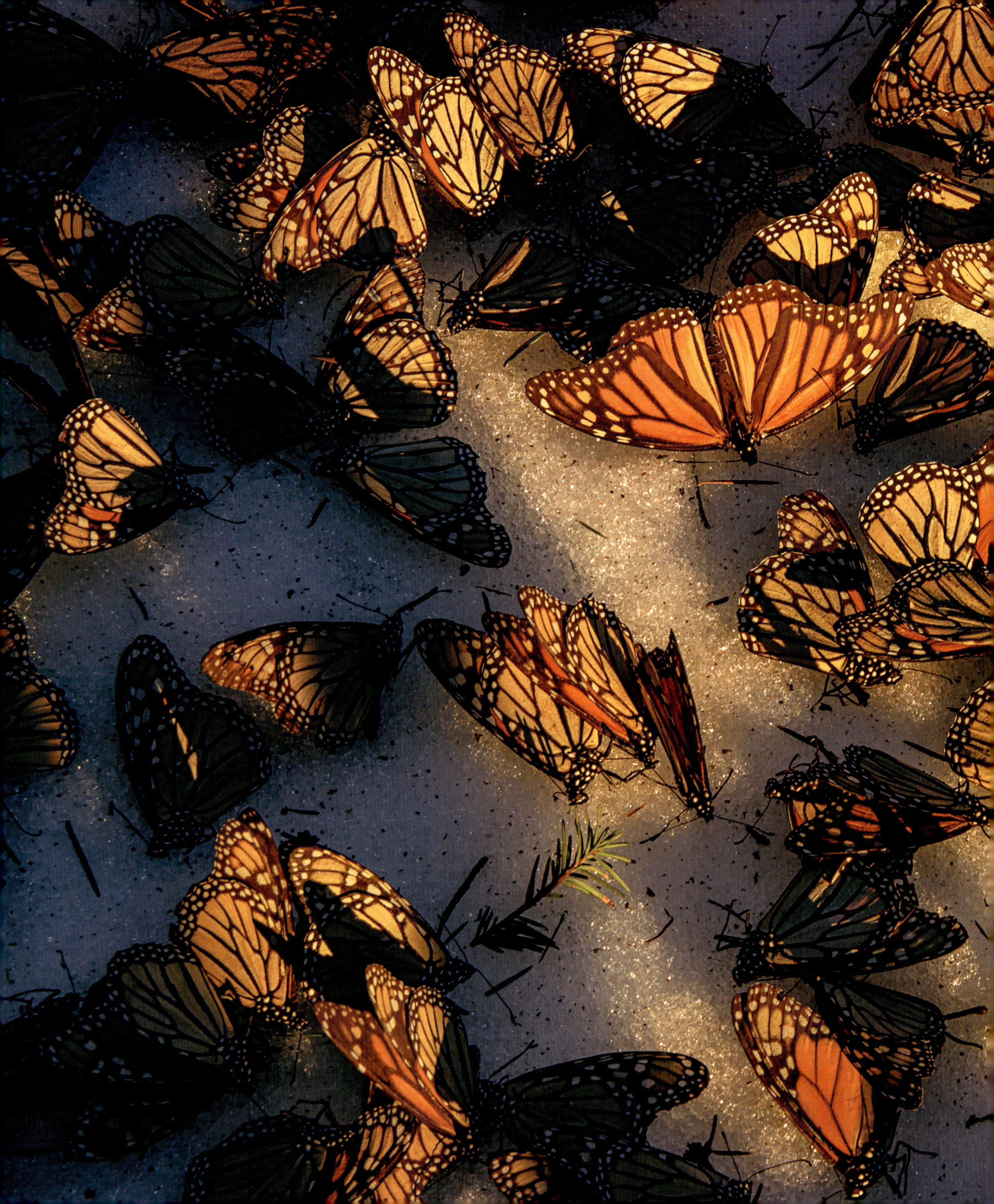

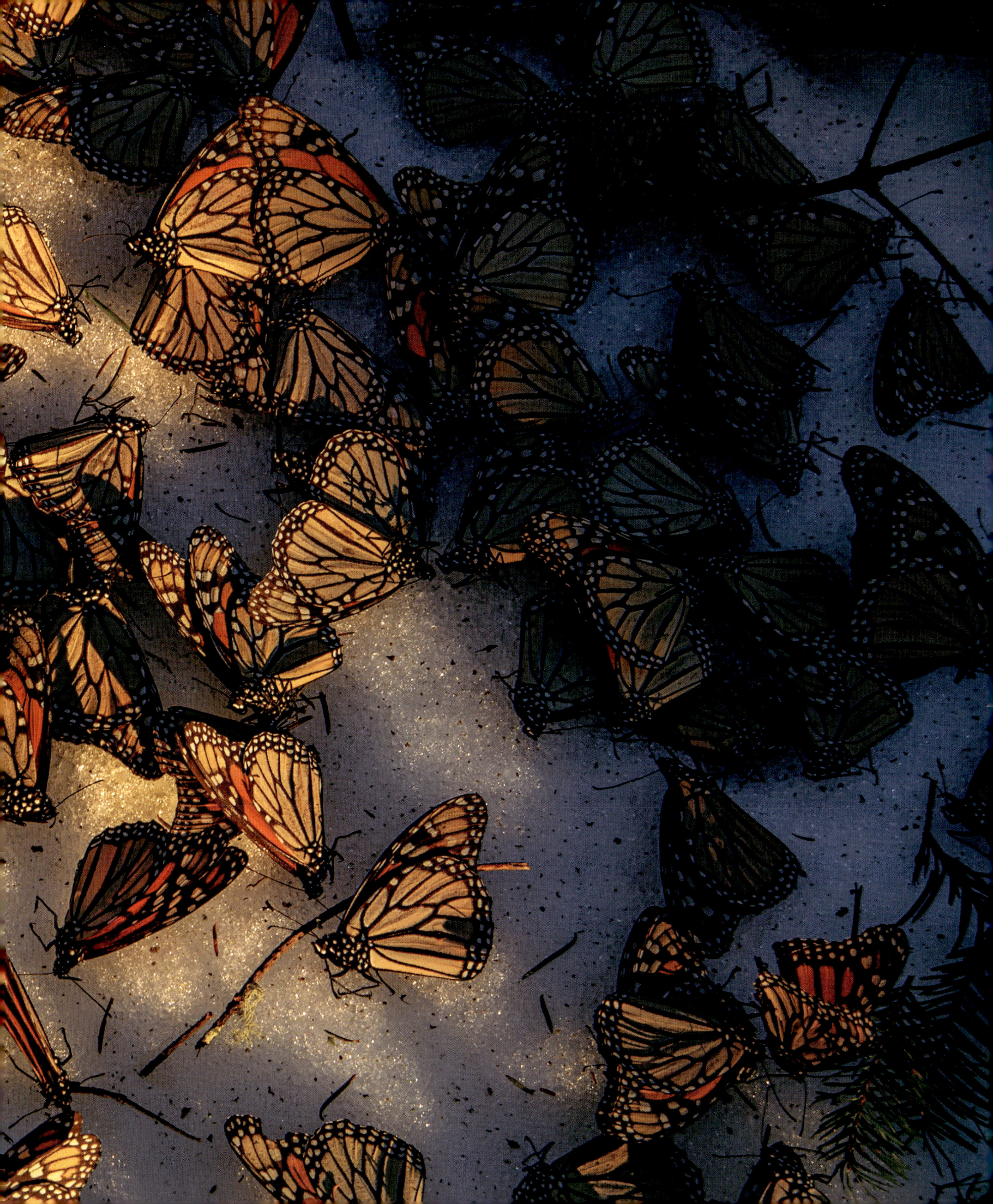

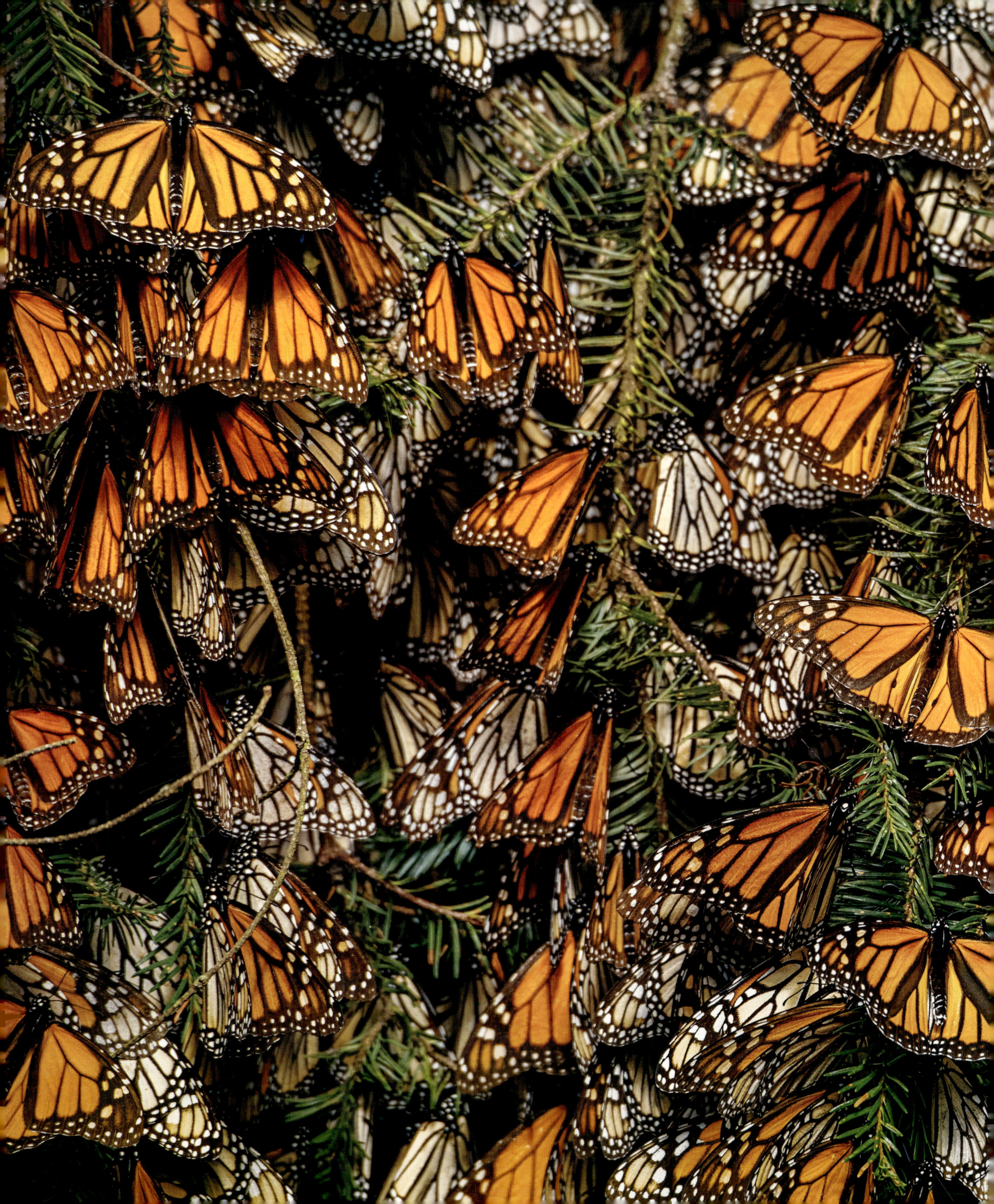

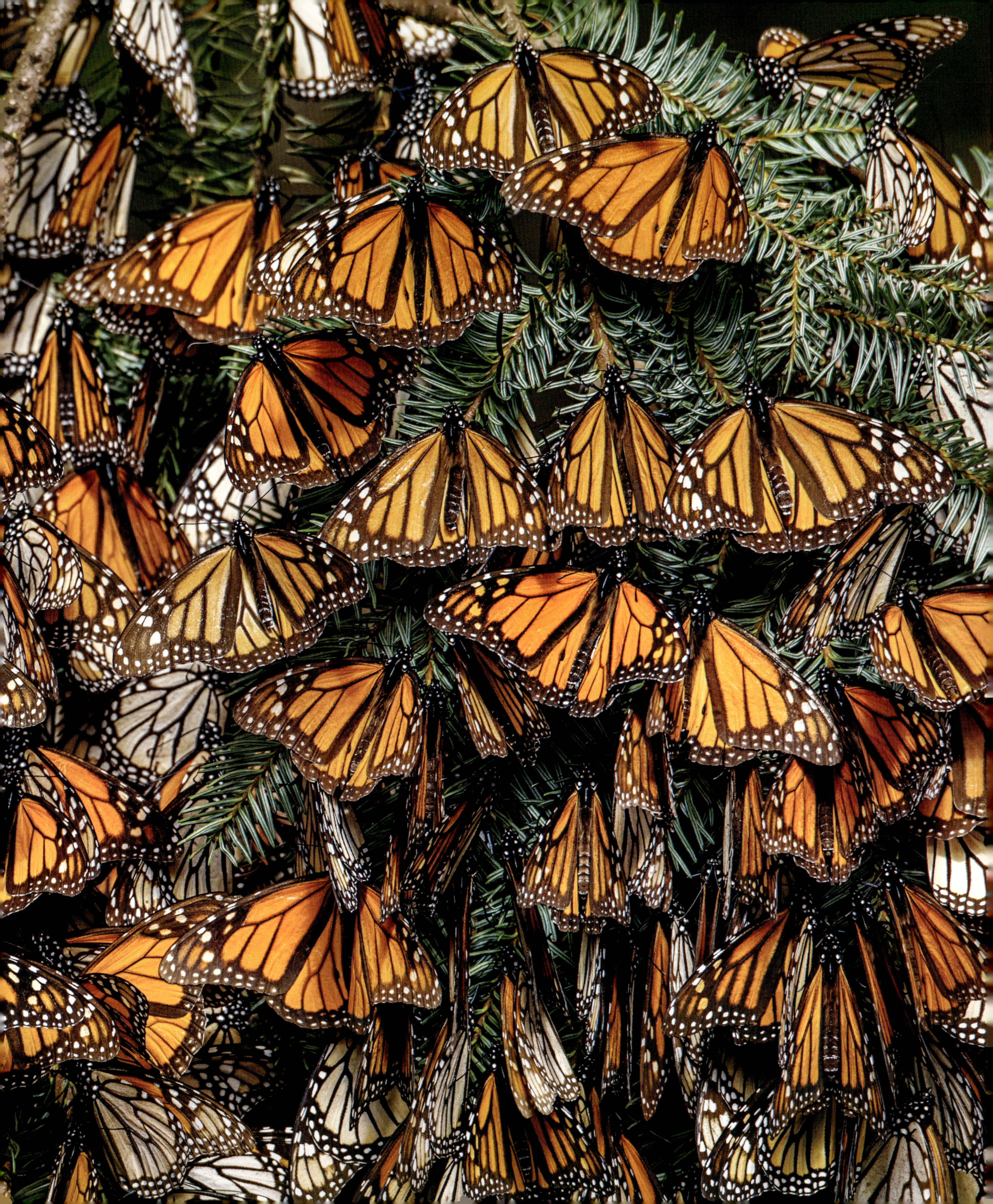

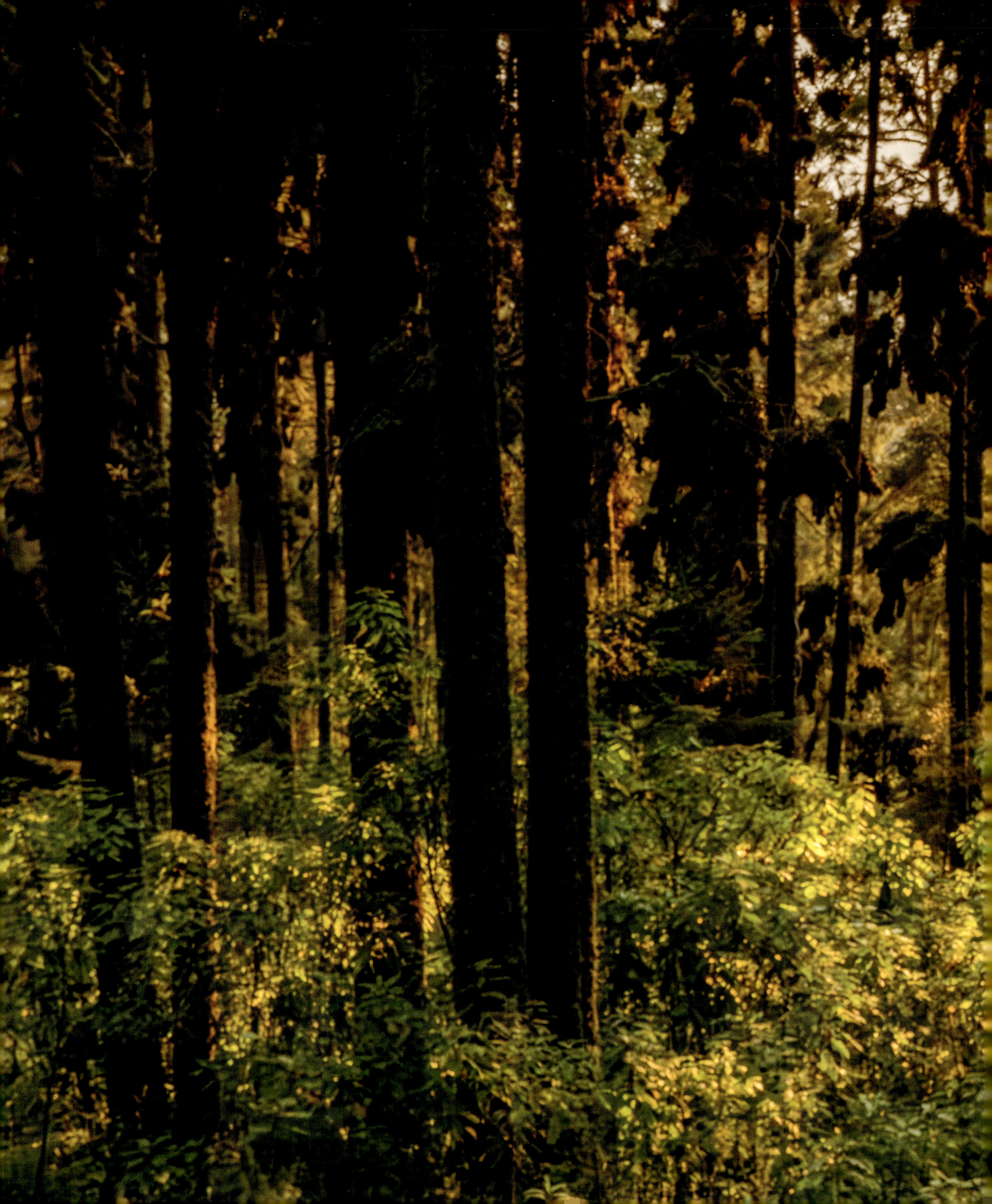

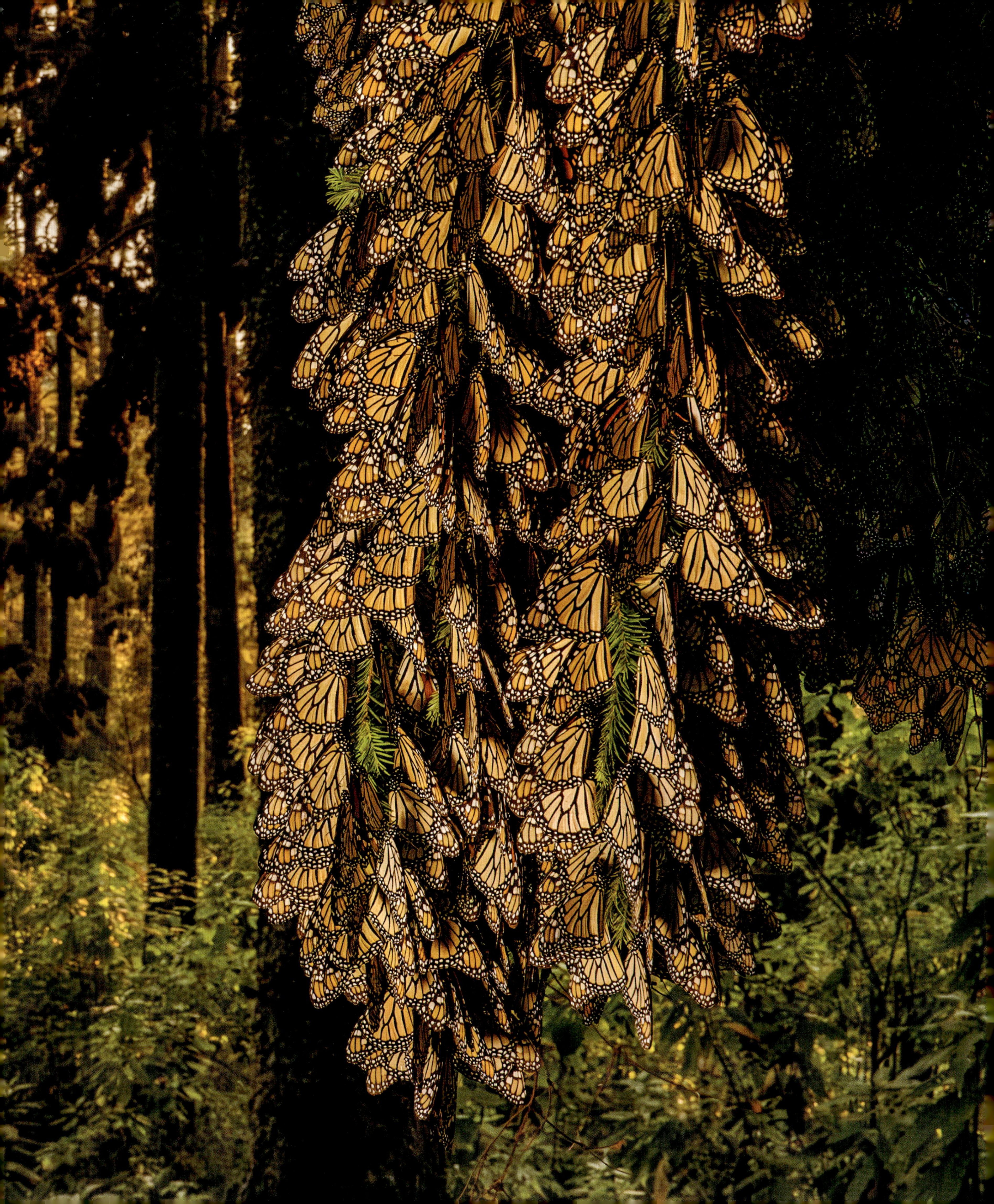

SPRING & SUMMER

A NEW CYCLE BEGINS

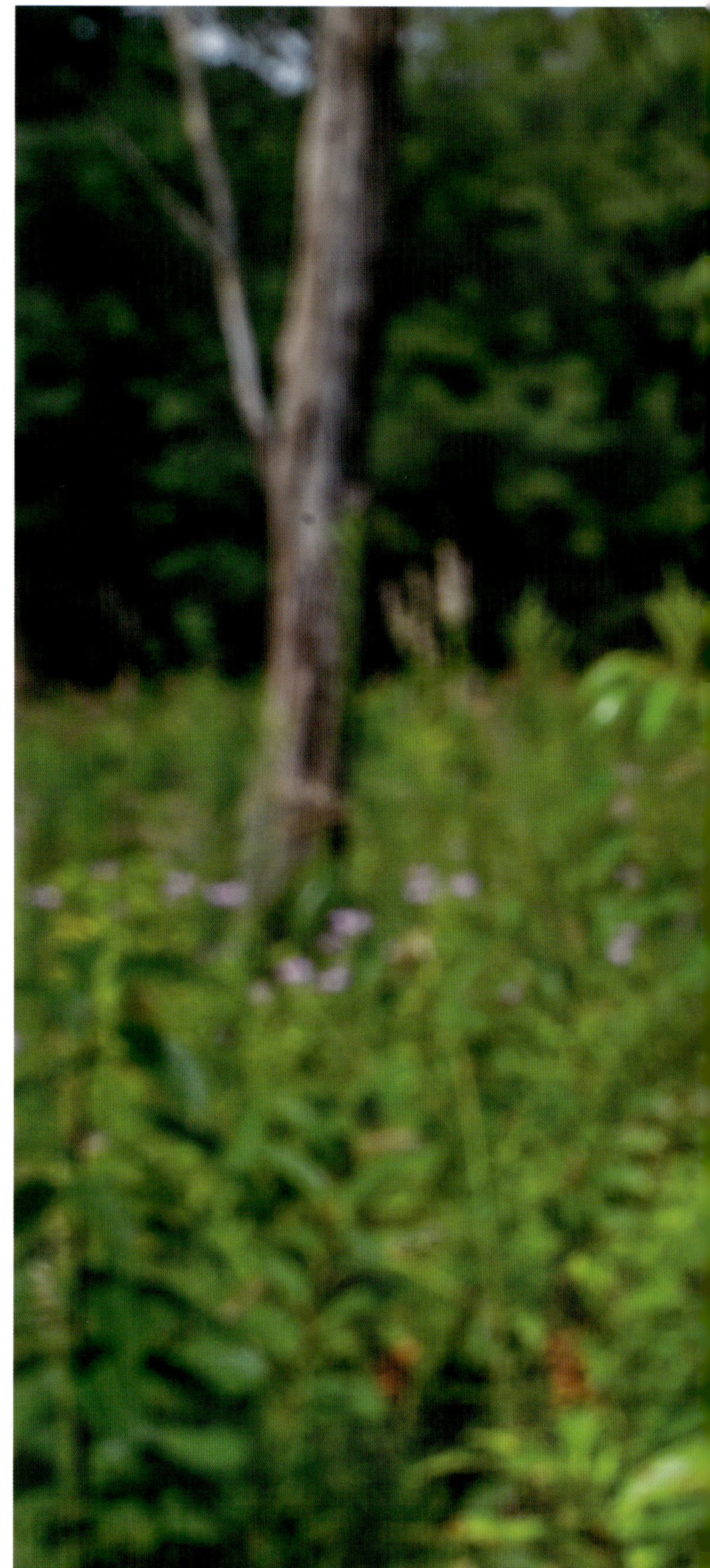

THE FOREST WAS EMPTY AND STILL. A faint breeze blew downhill, and a thin layer of dust covered the leaves of the bushes near the trails, where the footsteps of thousands of visitors had left marks that would take weeks to fade. Just days ago, this forest had been bursting with excitement—guides trying to keep their groups together, children pointing upward with wide eyes, muleteers tending to the animals that brought visitors to the spot where the monarch clusters hung. I had been busy, fluttering from one group of tourists to another, trying to capture their experience with my camera. But now everything was quiet, almost unrecognizable.

It was my first time visiting the monarch sanctuaries outside of the wintering season. Now it was spring, and the monarchs had gone, leaving behind a cemetery of those that had been too weak to make the trip north. The forest floor was covered with thousands of dead butterflies, battered and worn from the long winter. The whole place had a derelict air. As sad as it was, that dramatic scene marked the beginning of a new cycle that has been repeated for millions of years across North America. It's a cycle that includes one of the most epic and complex journeys in the animal world: a journey that is becoming increasingly dangerous; a journey that has puzzled scientists for decades—and still does.

In early spring, the monarchs that spent the winter in Mexico begin traveling back north to the prairies they left in the fall. They are searching for milkweed, a wild plant critical to their survival. Milkweed is their host species, where they will lay their eggs and die in one last gesture of endurance. Those new monarchs, born in the spring, are known as the first generation. The first and second generations are born and remain in the United States and Canada, each living just two to six weeks. By late summer and early fall, the third and fourth (and sometimes even fifth) generations emerge.

As the days grow shorter and cooler, these late-season monarchs enter a new phase of their life cycle, experiencing reproductive diapause: a hormonal shift that puts mating on hold, extends their lifespan, and triggers their migratory instinct. This so-called super generation can live up to nine months as they travel thousands of miles to their forebears' wintering

A female monarch, likely recently emerged, at the Monarch Butterfly Grady Tract in Madison, Wisconsin

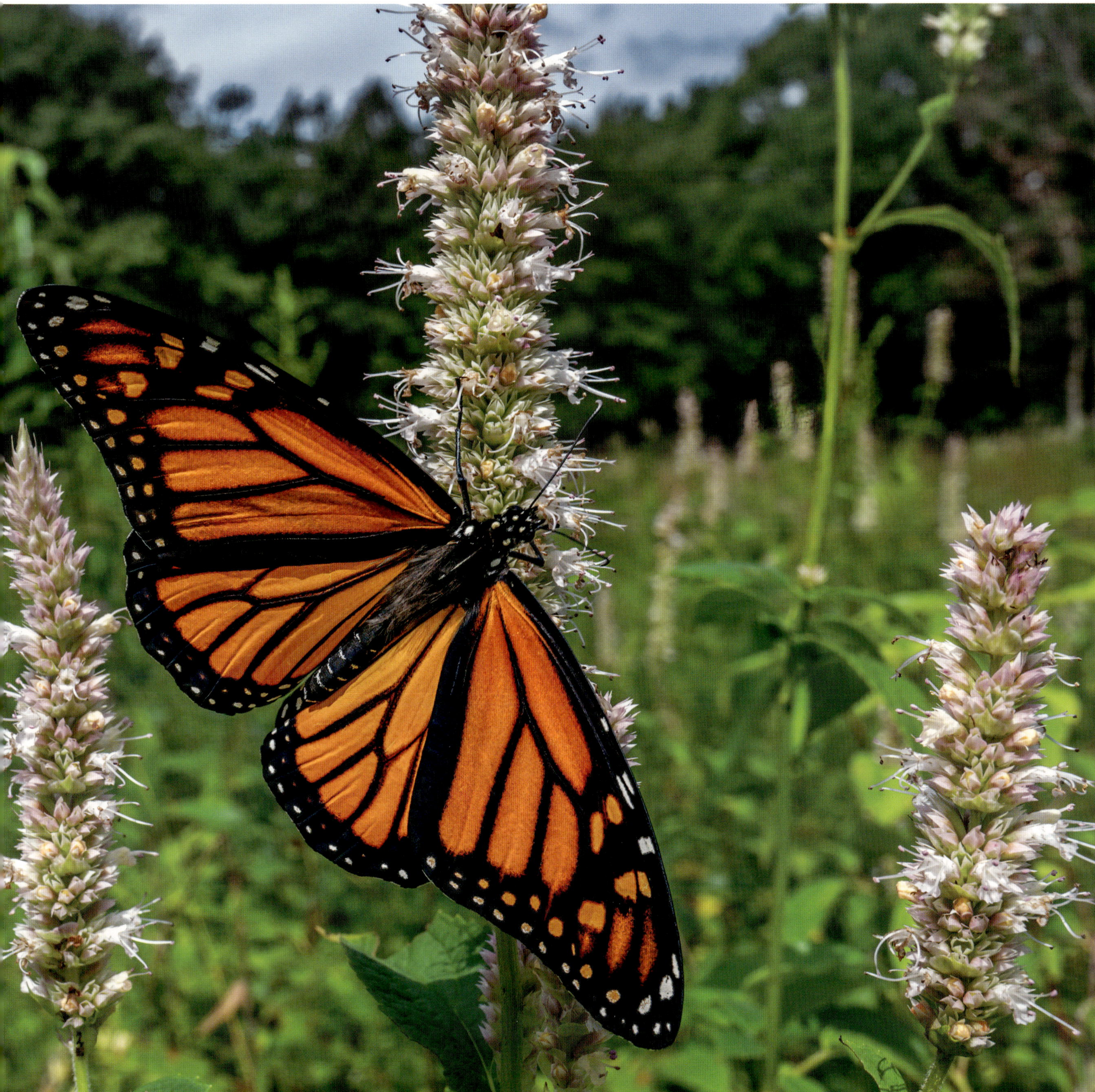

grounds in central Mexico—a place they have never seen before. They are the great-grandchildren of the butterflies that left in the spring. How they find their way to the very same patch of forest remains one of the great mysteries of the natural world.

FLYING NORTH, BUT ...

The spring migration of the monarchs is essentially invisible to the human eye. The butterflies spread out individually, nothing flashy to see. They leave not all at once but in gradual pulses, dispersing in increasingly smaller groups, to the point that it's rare to see any clusters after April or discern any pattern of directional flight by June.

I knew the migration was multigenerational, but like many others, I assumed it was a linear phenomenon: that wintering butterflies travel from Mexico to Texas; the first generation travels from Texas to, let's say, Kansas; the second generation from Kansas to Minnesota; and the third generation from Minnesota to Ontario. Then the fourth or fifth generation flies back to Mexico and starts the cycle again.

When I started researching more deeply, I realized that I had fallen into the same trap as well-intentioned textbooks, which oversimplify the monarchs' life cycle story into one flight north and another flight south. The real migration, I learned, is a much more organic dispersion, a complex treelike pattern with overlapping generations, multiple phases of localized spring migrations, and one long fall migration to Mexico. Some butterflies reach Texas and stay there, and their offspring don't migrate any farther. Others are restless fliers, driven north by instinct, reaching their final breeding range in Canada by late June—a time when most of their former breeding grounds have already been recolonized by new generations of monarchs.

A POCKET OF PRAIRIE

Well before humans became a driving force changing the landscapes of this planet, the spring migration of the monarchs happened naturally, unburdened by any barrier or complication aside from the random storm, flood, or drought. When the first European settlers arrived, the prairies stretched like an ocean across the heart of North America. The tallgrass prairie alone covered some 170 million acres from Manitoba to Texas and from Indiana to eastern Nebraska and Kansas, thick with grasses so tall they could hide a horse: big bluestem, Indian grass, and switchgrass. But after the steel plow broke the ground in the 1800s, the transformation was almost total. In just a few generations, nearly all of that prairie turned into farms and towns, suppressing fire and draining wetlands, both essential to the health of these ecosystems.

Despite this, milkweed still grew freely along roadsides, in field margins, and even inside cornfields in those prairie remnants. For millions of years, monarchs have depended on this plant. They lay their eggs on its leaves, and their caterpillars feed on its bitter sap. Until not so long ago, even after industrial farming was well established in the Midwest, milkweed was a common presence in the fields. It was tolerated. But that changed in the mid-1990s, when corn and soybeans genetically engineered to be resistant to the herbicide glyphosate took over. Farmers could spray entire fields with herbicide, wiping out every plant except the crop. The result was "clean" fields, free of weeds. This development made large-scale agriculture more efficient, but it was devastating for milkweed—and therefore for monarchs. In less than two decades, tens of millions of acres of monarch breeding habitat vanished.

In the summer of 2022, driving across the Midwest on my way to Iowa, I noticed how empty the fencerows and ditches looked. A generation ago, they would have been tangled with milkweed and wildflowers. Now they had been stripped bare. For monarchs, this change has been catastrophic. They return each spring, looking for milkweed, the only plant on which they can lay their eggs. Suddenly, the places that had supported their journey for centuries are gone. Where we see endless fields of corn and soy, migrating monarchs see a desert. This quiet transformation of the landscape, driven by herbicides and crop technology, has helped push one of the most remarkable migrations in the animal kingdom to the edge of collapse.

I wanted to visit Iowa because if there is one place in the United States to witness prairie loss, this is it. Once, the entire state was blanketed with tallgrass prairie. Now, less than one tenth of one percent survives. Iowa's story is a proxy for the story of prairies across the continent. In Cedar Falls, I met with Laura Jackson, director of the Tallgrass Prairie Center, an enthusiastic woman full of ideas, energy, and optimism—a rare trait in conservationists these days. She explained how Iowa's fertile soil, born from prairie, was now almost completely dedicated to growing corn and soybeans. For Laura, the story of Iowa is both a tragedy of loss and a chance to show how prairie can return, even in the middle of the most altered landscape in North America. She suggested that I visit some of the restored prairie sites.

The Tallgrass Prairie Center is spearheading an initiative called Science-based Trials of Rowcrops Integrated with Prairie Strips (STRIPS), which encourages a farmland conservation practice that protects soil and water while providing habitat for wildlife. The STRIPS team has been conducting research for more than a decade. Their work shows that strategically integrating small amounts of prairie among corn and soybean fields—in the form of in-field contour buffer strips and edge-of-field filter strips—can yield great benefits for soil, water, and biodiversity (pages 40–41). Even a modest strip of prairie within a field can slow erosion, filter runoff, and provide nectar and shelter for pollinators. Monarchs, truth be told, are not the most efficient pollinators, but they greatly benefit from these prairie restoration programs, too.

Another Tallgrass Prairie Center program, Iowa Roadside Management, was introduced in the mid-1980s. The idea was simple: Establish diverse stands of native plants along the edges of Iowa's more than 35,000 miles of county roads. In a state with so little public land available for conservation, those roadsides are one of the only places where prairie can come back. For monarchs and other pollinators, these ribbons of habitat are lifelines across an ocean of farmed landscape. A driver might see only weeds along the ditch, but a monarch sees a breeding and feeding haven.

I also connected with Peter Berthelsen, president of Conservation Blueprint, which has been pioneering prairie restoration in unlikely places: solar and wind energy developments. Instead of leaving the ground beneath solar panels bare or covered in tidy lawn grasses, his team seeds it with diverse native prairie mixes, turning industrial landscapes into pollinator habitat. On his recommendation, I visited a solar site in Anoka, Minnesota, managed by ENGIE, a French energy company with a growing investment in renewables. I spent an entire day there, chasing monarchs between the panels, trying to capture photographs that framed the feeding butterflies against the shining rows of glass. I failed to get the picture I wanted, but the message was clear: Even here, in the shade of solar arrays, prairie can return and offer monarchs what they need (pages 82 and 83).

Prairie today, I learned, is not some idyllic wilderness in the distant past. It's coming back in pockets—in a roadside ditch, in the corner of a soybean field, in a strip of restored grasses and flowers underneath the neat arrays of solar panels. Prairie here is not nostalgia; it's practical. It filters water, keeps soil in place, and helps farmers as much as it does butterflies and other pollinators.

A few days later, on a stifling hot morning in Ames, Iowa, I met Kelsey Fisher and Steve Bradbury, two scientists studying how monarchs move across the landscape. They had attached tiny radio transmitters to butterflies to track their paths through farm country. It felt magical to watch a monarch fly off with an antenna no bigger than a hair glued to its back. But the butterflies proved to be quite a challenge to photograph. It took me most of the morning and part of the afternoon to get just a couple of decent frames (see one on pages 62–63). What Fisher and Bradbury and their team confirmed was simple but important: In today's midwestern agricultural landscape, monarchs don't stay in one patch of milkweed. They keep moving, laying a few eggs here and a few there. Conservation, therefore, isn't just about planting milkweed anywhere—it's about putting it in the right places, close enough together that females can find the plants without flying too far. Creating stepping-stone habitats across the Midwest could make it easier for monarchs to reach their full potential and, in the long run, help grow the numbers that make it back to Mexico. Even in a landscape as altered as Iowa's, there are ways to weave back in enough pockets of prairie to keep the migration alive. And those pockets need to include the secret ingredient for success: milkweed.

MILKWEED & MONARCHS

One of the first books I read when I started digging deep into my monarch research was *Monarchs and Milkweed,* by Anurag Agrawal, a professor in the Department of Ecology and Evolutionary Biology at Cornell University. In his book, which I highly recommend, Agrawal explains the complex coevolution between monarchs and milkweed. The two species are locked in one of nature's classic evolutionary battles. Milkweed defends itself with bitter toxins, sticky latex, and tough fibers in its leaves and stem that make it nearly impossible for most insects to eat. Monarchs, however, have evolved to turn those defenses into an advantage. Their caterpillars chew the leaves carefully, to avoid the sticky latex, and they not only tolerate the toxins but

(Continued on page 34)

HOW THEY FIND THEIR WAY TO THE VERY SAME PATCH OF FOREST REMAINS ONE OF THE GREAT MYSTERIES OF THE NATURAL WORLD.

PAGES 32–33: Using drones, farm scientists and conservationists survey milkweed abundance near New Germany, Minnesota.

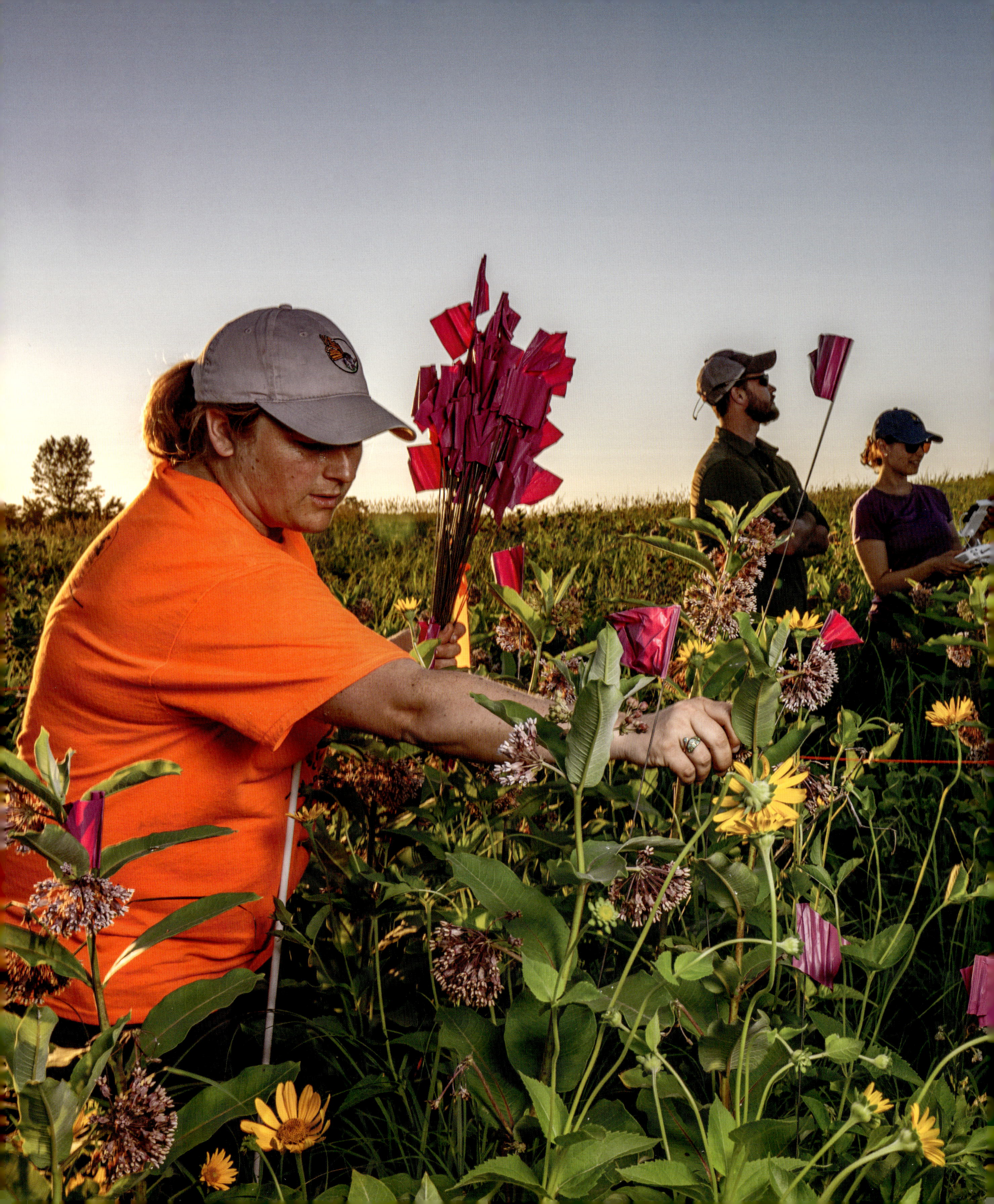

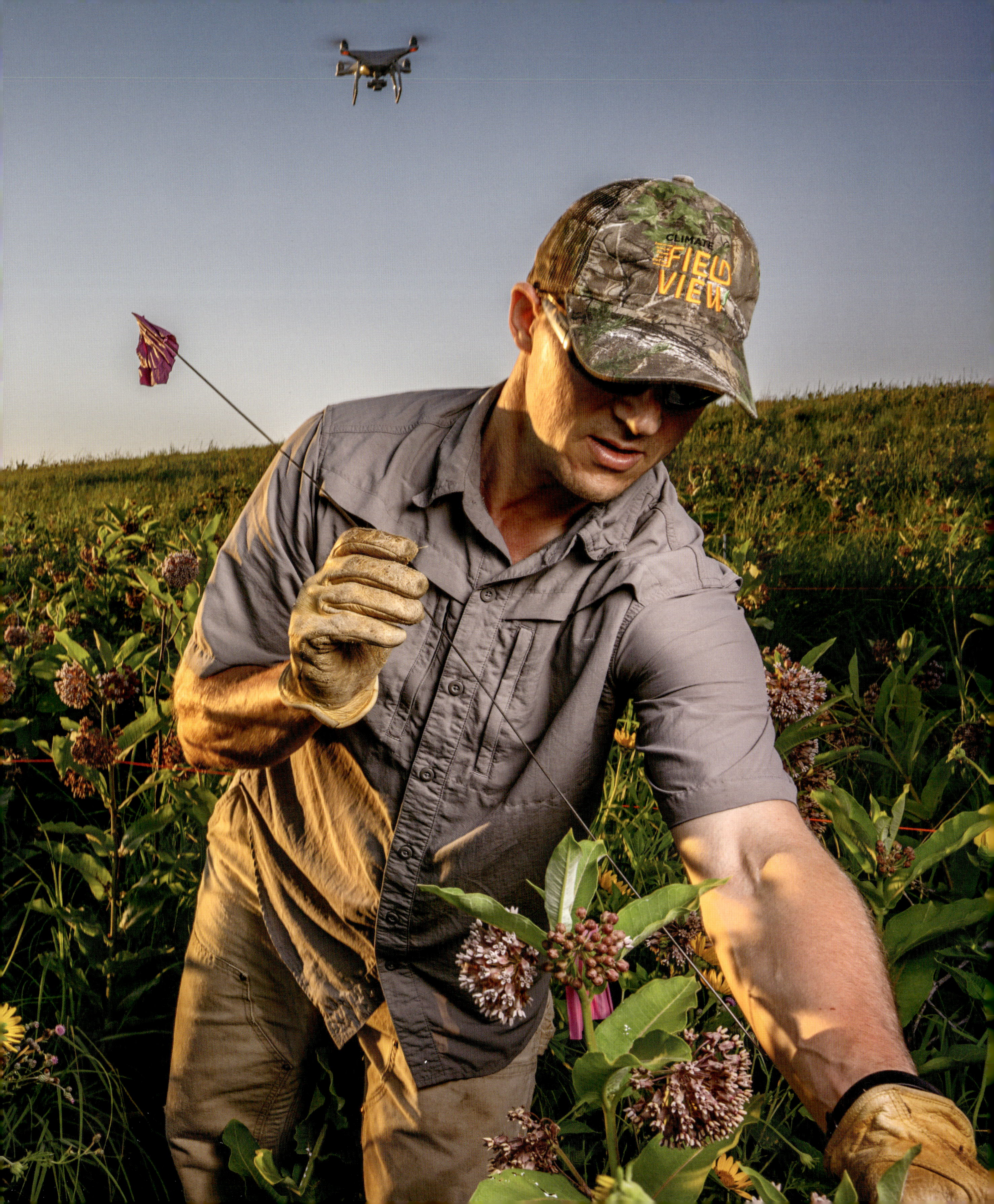
CLIMATE
FIELD
VIEW

I WAS LOOKING FOR WHAT ENTOMOLOGISTS CALL THE FIFTH INSTAR—THE FAT, ICONIC, ALIEN-LOOKING MONARCH CATERPILLAR.

(Continued from page 31)
even store them in their bodies, carrying the plant's poisons as their own protection throughout their adult lives as butterflies.

This tight link explains why monarchs cannot live without milkweed. Every egg is laid on a milkweed leaf, and every caterpillar depends on the leaves for food. The adult butterfly's survival is tied to that risky start. Agrawal shows how this relationship has driven both plants and butterflies to new extremes over millions of years. Milkweed keeps finding ways to fight back, and monarchs keep finding ways to counter it, in a coevolutionary arms race that has shaped one of the most remarkable migrations on Earth.

My early work includes some monarch caterpillar images (pages 52–53), but I wanted to go beyond them to get fresh, strong images that illustrate this complex relationship between monarchs and milkweed. Sometimes images just pop up in front of you. Sometimes you completely anticipate what you want—which was the case this time. I had already imagined the picture I wanted, with a set of storytelling elements: a caterpillar, the milkweed plant, and a corn or soybean field in the background.

I wandered across the Midwest, meeting monarch experts, photographing their science and conservation work, my head full of facts and data—but I hadn't seen a single caterpillar. The summer was proving very bad for monarchs. Scientists themselves weren't finding many eggs or larvae and were already anticipating a poor overwintering season.

To make things harder, I was looking for what entomologists call the fifth instar—the fat, iconic, alien-looking monarch caterpillar. Monarch caterpillars grow in five stages, each a different instar. After hatching, the tiny first instar—smaller than the nail on my little finger—feeds on milkweed then molts, shedding its skin to grow. This happens four times. Each instar looks a little different—gaining stripes, tentacles, and size—until by the fifth instar the caterpillar is large, boldly patterned, and ready to form its chrysalis. As I sought out the image I had in my mind, I would sometimes find the kind of caterpillar I wanted to photograph, but it would be too small or in the wrong place. I had already tried making some macro images, which are not really my strength, and I wasn't happy. I was starting to get nervous.

By this point I had spent almost seven weeks on the road, and I was missing my family, who were back home in Spain. I knew I needed to get home to them, but I also needed that picture. The hardest part of my otherwise dream job as a photographer is trying to find a balance with my family life. I have two daughters, Candela and Lola, and a wonderful wife, Fátima, and we are a tight-knit family; we really don't like being apart, and we constantly look for excuses to be together while I'm working. This time it wasn't just an excuse; it was a necessity. I wanted my daughters, then ages six and four, to help me search for caterpillars. Their restless, curious eyes are perfect for the task. And we would all enjoy the much needed time together.

So they traveled all the way from Spain and joined me in Madison, Wisconsin, where I introduced them to Karen Oberhauser, one of the leading experts on monarch science. She has always been kind and generous, a great help for my project, and I wanted my daughters to meet her, to be inspired by her dedication and persistence. For 30 years, Karen has led the Monarch Larva Monitoring Project, one of North America's longest-running citizen science programs, training volunteers to track monarch eggs and caterpillars in order to monitor populations and habitat health. If anyone could help us find caterpillars, it would be Karen.

But our days in Madison weren't successful—it was likely too late in the season—so we decided to head north to the Ney Nature Center in Henderson, Minnesota, a small nature reserve with short, accessible hiking trails through prime monarch-breeding habitat. We started the search early, full of energy, but as the hours passed and the humid July heat grew, there were no caterpillars in sight, and the fun turned into a tedious task. My daughters were cranky with disappointment (and probably hangry—my new favorite expression in English since I became a parent). We took a lunch break in the shade to reset the mood.

We resumed the search in the cooler hours of the afternoon, when the light was already softer. We split into search parties, and soon I heard cries of excitement. They had found a caterpillar at last, and in the perfect location. It was a beautiful, fat fifth instar at the edge of a cornfield. I prepared my gear. I knew it would be a long shoot because I wanted the caterpillar to crawl under a leaf, like they do just before nightfall. So I said good night to the girls, who headed back to our lodging with my wife, and I stayed there until well after sunset, testing every possible lighting combination. Finally, the caterpillar slid under a milkweed leaf, and I got the image I had envisioned (pages 66–67).

I am often asked about the color palette of my monarch work. My photos from Mexico's overwintering sanctuaries are colorful and warm. But when I set out to document the work of scientists and conservationists along the migratory route, I had to face the reality that most of their work happens at noon, in harsh light. I began using artificial lights to control the palette, and eventually I found the right balance between natural and artificial sources, creating a contrast of cold and warm tones. For this caterpillar image, I had always envisioned a mix of twilight's soft blue sky—my wife calls this "Magritte's sky," recalling the work of the Belgian painter—and a warm glow to emphasize the texture and color of the milkweed.

The next morning, I showed the images to Candela and Lola. They were thrilled and asked me a million questions. Experiences like this keep us together as a family and reinforce my daughters' love of the outdoors and their sense of belonging. A part of my work is now theirs.

DRONE MONITORING

Given its importance, milkweed is driving unprecedented innovation and alliances between farmers and conservationists—something I did not expect at first. But during my journey in search of monarch photographs, I had witnessed one of the most interesting developments in this area.

A drone was flying along established transect lines, buzzing over the sound of cicadas on a stuffy summer afternoon in New Germany, a charming rural town in central Minnesota. The small but agile land crew worked fast, flagging milkweed stems as Christine Chase, the pilot, kept an eye on the drone above. It was hot, and I was struggling to get the right shot. The sun was still high, the light too harsh, and I needed to use fill flash to lighten the shadows. There were too many moving pieces to manage on my own. My assistant had left the day before, and so I had recruited Peter Sanderson—Christine's retired father—to help.

This shoot was the culmination of a long process that had started months earlier, when I first reached out to Wendy Caldwell, the executive director of Monarch Joint Venture—a small, influential organization in the monarch conservation world. I had read about its effort to monitor milkweed using drones and artificial intelligence, and I wanted to see it firsthand. Knowing how much milkweed is left, and where, is the first step in any serious monarch recovery plan, yet the vast expanses of farmland and prairie in the Midwest make this a daunting task.

On the ground, Wendy and the rest of the land crew moved through the grass, marking stems one by one by sticking little purple flags in the soil. Christine and her co-pilot, Drew Smith, steered the drone in steady parallel lines as its camera recorded a patchwork of images that would later be stitched into a seamless map. Once you learn to spot them, the milkweed plants stand out—tall and sturdy, almost defiant—but at first the plants are easy to miss, drowned out by the chaos of summer prairie vegetation. The flags placed by their team provided a visible reference for the drones flying above. The operating algorithm links the flags with a certain shade of green, distinguishing milkweed stems from the rest of the vegetation, a task that would take endless hours if done entirely on foot. Here, the technology and the fieldwork depended on each other: the drone doing the heavy lifting, the crew on the ground making sure there were no mistakes.

Slowly, the rhythm of the shoot settled in. Peter, my impromptu assistant, was doing a great job with the lights. The drone passed overhead as the flags multiplied across the field, and I began to see the picture emerging. Soon, the scene turned dramatic, as late afternoon's golden light swept through the prairie. It was cinematic: the geometry of the crew, the drone in the air, the cadence of their movements, busy hands working plant by plant, all moving toward something bigger (pages 32–33). Every flagged stem was another data point in a much larger story, helping us understand what remains of the milkweed across the Midwest and whether it is enough to sustain the monarch migration.

Milkweed doesn't look like much, and yet it carries the weight of an entire species on its leaves. Conservation often begins with the humblest of things.

SEEDS OF RENEWAL

Things don't always go smoothly on photo shoots. No matter what the plan is, the first few days are challenging. There's this lingering fear that you're going to fail, no matter how much you've researched, no matter how carefully you've prepared.

I landed in Dallas–Fort Worth, having left Spain the day before, and met up with Thor Morales, my assistant and videographer.

(Continued on page 38)

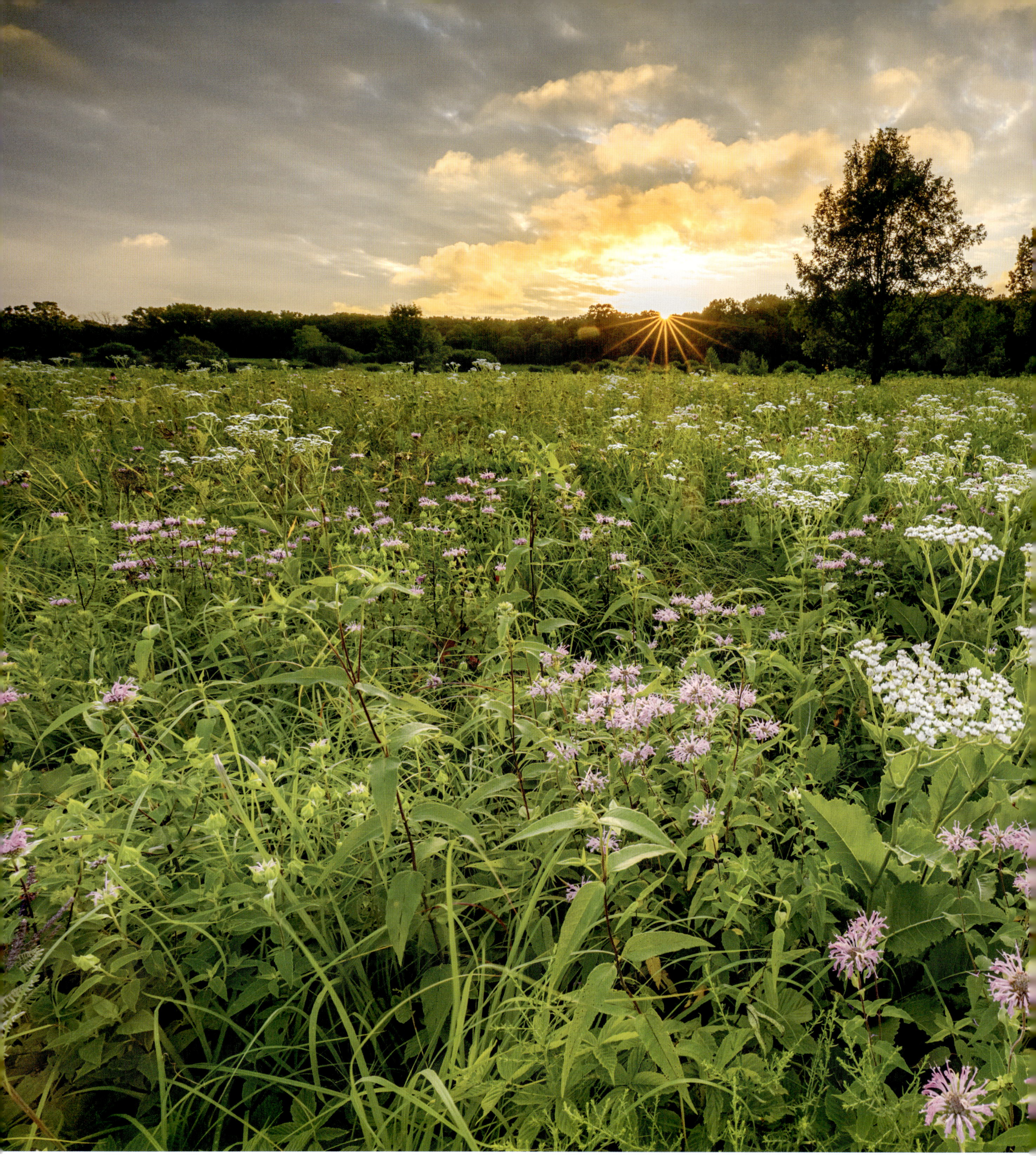

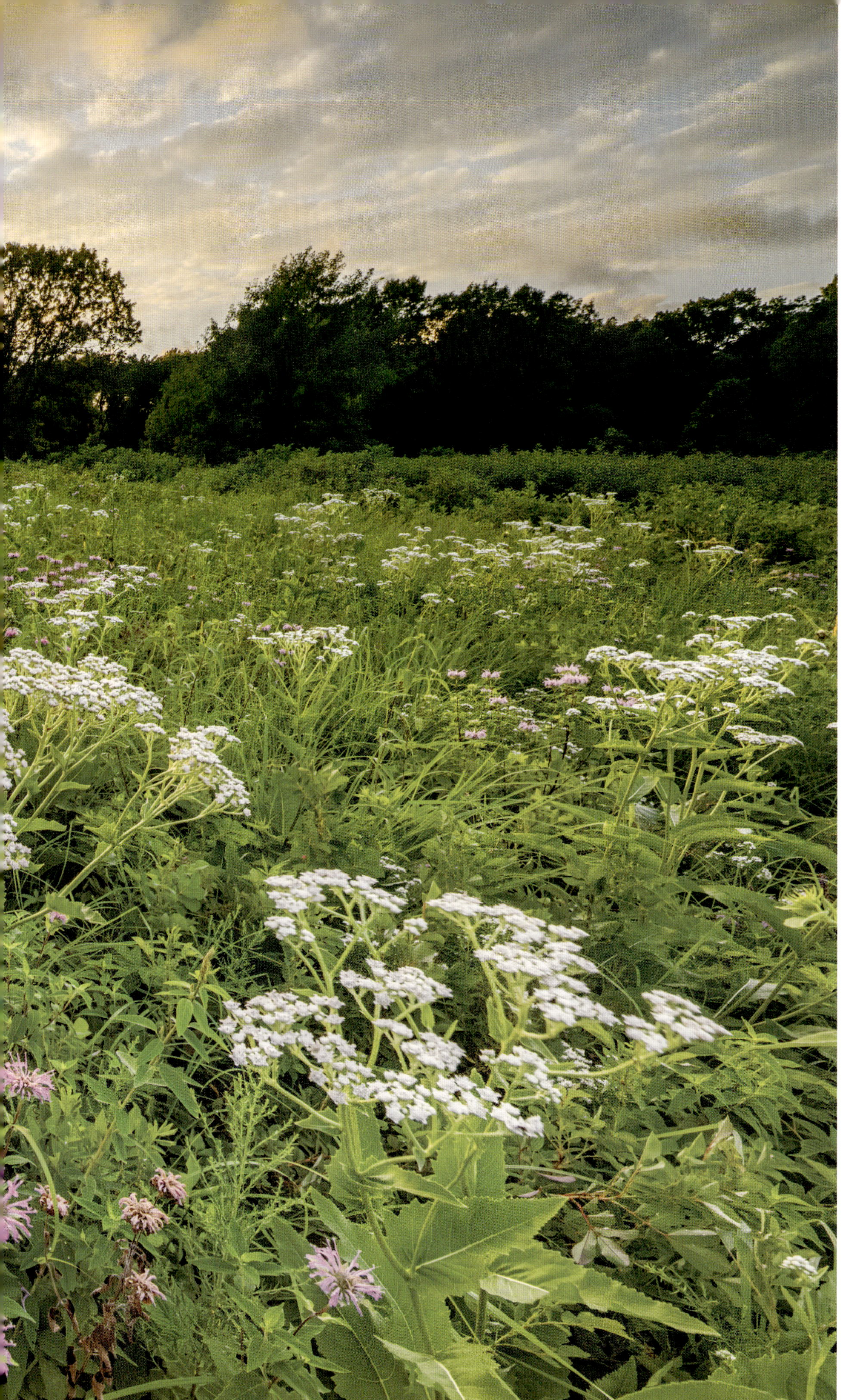

Sunset gleams over Greene Prairie, a restored tallgrass stand at the University of Wisconsin–Madison Arboretum.

(Continued from page 35)
We drove four hours north to Leonard, Oklahoma, to meet Jane Breckinridge, a citizen of the Muscogee (Creek) Nation, on her family farm. In 2017, Jane cofounded the Tribal Alliance for Pollinators (TAP), a Native-led nonprofit organization that uses traditional ecological knowledge to restore native plants and pollinator habitat on tribal lands. TAP's work shows how ecosystem recovery can also mean cultural recovery.

Jane and her husband, David Bohlken, hosted us at their beautiful farm (pages 48–49), and that first night of real sleep on location helped me reset. The next morning, Jane took us to meet her team at an abandoned golf course in Okmulgee where they were planting native species as part of TAP's restoration program. By the time we arrived, greeted everyone, and got to work, it was nearly noon. The July sun was brutal, the light harsh, the air heavy with moisture. Thor started setting up lights and diffusers—gear I hadn't expected to use so early, though it would soon prove essential for much of the assignment.

I try to keep myself nimble, always taking pictures—at home with my family, outdoors, anywhere—so I don't feel rusty when a project starts. But that morning nothing went smoothly. I was sweating, missing angles. The lighting was too much. Then two young men, Brandon Gibson and Collin Spriggs (from the Navajo and Cherokee Nations, respectively), began planting along the shore of a small pond. I adjusted the lights, cleaned up my background, and suddenly everything clicked. As if a switch had been flipped, the images started to work (pages 44–45). The heat, the jet lag, the frustration—gone. I reached that rare creative flow on the very first day. Unheard of.

After the team wrapped up and we had said our goodbyes, Jane took us to visit the official Muscogee Nation facilities: the council building, the college, and more. It was my first time in Native American territory, and it was eye-opening. It is easy to forget how deeply the history of colonization and nature conservation are intertwined. Two worldviews clashed centuries ago—those of settlers and Native peoples—and the work of reconciliation continues to this day. Jane's story, and that of her family, embodies this process. It will never be easy, but it is necessary. Jane was one of the nicest people I worked with as I sought ways to document the monarch migration, and I will always be thankful for how generous, attentive, and mindful she was. Those first few days with her and her team set my whole trip on the right path.

SUMMER MAINTENANCE

After a few long and intense summer weeks in the prairies of the United States, discovering new angles and refining others about the spring migration of monarchs, I headed back to where it all started: the forests of Mexico. We were driving up the bumpy road in a stake-bed truck on a cool July morning. The air was filled with the smells of moss and moisture. After years of visiting during the winter months, it felt strange and exciting to be here during the rainy summer season.

Our group, which included local community members and graduate students, was on its way to visit one of the experimental sites of Cuauhtémoc Sáenz Romero, a forest scientist in Michoacán who studies how climate change is reshaping the high-elevation oyamel fir forests where monarch butterflies spend the winter. As temperatures rise, these trees are struggling to survive in the places where they have stood for centuries. Cuauhtémoc's response has been to experiment with "assisted migration": The idea is to use seeds from oyamel populations at the upper limit of their elevation range to plant young fir seedlings even higher up the mountains—in places where the climate of tomorrow is already taking shape today.

After a couple of hours of bumps, we stopped where the track ended, at a clearing in the forest. It took me a while to realize that we had reached the iconic Llano de los Conejos (Plain of the Rabbits), a meadow in El Rosario Sanctuary, the largest of the forest conservation areas that make up Mexico's Monarch Butterfly Biosphere Reserve. One must almost always cross through here to reach the monarch colonies by foot. The usually sun-scorched and dusty plain was today covered in a carpet of grass, bright green from the recent rains. With one eye on the dark clouds, black with rain, and the other on my camera bags, I started preparing my gear for the photo shoot. I wanted to get an image of another innovation in Cuauhtémoc's project: the use of fast-growing *Baccharis* nurse plants (a hardy native shrub in the aster family) to provide shade for the oyamel saplings and increase their survival rate. This technique has proved to be both cheap and extremely effective.

I must admit that I was skeptical at first. After photographing hundreds of reforestation sites in Mexico and revisiting them years later, I had a very strong opinion against reforestation as a useful restoration strategy. I had the feeling it had become a bit of a corporate or government greenwashing ploy. But the numbers in Cuauhtémoc's scientific papers had gotten my attention.

Cuauhtémoc's doctorate students started distributing tools and allocating work sites to each member of the crew. We were joined by some *ejidatarios*—members of Ejido El Rosario, the community that owns this part of the forest. Beyond his technical innovations, Cuauhtémoc's work is based on collaboration. He partners with local communities and conservation groups to weave science into practical restoration. By tracking how transplanted firs grow and adapt, and by involving people who live around the monarch reserves, he is testing a way to secure the habitat the butterflies—and the livelihood of nearby communities—depend on. It's a long but necessary process, aimed at ensuring that forest, butterflies, and people have a future in a changing climate.

The rain had almost arrived, and I was growing anxious. There was too much chaos, too many things going on to concentrate on anything properly. Finally I saw José Humberto García Miranda, one of the *ejidatarios,* moving to a new site that had it all: the clean background, the beautifully arranged *Baccharis* species, and even the barbed wire, which added some context to the experiment, as it is used to protect the saplings from the grazing cattle that roam these pastures during the summer. José Humberto worked fast, with the precise movements of someone who has done this task a thousand times. I asked if it was okay for me to hang out with him for a while to photograph his work, and he reluctantly accepted. I backed off for a few minutes while I discussed the lighting strategy with Ganesh

Marín—my longtime friend, fellow National Geographic Explorer, and last-minute photo assistant for this trip. I knew I had only a few minutes left before the rain, so I shot a couple of test exposures and looked at them with Ganesh. We quickly fell into the intense concentration of the process, where you frame and shoot constantly, keeping an eye on the subject. The dark sky added a lot of drama. The *Baccharis* nurse plant was standing out, and the oyamel saplings that José Humberto was planting were shining—little bright green lights of hope for the monarchs (pages 88–89). After a few more frames, we felt the first drops of rain and knew it was time to pack up and drive back before the road became impossible to navigate.

On the way back, bumping along the track, I had a nice chat with Eligio García Serrano. Eligio is the executive director of Neovolcanic Axis Conservation Fund (FOCEN), and he coordinates the Monarch Fund (MF), a key financial tool for conservation in the Monarch Butterfly Biosphere Reserve. MF provides long-term financial support to landowners and communities to protect the critical overwintering forests. Over the years, he has become my trusted adviser, always providing insights and knowledge from the region. Few people understand the local communities as well as he does, and the success of that day's shoot had a lot to do with his always humble and discreet, but extremely effective, influence.

That same afternoon we drove to the training center of Alternare, a small but successful nonprofit also dedicated to monarch conservation. For more than 20 years, its facility on the outskirts of the Monarch Butterfly Biosphere Reserve has been training local communities in economic activities other than logging that provide livelihoods, such as organic food production, water-saving technologies, and reforestation, thus restoring and protecting the landscape. Its training program has an interesting twist: Young leaders from the region attend an intensive eight-week course, supported by scholarships and delivered in Alternare's facility. When they finish, they become certified trainers and return to their own communities, where they teach others: an exponential win-win for the future conservation leaders of the region.

The next morning we joined a community reforestation project in Macho de Agua, a small locality near Zitácuaro, Michoacán. We were welcomed by Luz Pérez Flores, a young woman with a big smile and endless energy. She had attended one of Alternare's training courses a few years earlier and was now working as a trainer herself, leading reforestations like this one as well as many other projects. We arrived at a steep, treeless plot owned by Doña Cayetana Estrada, a strong middle-aged woman with a determined stare who had invited most of her relatives and neighbors to participate. Soon, the event turned almost festive, with up to four generations working side by side, hands in the soil (pages 90–91). I couldn't keep up with photographing all the different scenes unfolding at once, but at the end of the day I was pleased because I had witnessed, firsthand, what a tight-knit community can achieve when there are both the means and the will to act. The community was preparing to welcome the monarchs, even if they still had to wait a few months.

In the meantime, the butterflies themselves were about to begin the next exciting stage of their complex life cycle.

THE OYAMEL SAPLINGS THAT JOSÉ HUMBERTO WAS PLANTING WERE SHINING—LITTLE BRIGHT GREEN LIGHTS OF HOPE FOR THE MONARCHS.

PAGES 40–41: Curving with the contours of soybean fields, restored native prairie strips in Rowley, Iowa, protect the soil, conserve water, and provide habitat for pollinators.

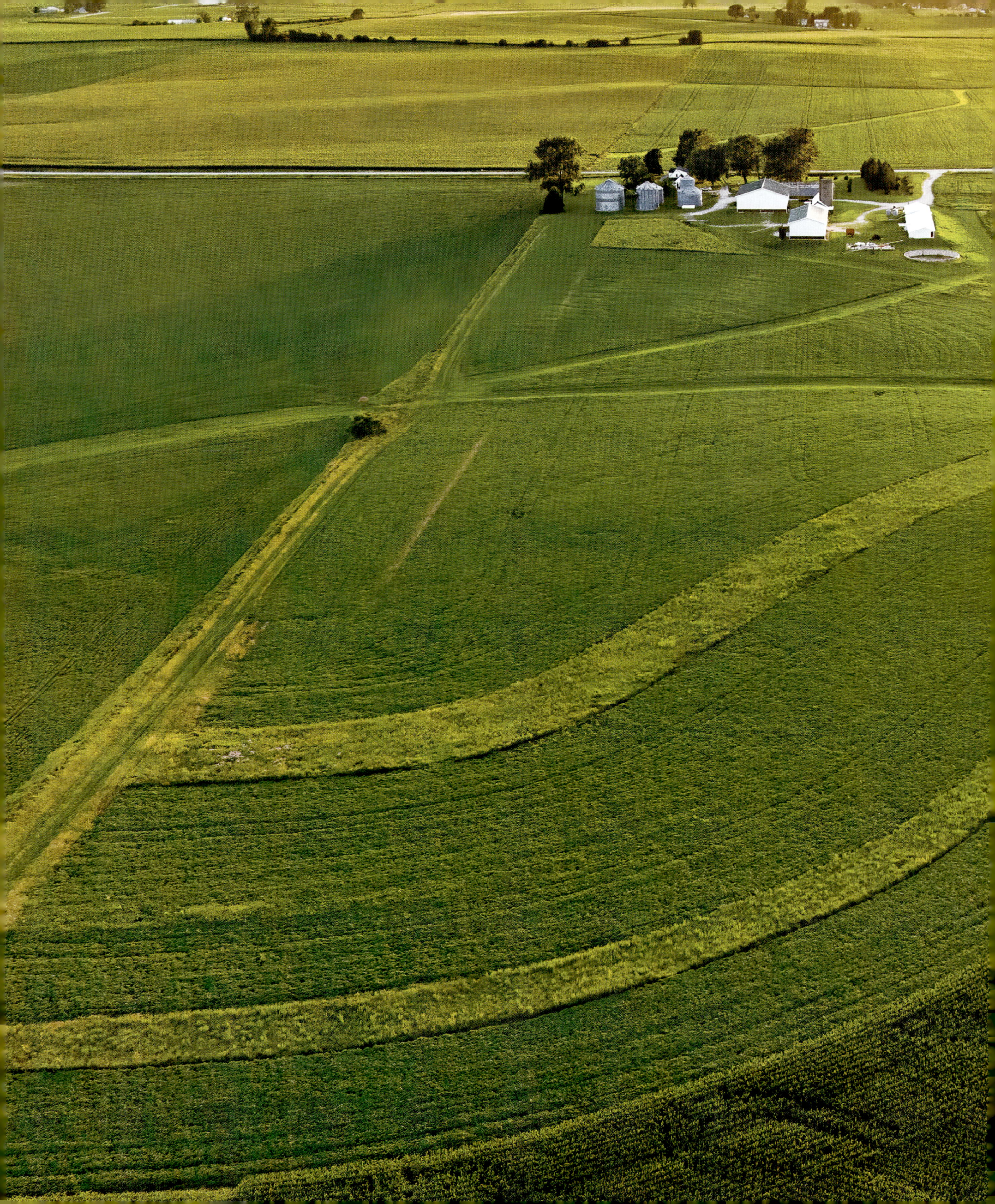

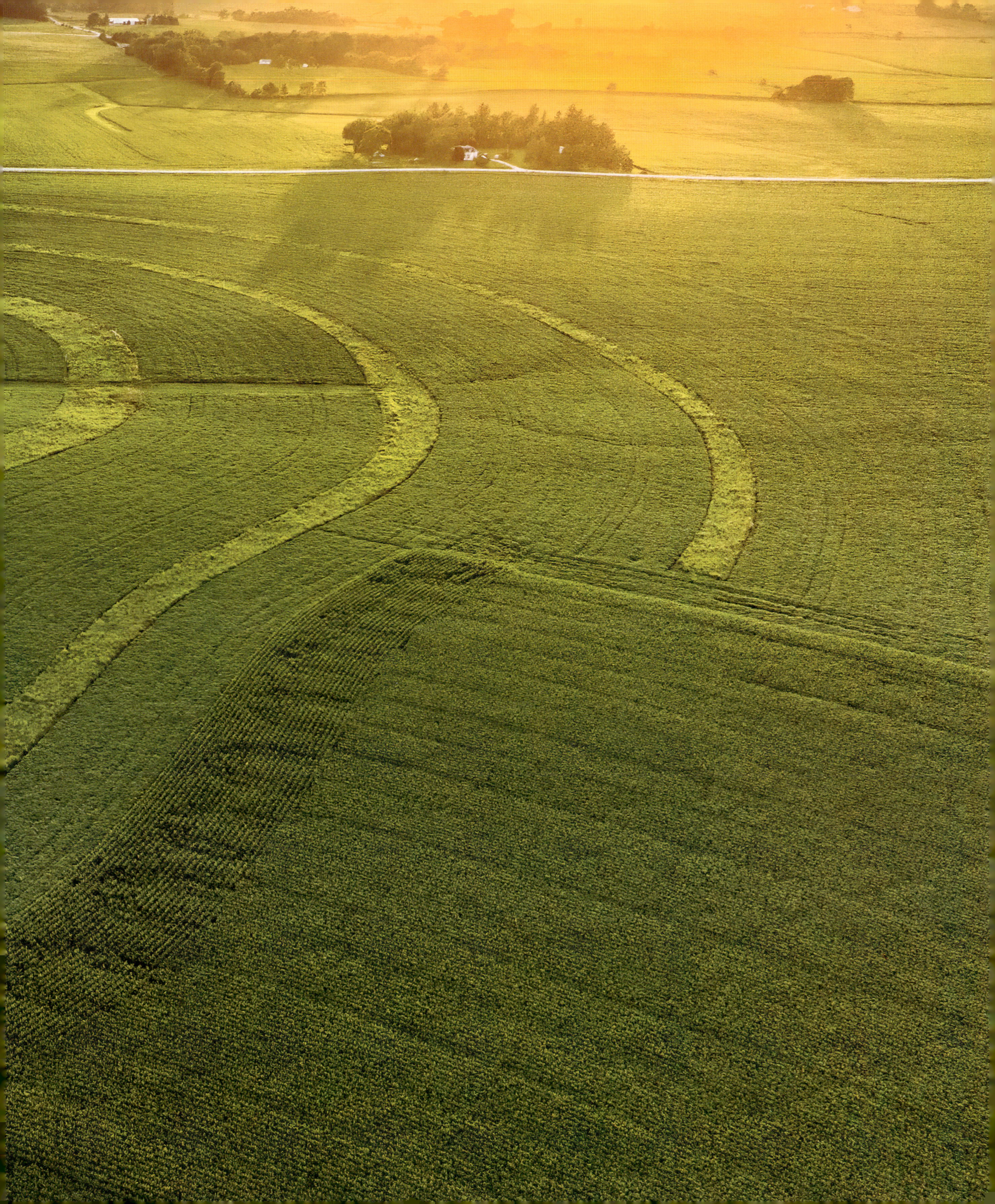

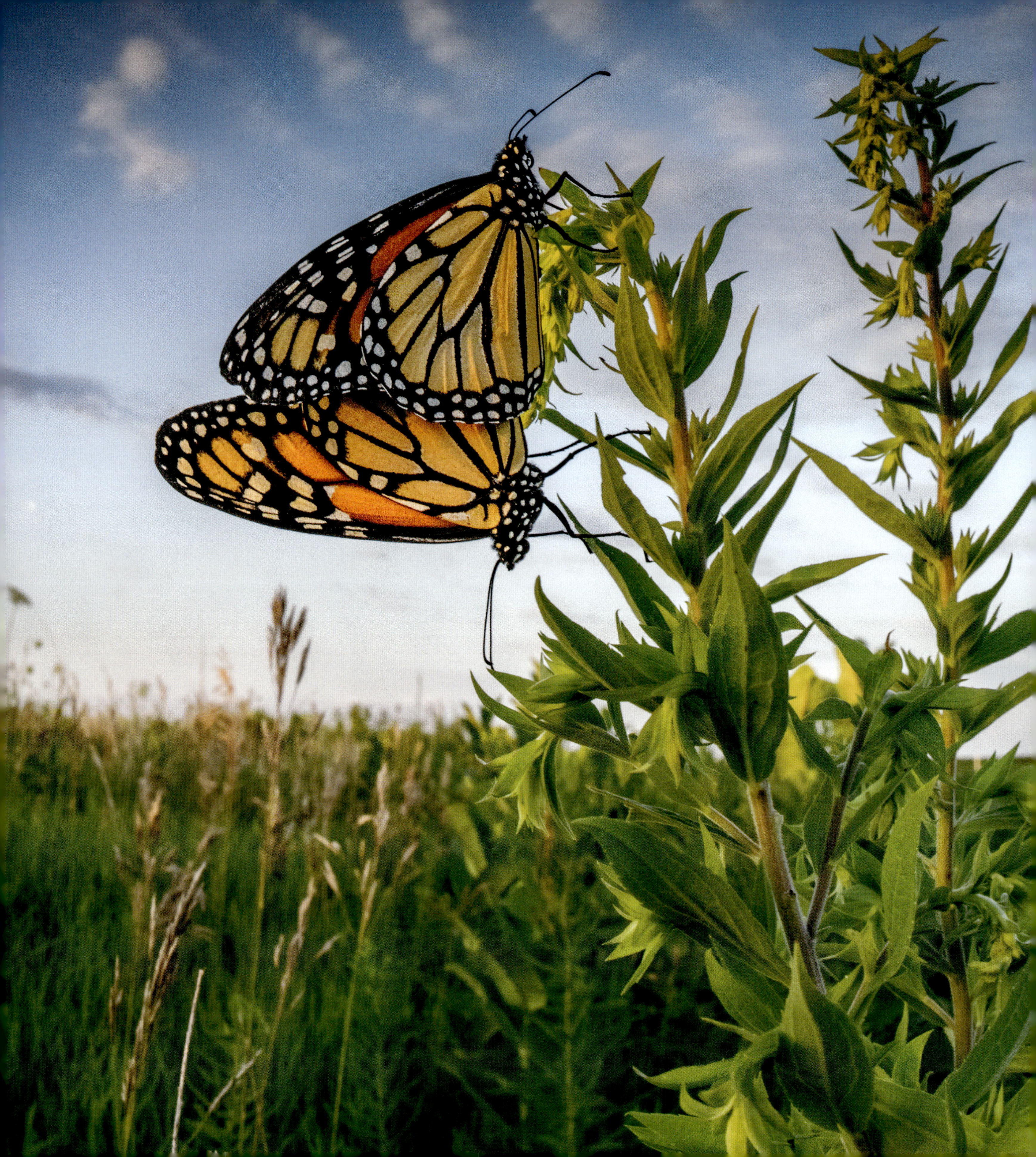

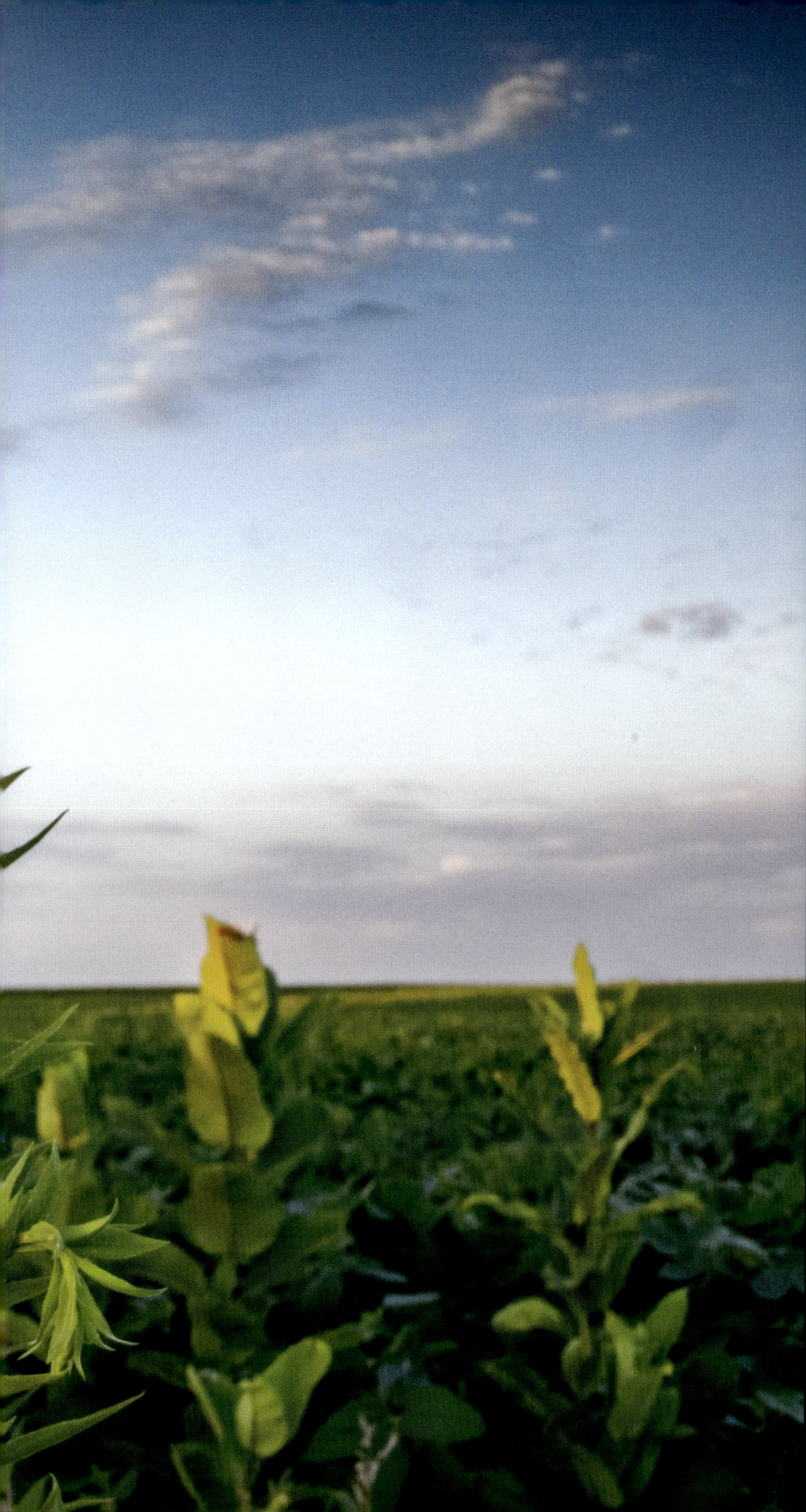

MATING MONARCHS

THE AFTERNOON LIGHT bathed Dick Sloan's farm in Rowley, Iowa, and the prairie strips stood out like an art deco garden among the endless corn and soybean fields. In Iowa, where so little—less than one percent—of the original prairie remains, these rows of native plants are introduced to protect the soil, conserve water, and provide habitat for pollinators, including monarch butterflies.

I had spent an entire summer chasing monarchs across the Midwest, but I was still missing a photo of what might be their most important behavior. It was my last day in the United States, and I had almost given up. The drone was flying high, the light was beautiful, and suddenly—out of the corner of my eye—I saw movement. It was a pair of monarchs, flying, attached in what's known as aerial courtship. I couldn't believe my luck. My heart raced. I grabbed my camera bag and ran after them.

I had to follow the fast-flying pair of butterflies across dozens of neatly planted rows of soybeans without stepping on the crop. I tracked them through my binoculars until they finally stopped. They landed on top of an evening primrose in one of the prairie strips, and they began to mate. The sun was low, the light golden but fading. Mating can last for up to 16 hours. I knew that if I approached them carefully, I could still make it in time. It was now or never. I slowly positioned myself, thinking about every possible variable: the focus distance, the background, the changing light.

I was getting closer, so I grabbed the first few frames with a telephoto. They were nice but boring, a bit small in the frame, with no context at all. One step at a time, I approached, almost forgetting to breathe. I tried again, but I wasn't capturing the complexity of the whole scene with a telephoto. I needed to switch to a wide angle to add context. I moved a bit closer, carefully testing how close they would let me get. Every step made me anxious, picturing the butterflies flying away forever. But they seemed calm, perched and still, their abdomens attached to one another, their wings folded. I took one step more and, to my horror, the male startled and beat his wings. I froze. But they remained there.

The seconds turned into minutes as I continued to slowly approach, grabbing some frames, closer each time. The pink glow of the clouds in the background acted as a giant reflector, and the light was soft and beautiful. One inch at a time, I got closer, and finally, I got the shot that I had dreamed of: a frame that told the story in full circle.

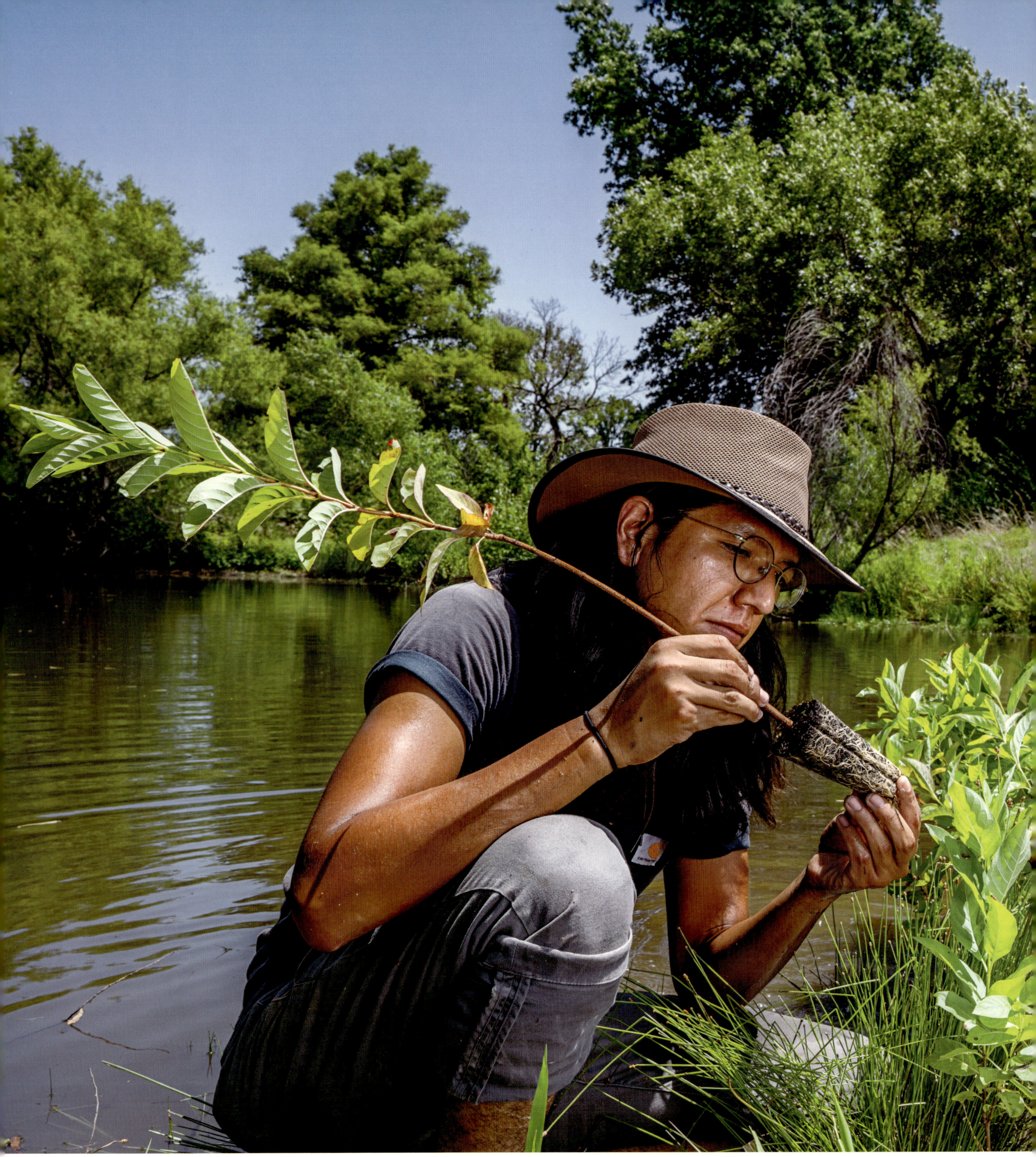

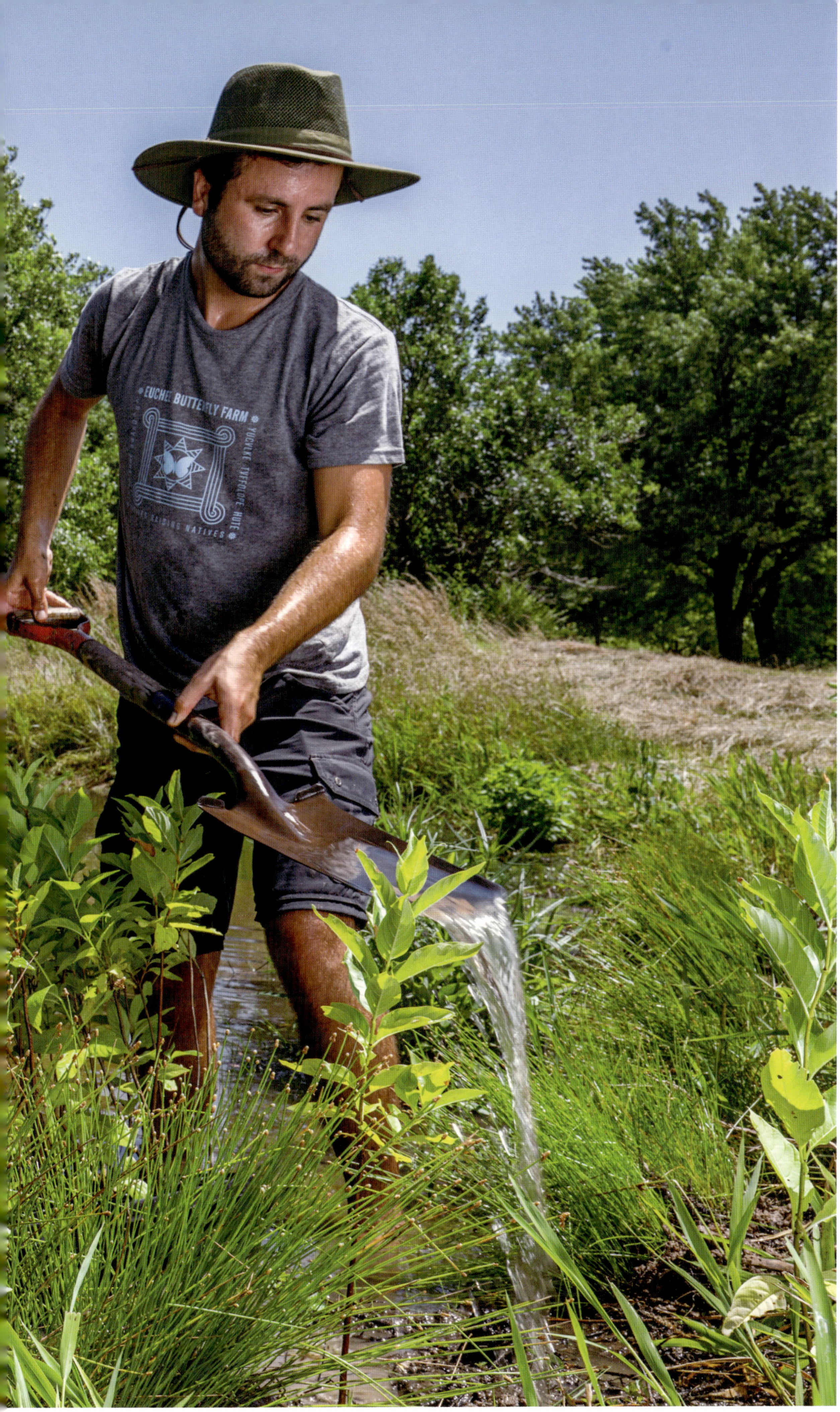

Brandon Gibson (left, of the Navajo Nation) and Collin Spriggs (right, of the Cherokee Nation) plant native species on a former golf course on Muscogee Nation lands in Okmulgee, Oklahoma, as part of a Tribal Alliance for Pollinators program in which Monarch Watch and Native nations partner to restore habitat.

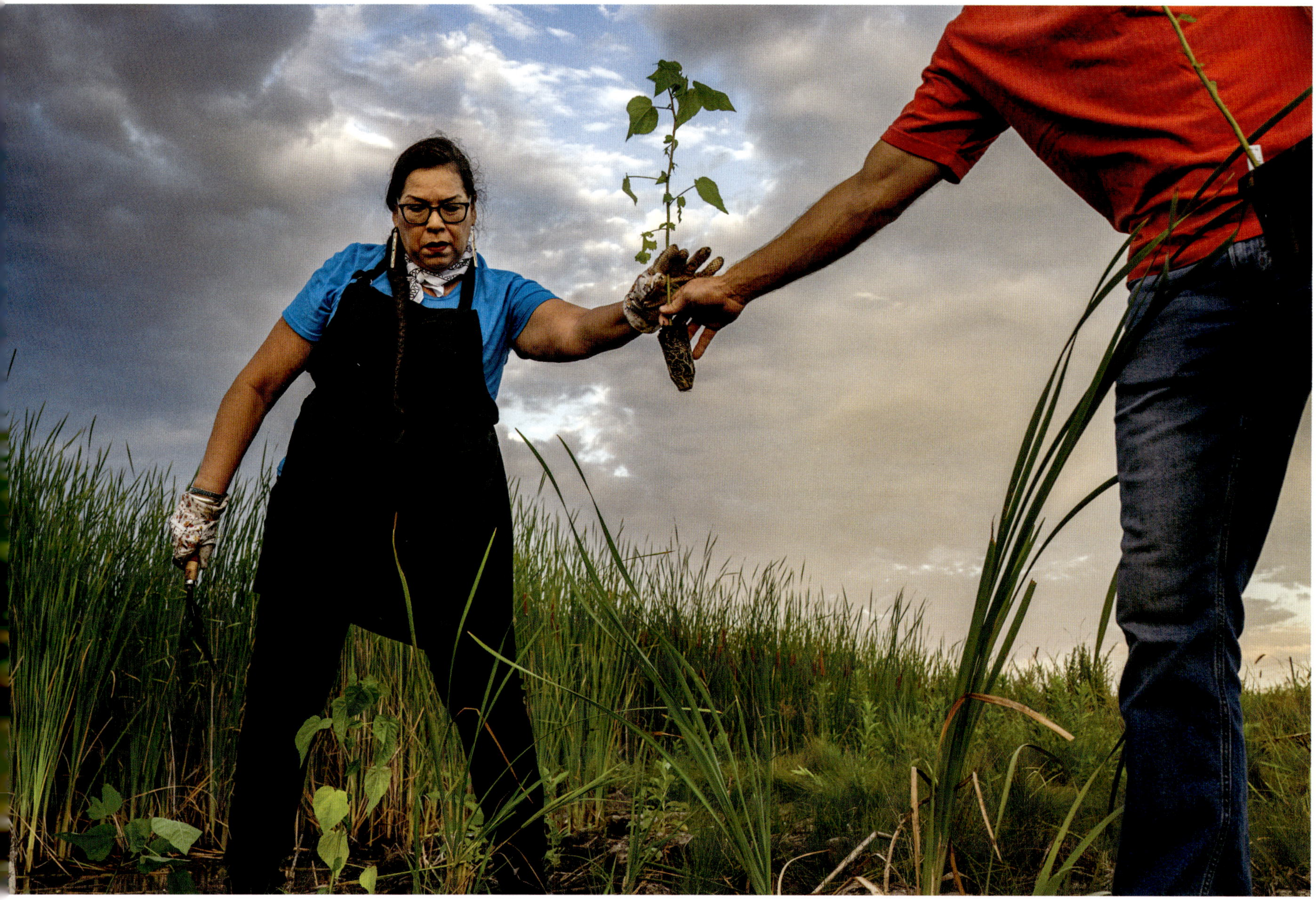

Margaret Sisk, an Osage Nation member, plants native species at Harvest Land Farm in Pawhuska, Oklahoma, part of an Osage program to restore traditional food systems and pollinator habitats.

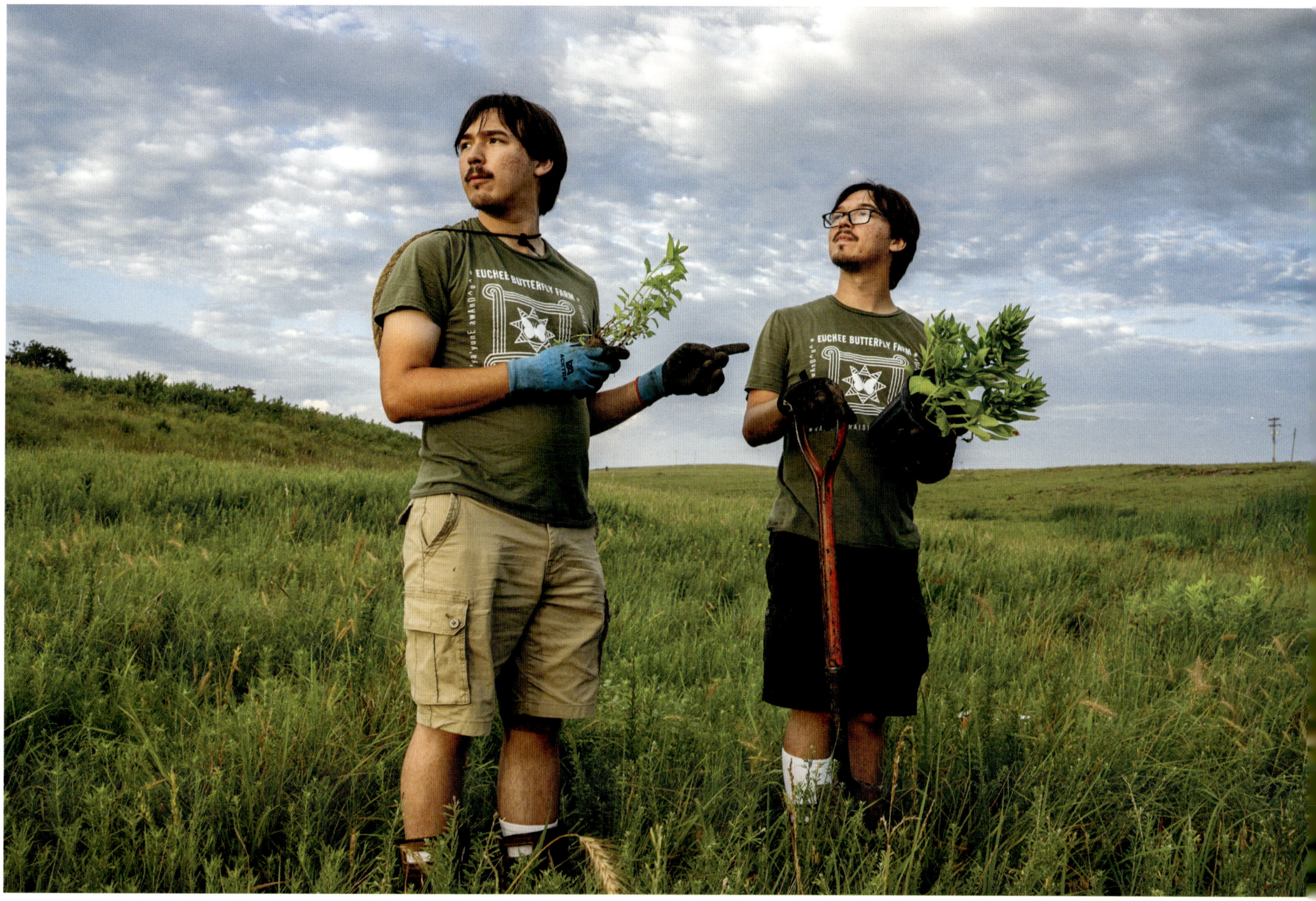

Brothers Bryce and Beau Whetstine take part in the Osage Nation program by planting native species.

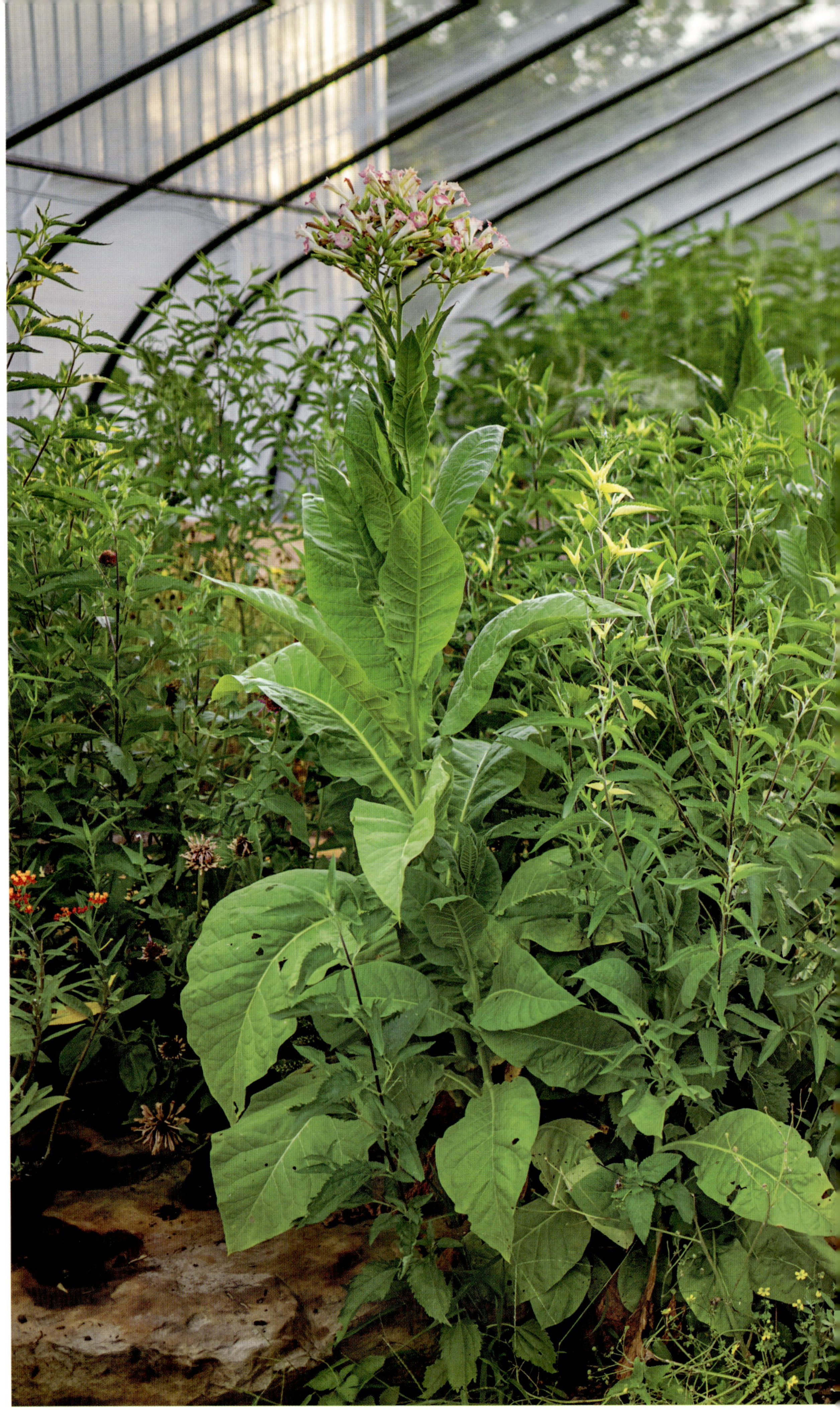

Jane Breckinridge and her husband, David Bohlken, founded the Euchee Butterfly Farm in Bixby, Oklahoma. She leads the Tribal Alliance for Pollinators, guiding Native nations in strengthening ecological and cultural resilience across tribal lands.

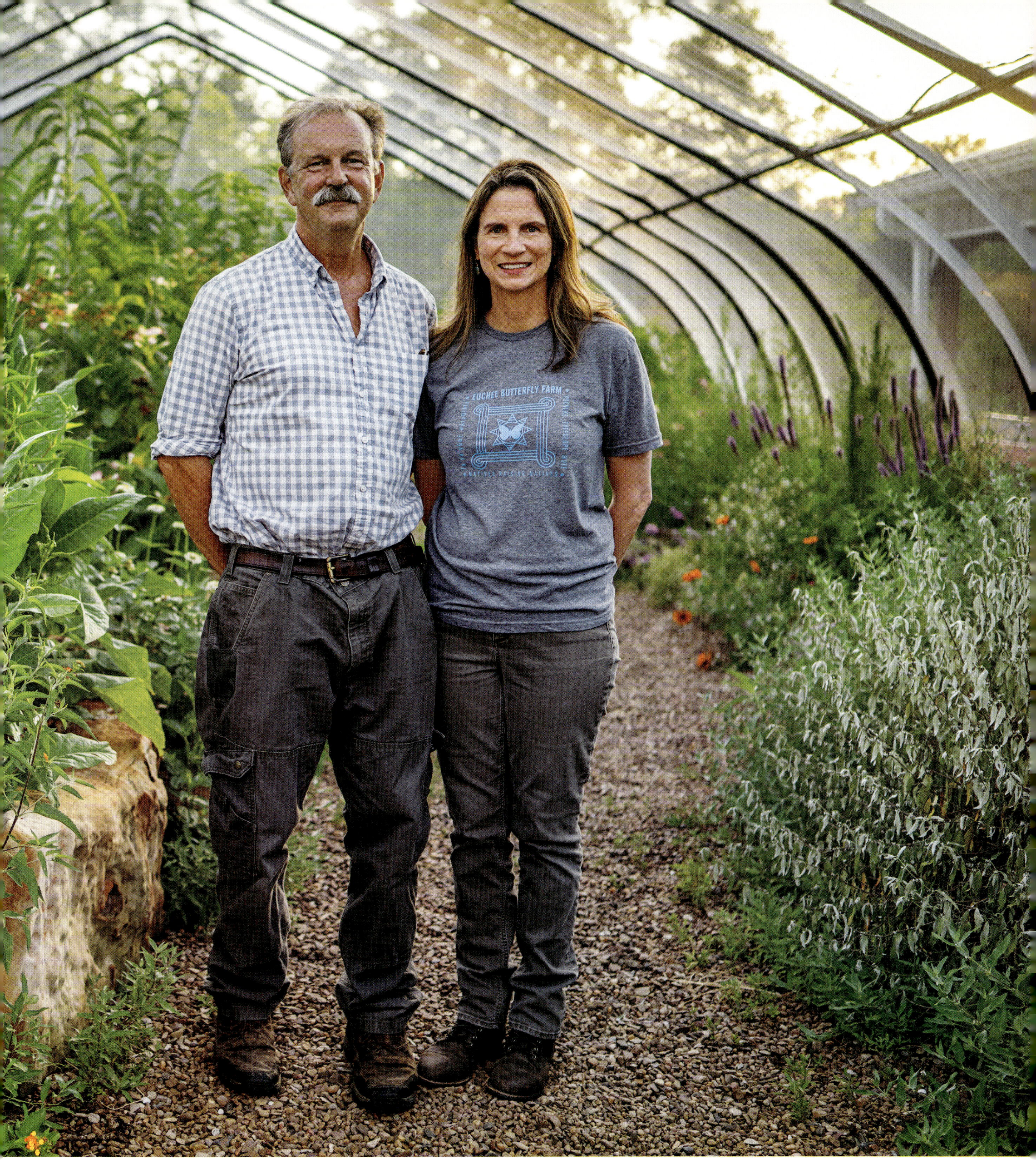
EUCHEE BUTTERFLY FARM

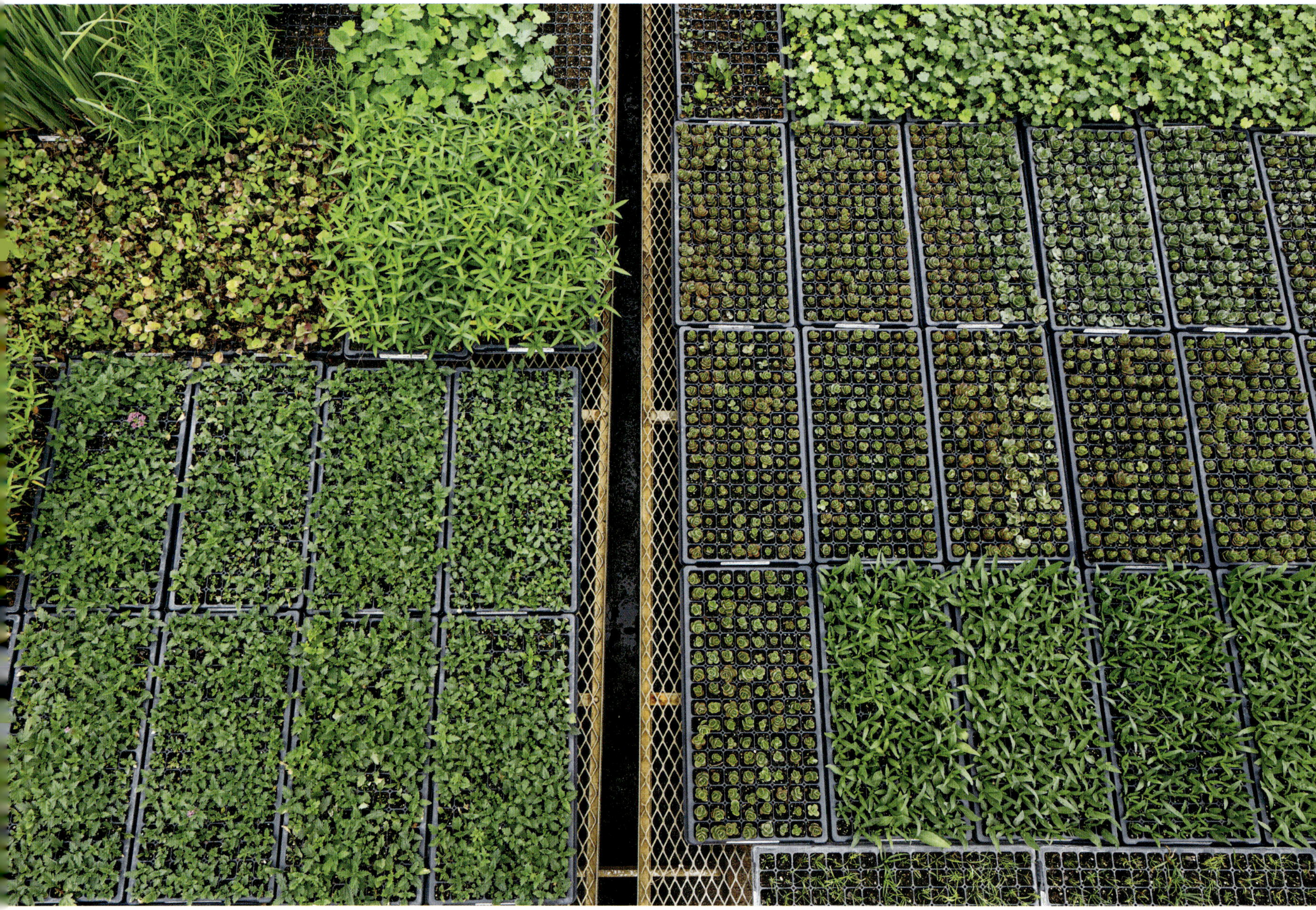

ABOVE AND OPPOSITE: Native prairie plants grow in a nursery in Baldwin City, Kansas, operated by Resource Environmental Solutions, the largest ecological restoration company based in the United States.

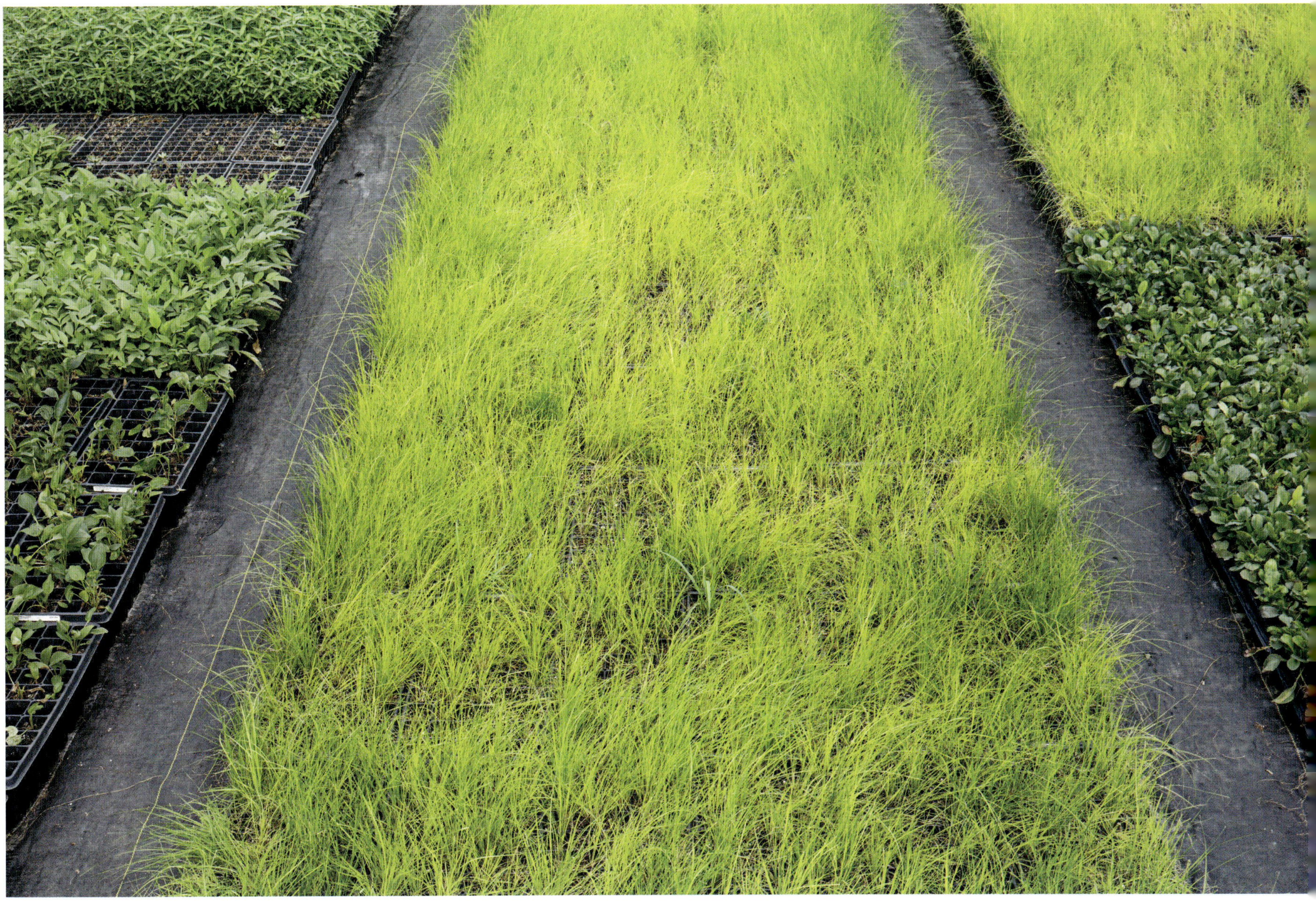

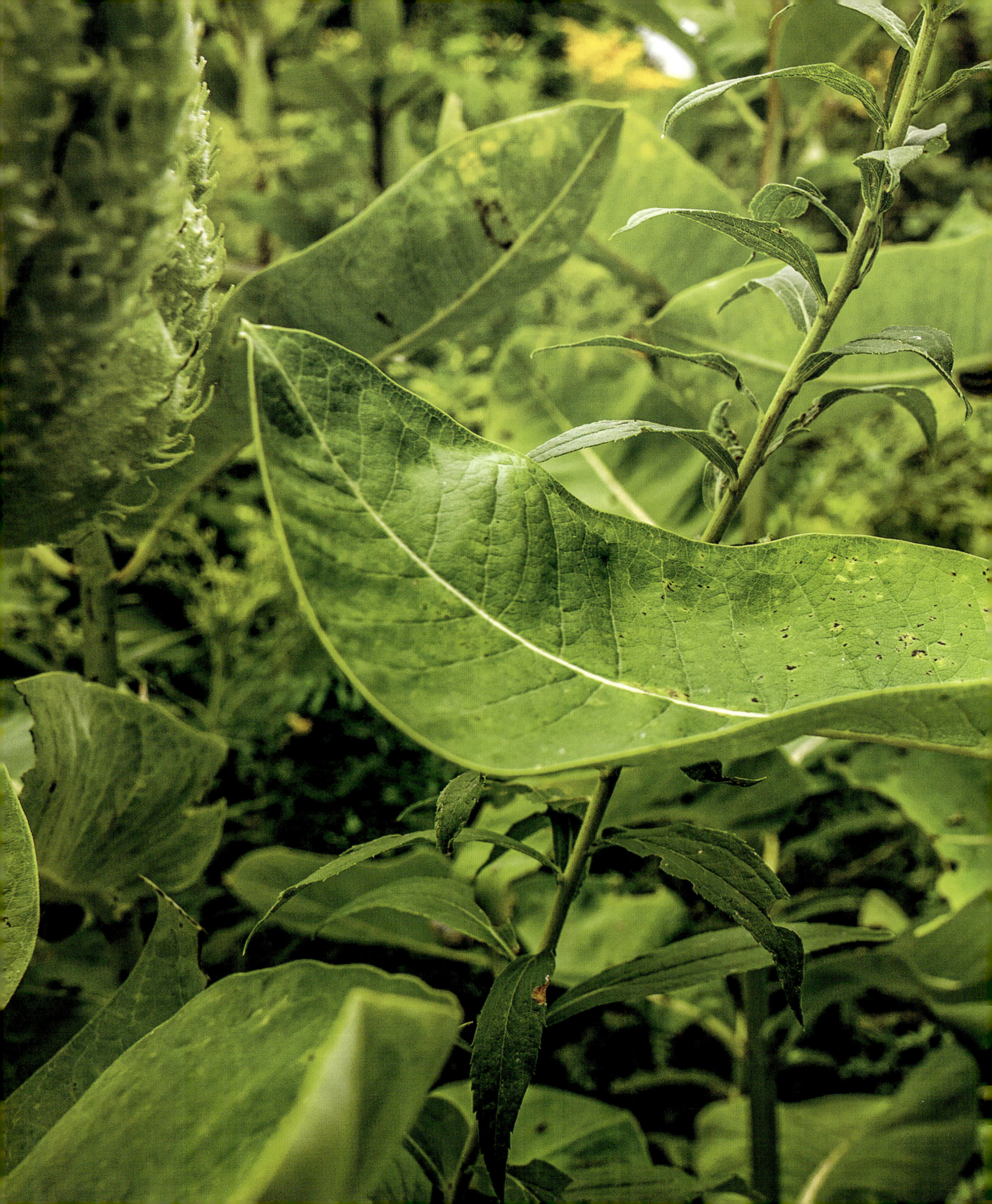

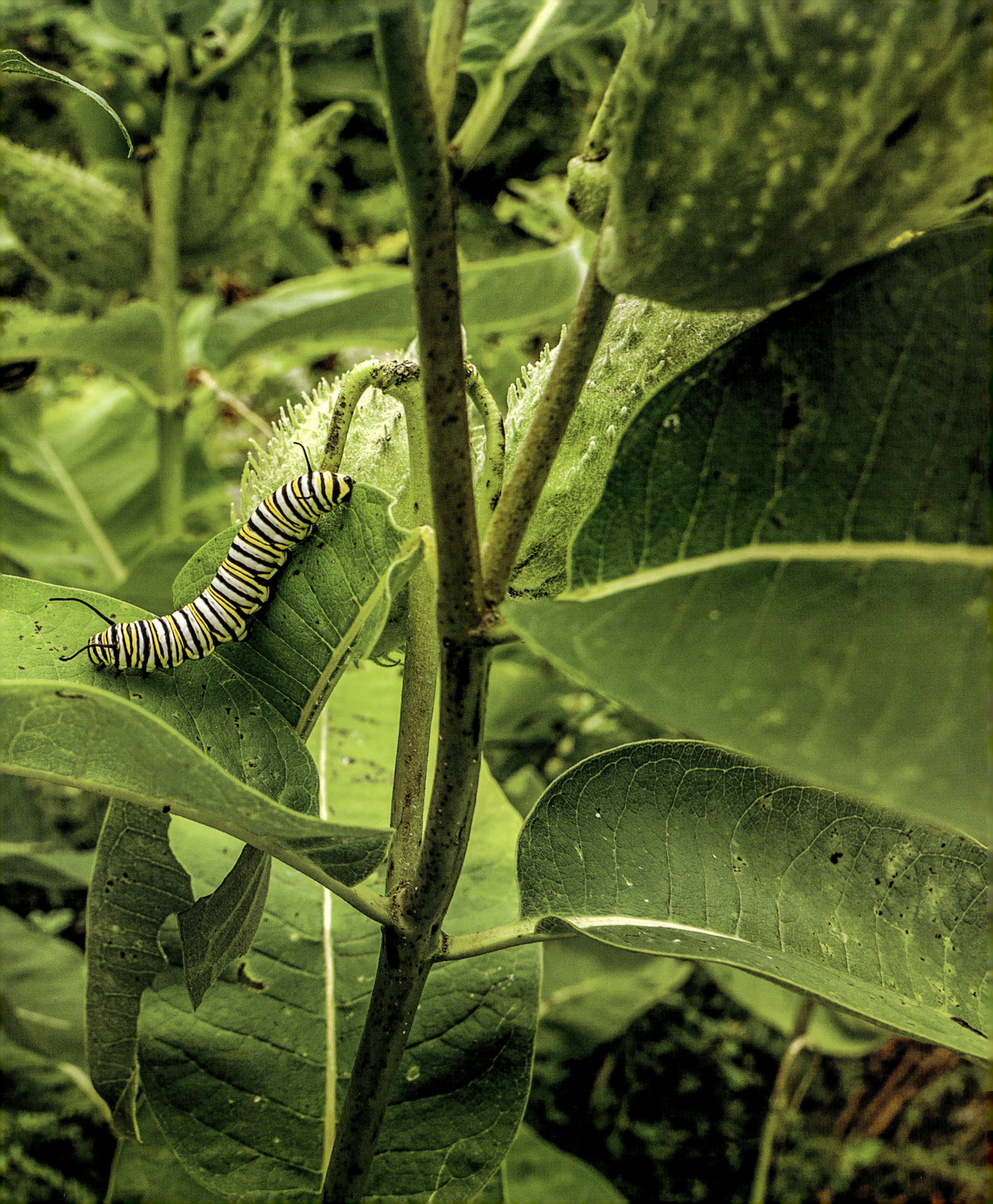

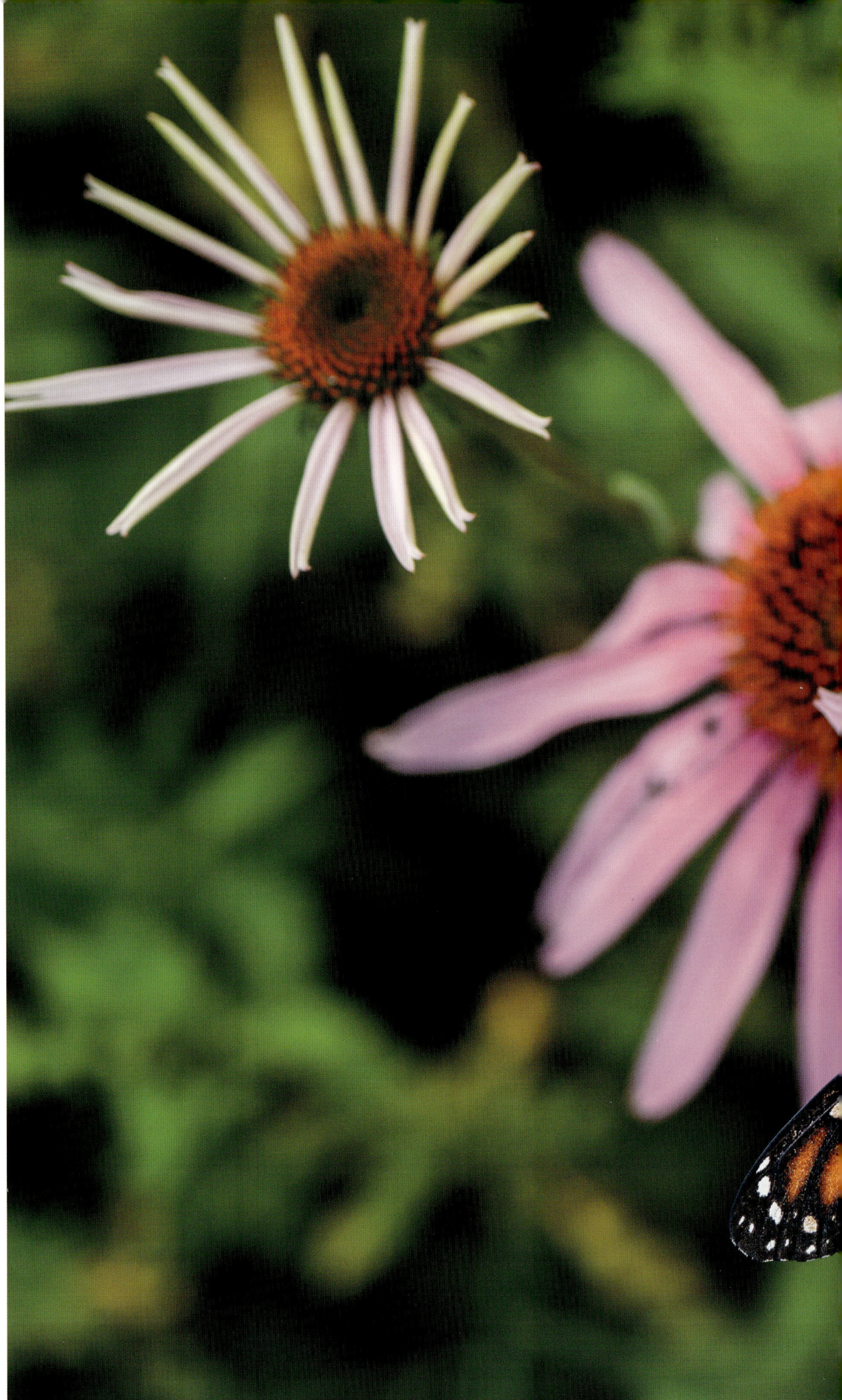

A male monarch butterfly seeks nectar from a purple coneflower (*Echinacea* sp.) at Prairie Oaks in Belle Plaine, Minnesota.

PAGES 52–53: A monarch caterpillar munches its way through a milkweed patch at Tommy Thompson Park, a reclaimed urban wilderness along the shore of Lake Ontario in Toronto.

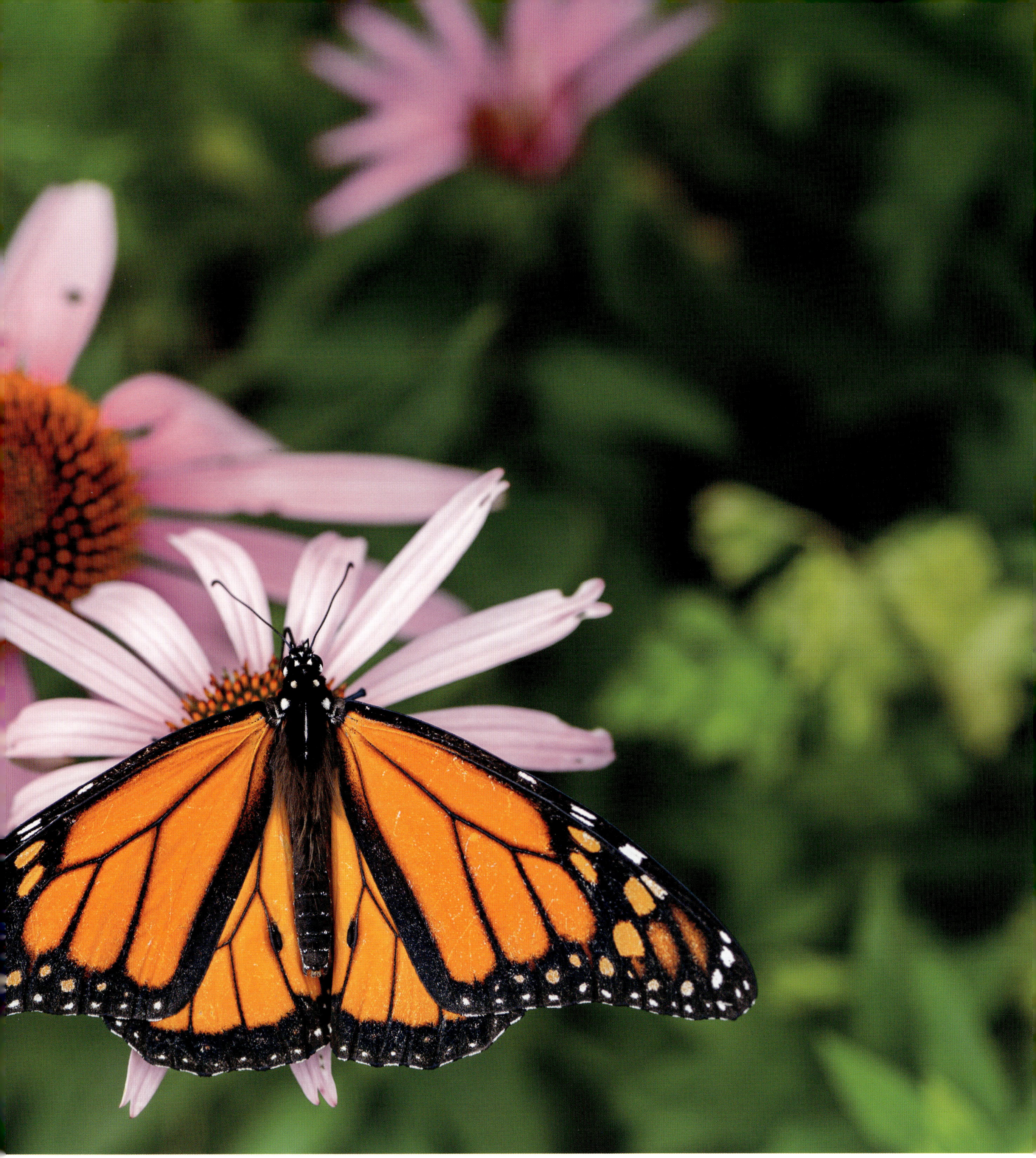

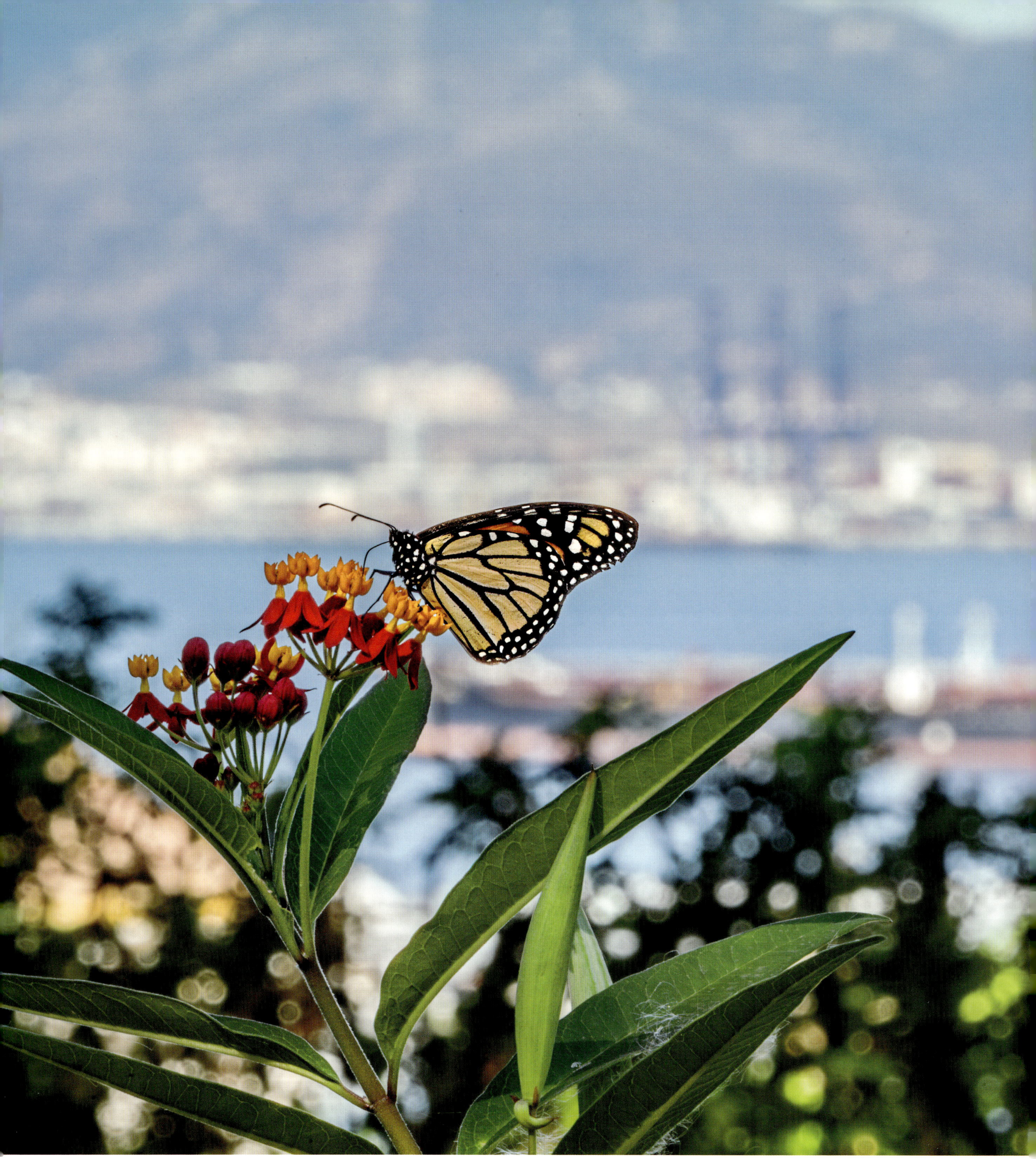

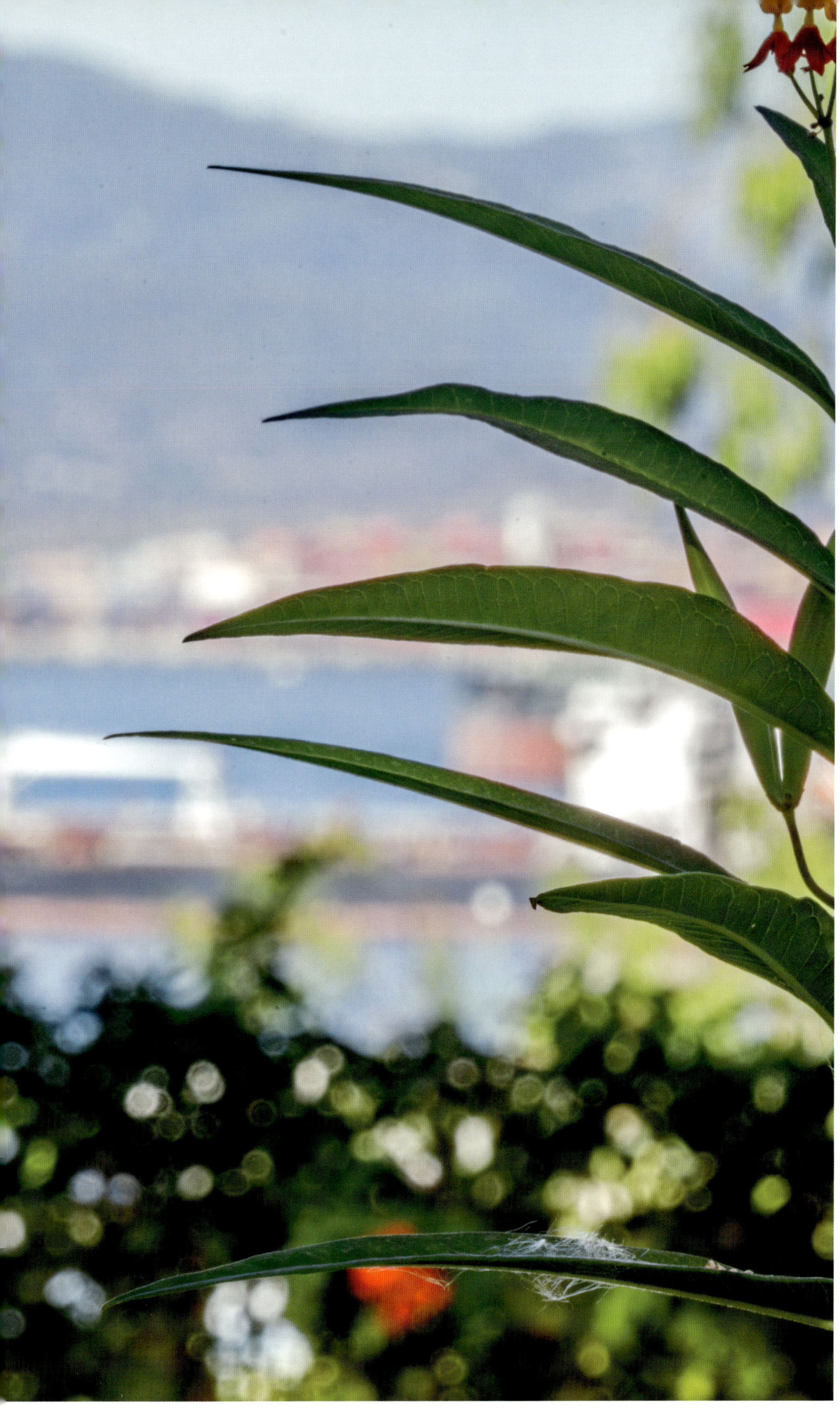

A female monarch feeds on tropical milkweed *(Asclepias curassavica)* at the Gibraltar Botanic Gardens, with the Bay of Algeciras and the Mediterranean Sea in view. No native milkweeds grow on the Iberian Peninsula, but introduced species such as this have enabled small resident monarch populations to persist.

Karen Oberhauser (right), founder of the Monarch Larva Monitoring Project, and volunteer Beth Whitaker inspect common milkweed for monarch eggs and caterpillars in Madison, Wisconsin.

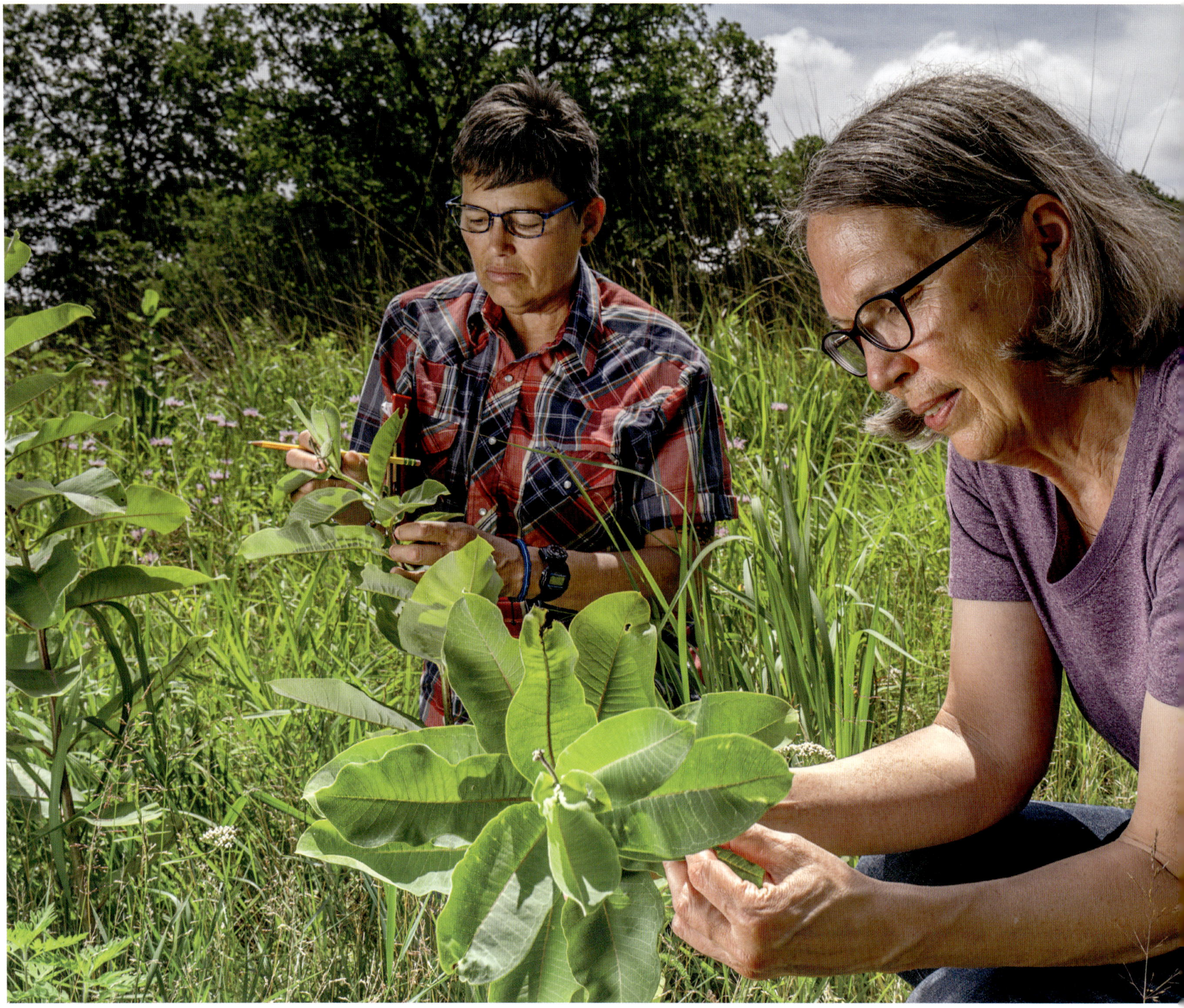

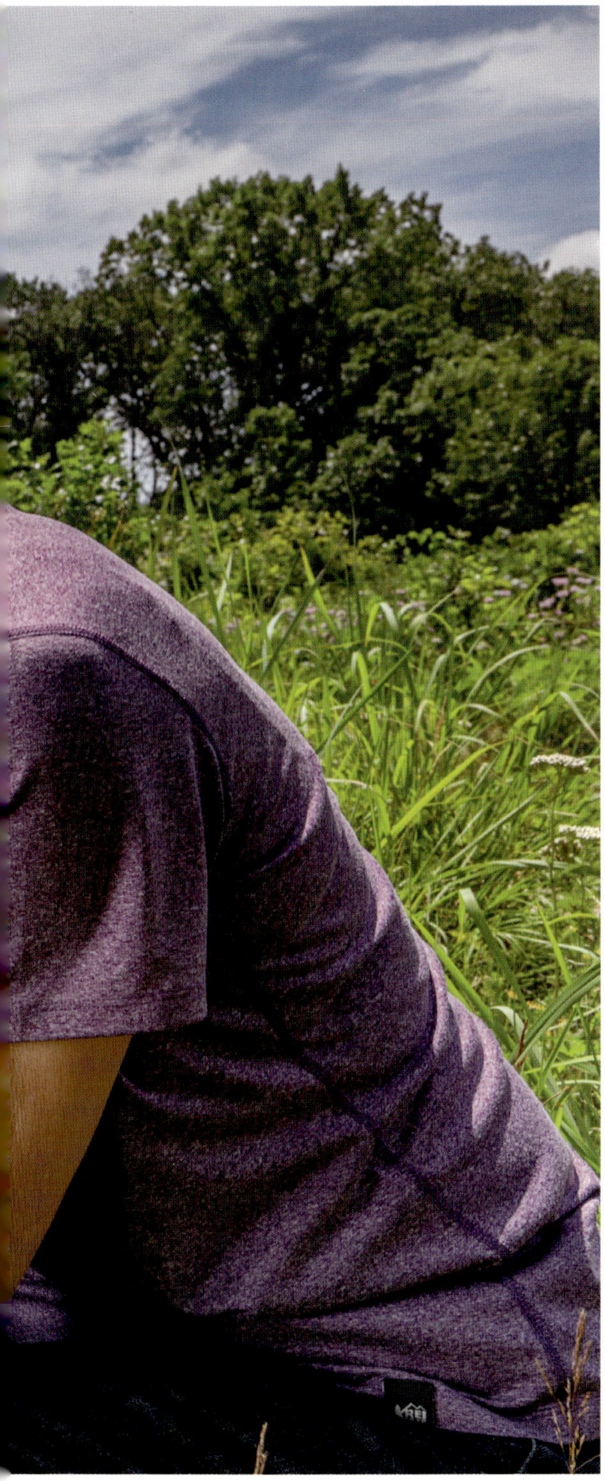

FIELD NOTES FROM

KAREN OBERHAUSER

DRIVEN BY CUES FROM THE SUN, monarchs leave their Mexican overwintering sites beginning in early March and fly northward. After spending the first several months of their adult lives in a state of reproductive dormancy, they break that dormancy before leaving their winter home. Mating and egg laying continue as they migrate north and encounter milkweed along the way.

During summer, monarchs in the eastern migratory population fan out to cover a large area across the northeastern quarter of the United States and southeastern Canada. Three generations are born from May through September, before the final generation begins the migration southward in late summer. Nonmigratory individuals live two to six weeks as adults, mating and laying eggs for all but the first few days.

Female monarchs probably lay between 400 and 600 eggs during their lifetime on most of the more than 100 North American milkweed species. The caterpillars consume leaves from these host plants. They grow for nine to 15 days, molting—or shedding their skin—five times, from the tiny first instar caterpillars that emerge from eggs the size of a pinhead to the large fifth instars about 2,000 times heavier and almost the size of your little finger. The fifth and final molt, also called pupation, results in a jade-colored pupa dotted with gold spots.

Monarchs gain some protection from predators throughout their lives by sequestering toxins from the milkweed they consumed as larvae. But this protection is far from perfect; only about 5 percent of eggs survive to adulthood. Most monarch caterpillars are eaten by invertebrate predators—like ants, spiders, wasps, and true bugs—before they even pupate.

For over four decades, my students and I carried out research designed to clarify monarchs' annual migratory cycle, especially the spring and summer portions. We worked in Mexico, studying the changes that occur before monarchs head north, and in Minnesota,

Wisconsin, and other U.S. states studying their host plants, mating and egg production, caterpillar growth, and predators and parasites.

While we used traditional lab and field techniques, we also engaged hundreds of citizen science volunteers. In 1996, led by graduate student Michelle Prysby, we developed the Monarch Larva Monitoring Project (MLMP), designed to track monarchs' spring and summer breeding. MLMP volunteers monitor milkweed patches in gardens, along roadsides, and in native and restored prairies, searching the plants for eggs and caterpillars. Their data allow us to track monarch densities and survival—making comparisons from year to year, across site types, and across locations.

I've also worked closely with Journey North, a citizen science project founded in 1995 by Elizabeth Howard and now run by Monarch Joint Venture. Journey North volunteers submit reports of the first adult monarch they see in a year, of eggs and larvae, and of milkweed—essentially any sighting of monarchs or milkweed is fair game. And Project Monarch Health volunteers use a simple, nondestructive technique to assess infection by a monarch parasite with a tongue twister of a name: *Ophryocystis elektroscirrha,* or OE for short. OE is a single-celled protozoan whose life cycle intersects with the monarch's. Eaten by the larva, it grows inside the developing pupa. Depending on the level of infection, the parasite kills or weakens the monarch. Project Monarch Health builds on research conducted in the 1990s in my lab by graduate student Sonia Altizer, who now runs it out of her own lab at the University of Georgia. With these and other monarch citizen science projects, we're learning how monarchs are doing and how we can best help them.

During their spring and summer breeding period, monarchs require intact habitat that includes milkweed for caterpillars and nectar for adults. The nectar can come from milkweed, but also from dozens of other flower species; what's important is that it's available throughout the entire time that monarchs are present. Weather also has a big impact on monarch numbers. Unfortunately, both the availability of quality habitat and the weather are changing in ways that present risks for monarchs, and much of my recent work focuses on how science can inform actions to mitigate these risks.

In collaboration with John Pleasants at Iowa State University and many other scientists, I've shown that a key cause of monarch decline in the early 21st century was the loss of breeding habitat in the upper Midwest due to the adoption of genetically modified row crops, mainly corn and soybeans. Throughout much of the 20th century, milkweed was common in row crop fields, and monarchs used these plants for egg laying. In fact, most monarchs that migrated to Mexico then probably originated in agricultural fields. However, beginning in the late 1990s, farmers began growing genetically modified crops: first, crops specifically resistant to the herbicide glyphosate, and then a series of crops resistant to other herbicides. These developments allowed farmers to treat their fields with herbicides after plants emerged, and they killed everything but the genetically modified crops. Milkweed disappeared from corn and soybean fields as adoption rates increased, reaching 90 percent by 2014.

I've also worked with several teams of scientists to use citizen science data to document strong associations between monarch numbers and weather conditions. We've found that hot and dry or cold and wet conditions result in fewer monarchs. Associations between weather and monarch numbers could be the result of direct impacts on monarchs themselves or impacts on plant quality and quantity. Insecticide use, in agriculture but also in lawns and parks, also poses problems for monarchs. Research that I and others have done has shown that many insecticides kill monarchs at concentrations used in the field.

As a conservation biologist, I'm concerned with not only why monarch numbers are declining but also what we can do to help. Monarchs' basic habitat needs are clear; increasing the amount and quality of habitat will result in more robust populations. Habitat sets an upper limit on how many monarchs are produced in a given year, and weather conditions determine if that limit is achieved. Widespread habitat makes it easier for females to find host plants for egg laying and flowers for nectar, enabling them to lay more eggs and live longer.

I often say that my field, conservation biology, is a science of hope, and there is reason to be hopeful about the eastern North American monarch population. The fact that monarchs cover a large area during the breeding and migratory phases of their annual cycle makes them more resilient than species with smaller ranges. Because female monarchs can lay large numbers of eggs, the population has the potential to rebound relatively quickly when weather conditions are good.

Perhaps most important, people who care can make a difference for monarchs, through our continued monitoring of the population, through our continent-wide efforts to protect, restore, and enhance their habitat, and by convincing as many people as possible to join us in caring.

Karen Oberhauser is a conservation biologist who has been studying monarch butterfly ecology for more than 40 years, developing projects that engage the public.

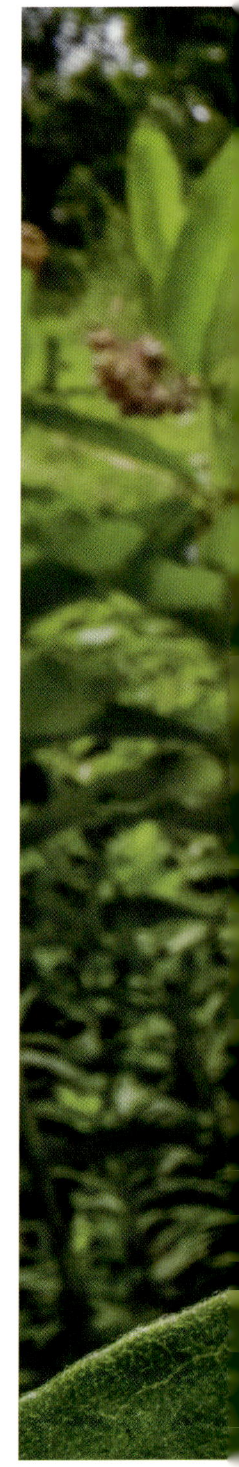

As volunteers with the Monarch Larva Monitoring Project inspect other plants, a monarch caterpillar curves along on a milkweed leaf—a tiny sign of success.

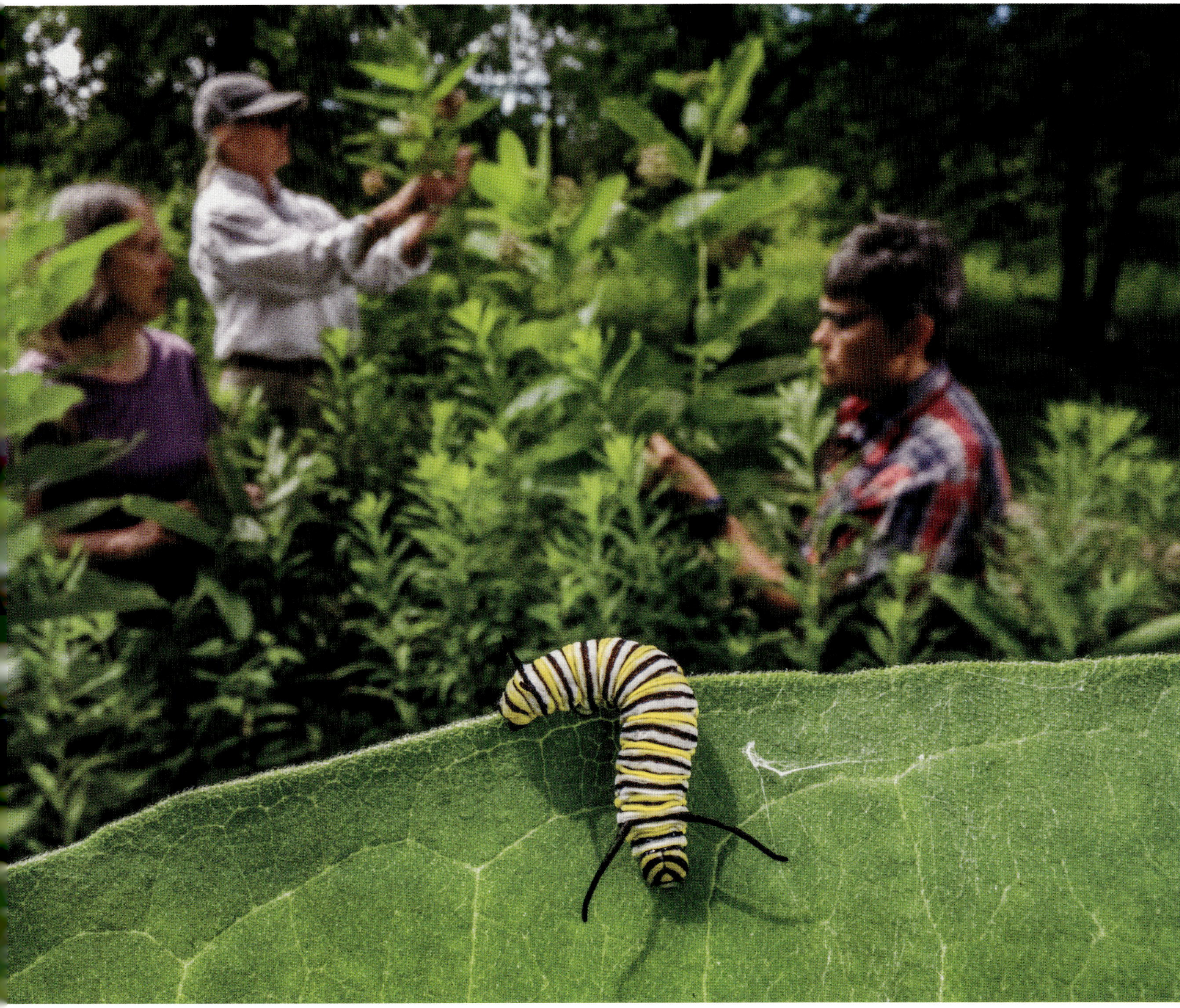

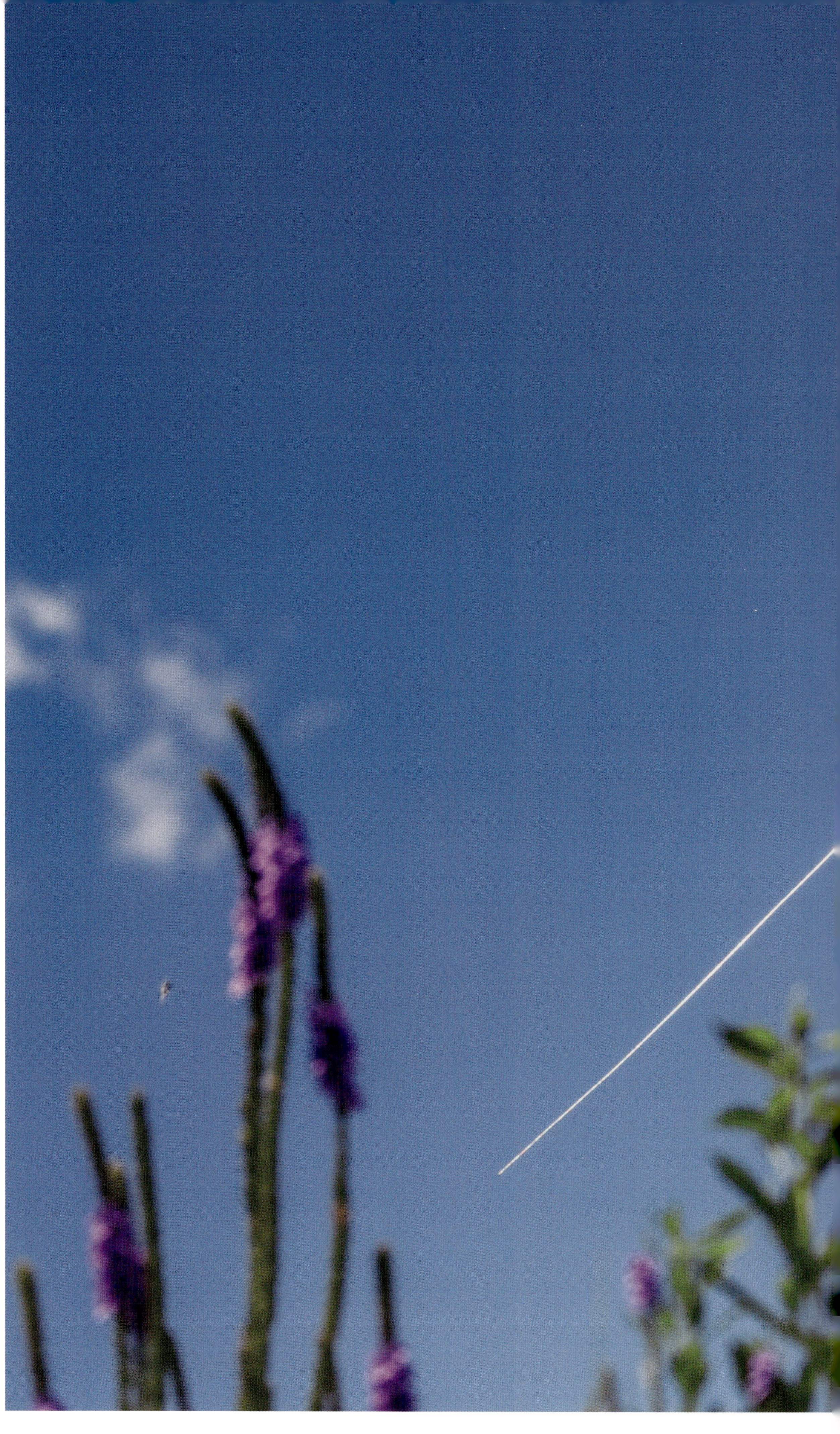

A female monarch forages for nectar near Ames, Iowa. The radio transmitter attached to her helps researchers study flight patterns and navigation, providing data that could guide future habitat restoration.

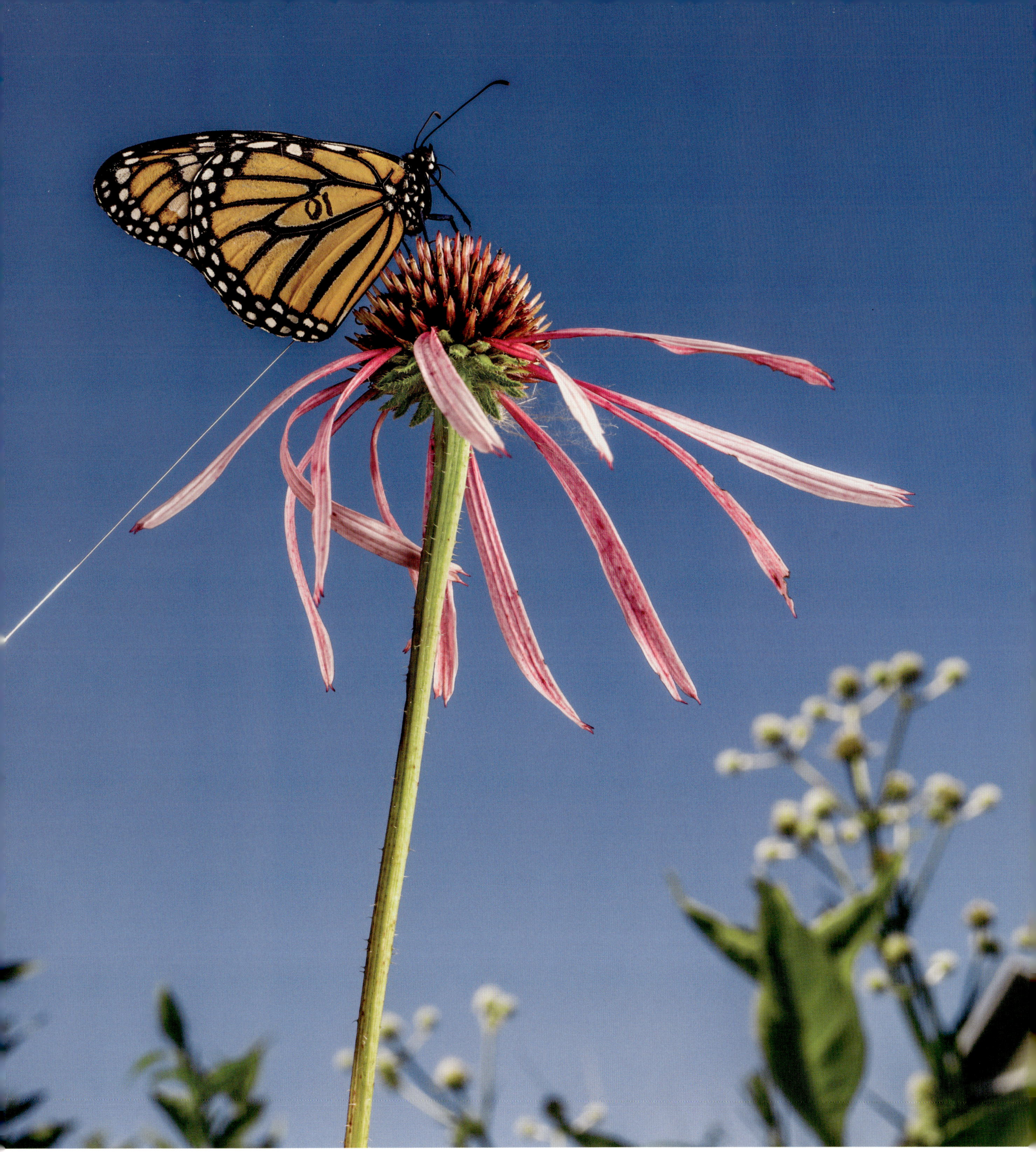
01

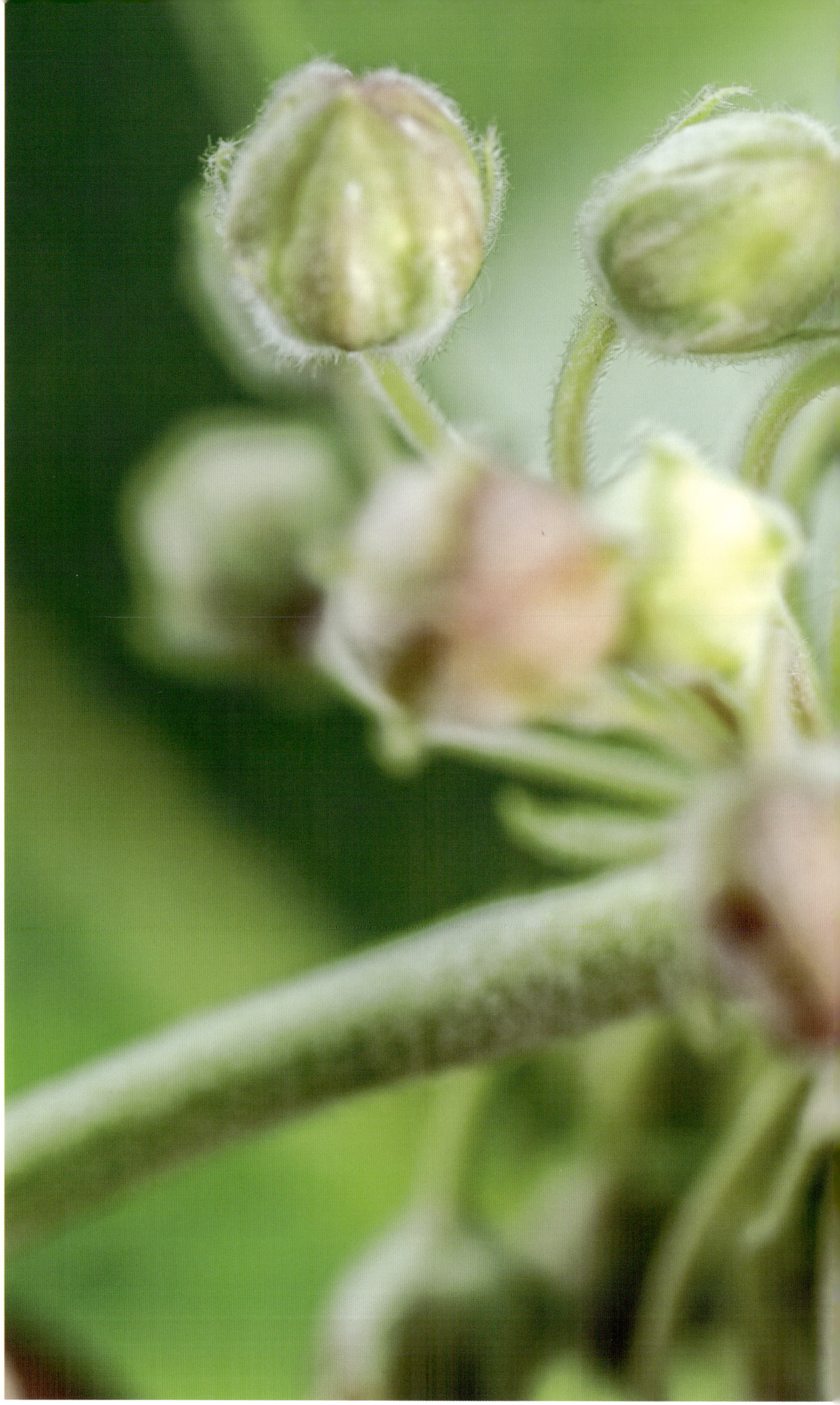

A monarch caterpillar rests on a milkweed flower just beginning to bloom.

PAGES 66–67: Gripping the leaf's central vein, a monarch caterpillar rests on a milkweed plant in Henderson, Minnesota.

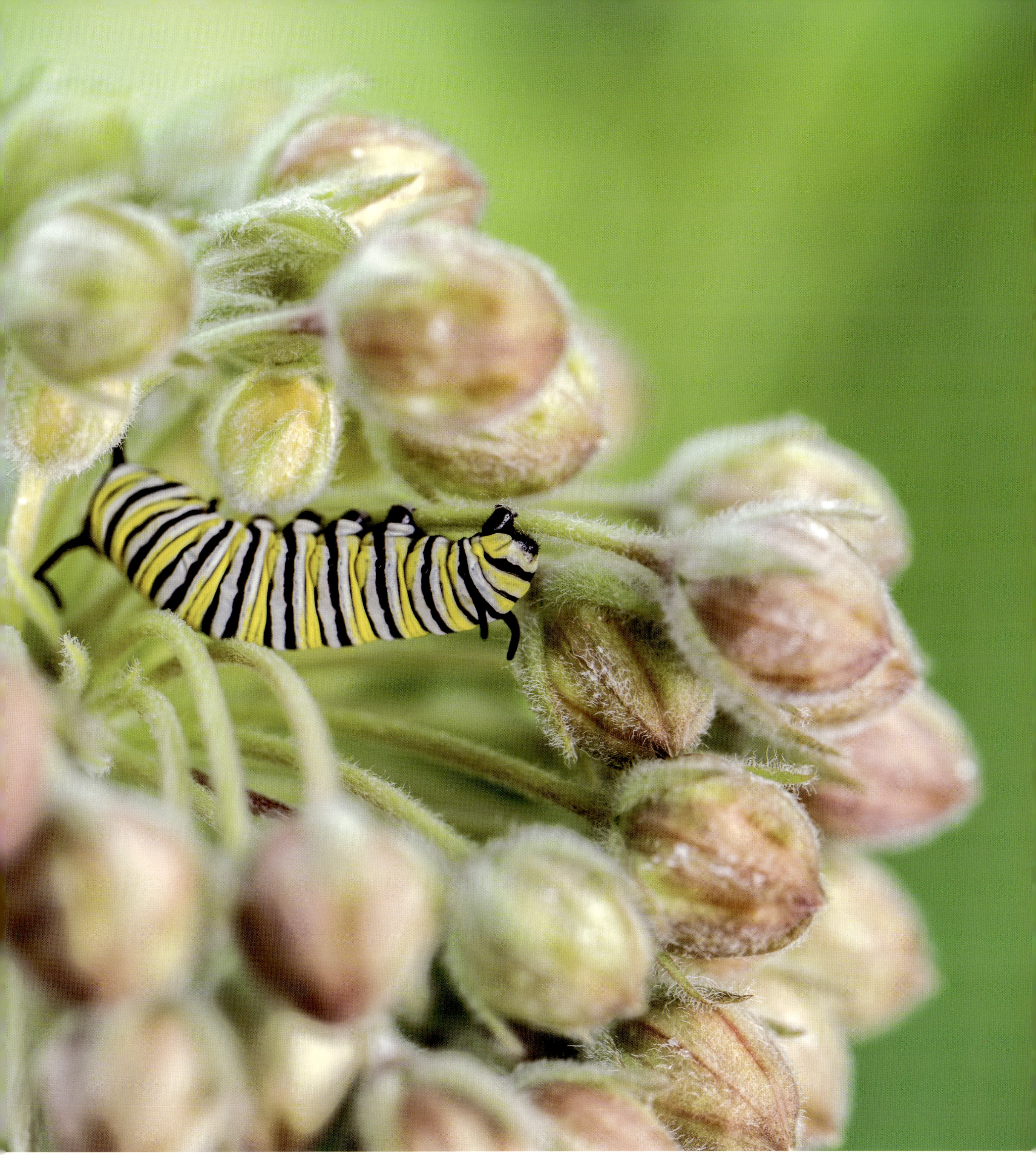

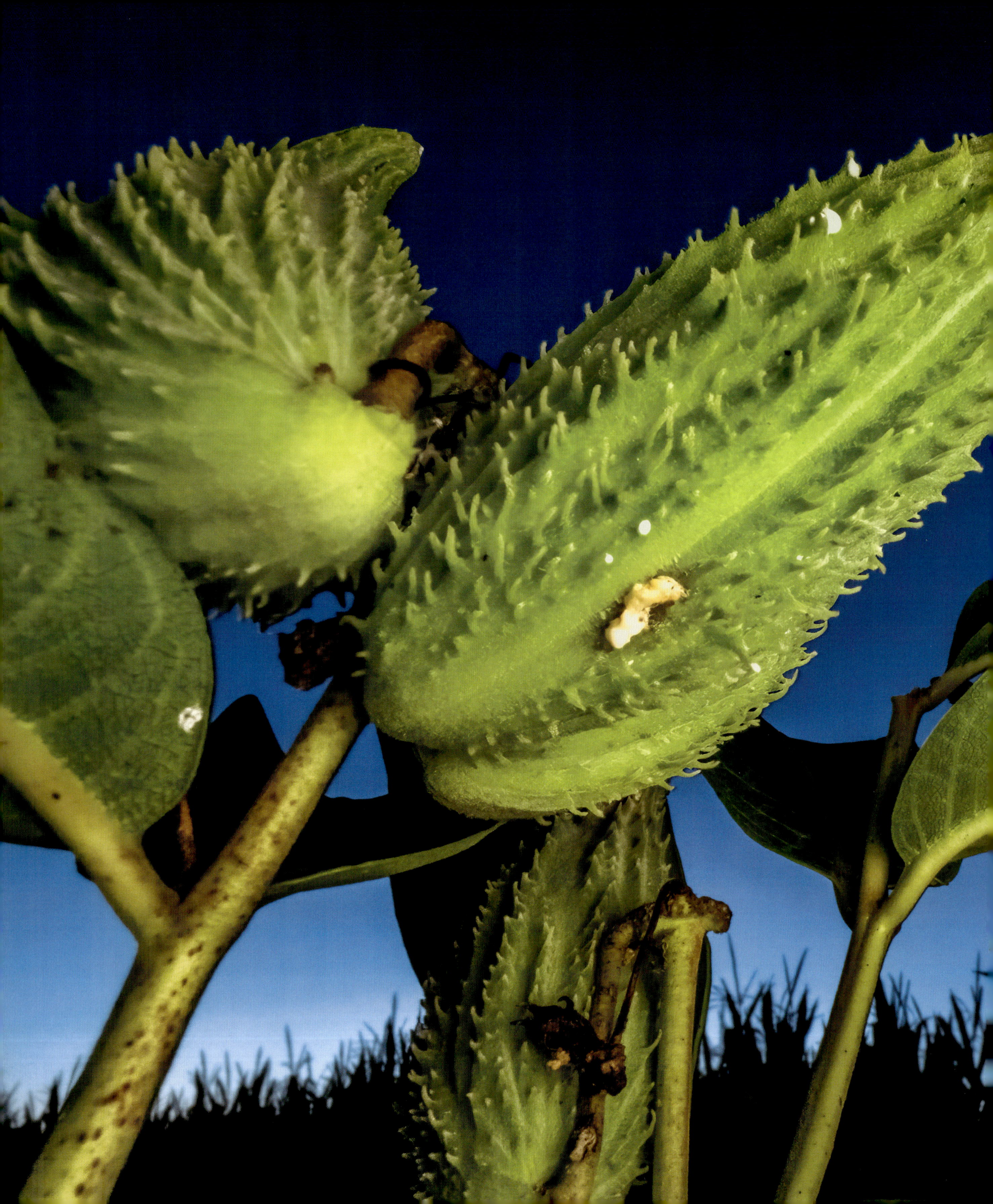

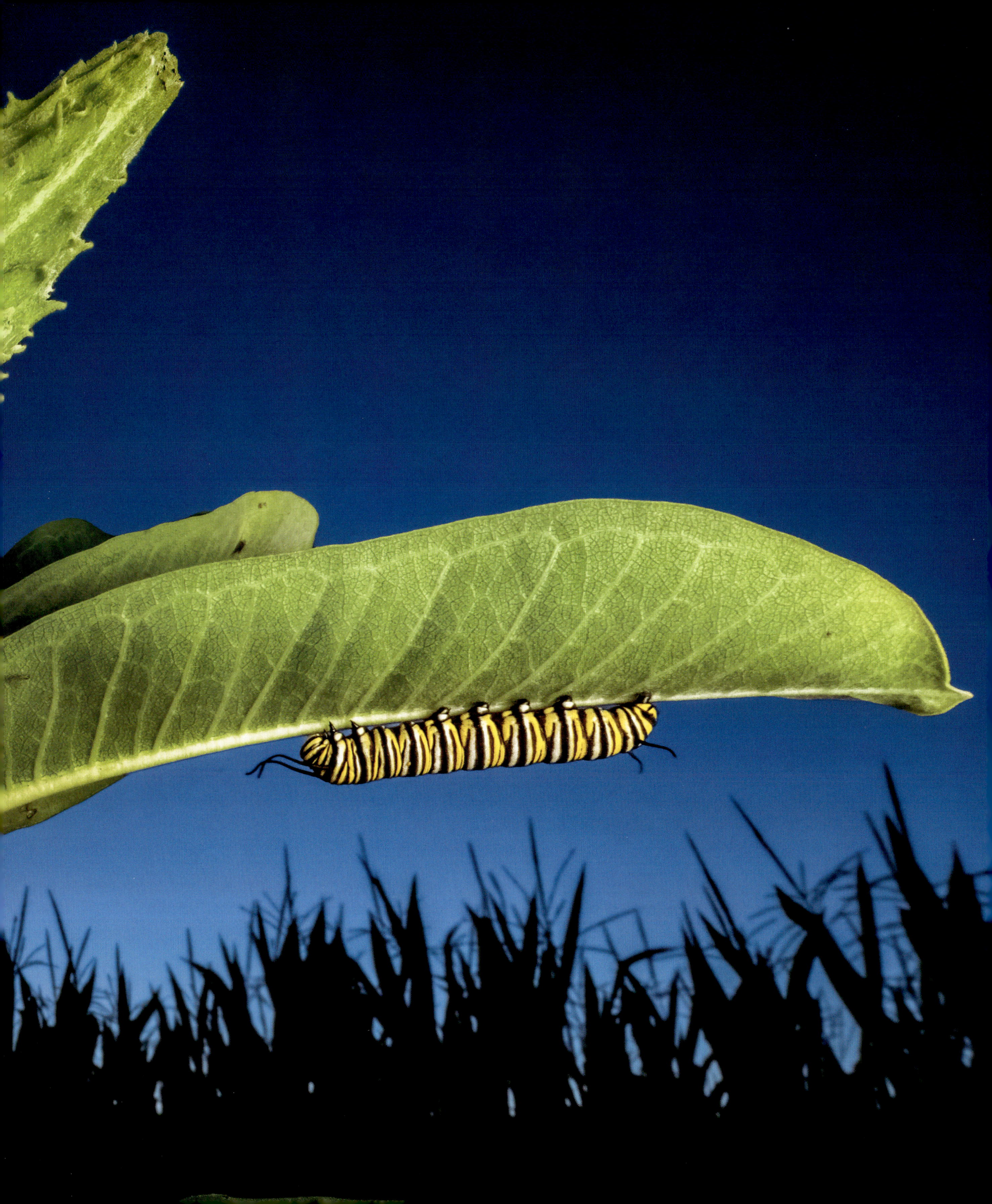

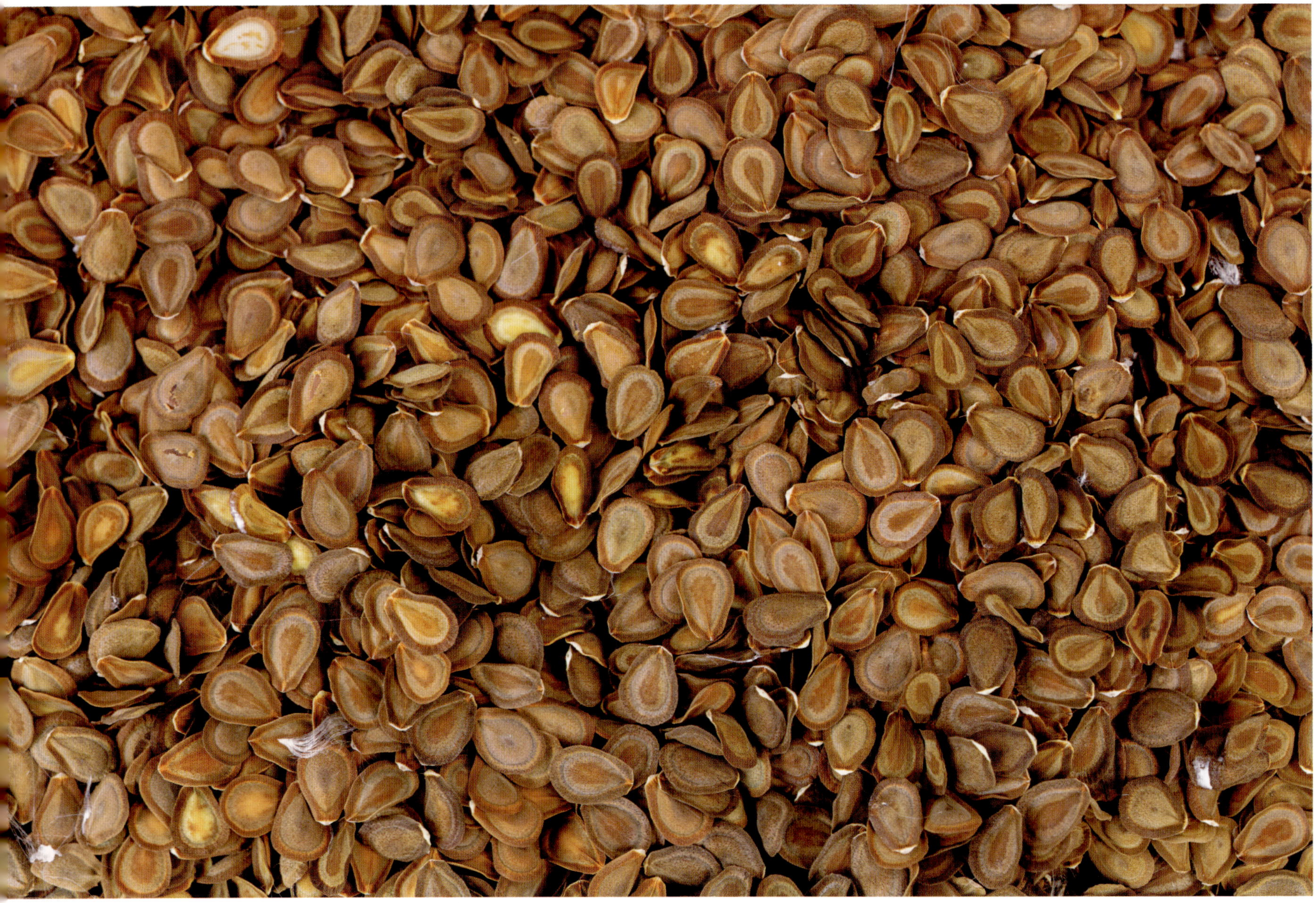

Volunteers collect milkweed seeds to support restoration efforts across North America. The seeds are cleaned, stored, and distributed to reestablish milkweed stands throughout the monarchs' breeding range.

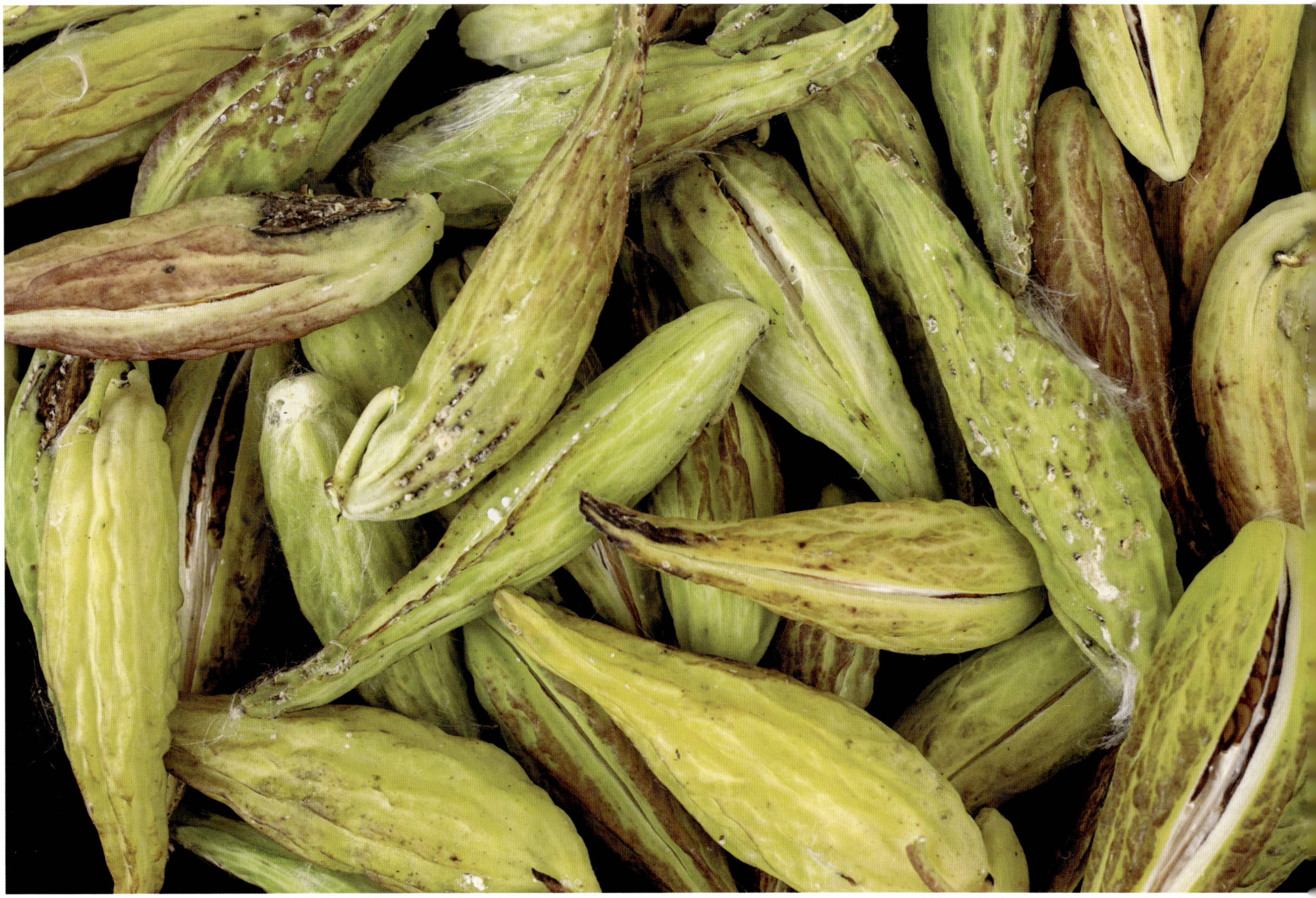

Mature milkweed pods split open and release silky seeds ready for collection. Each fall, Monarch Watch volunteers harvest these pods to expand vital monarch breeding and migration habitat across North America.

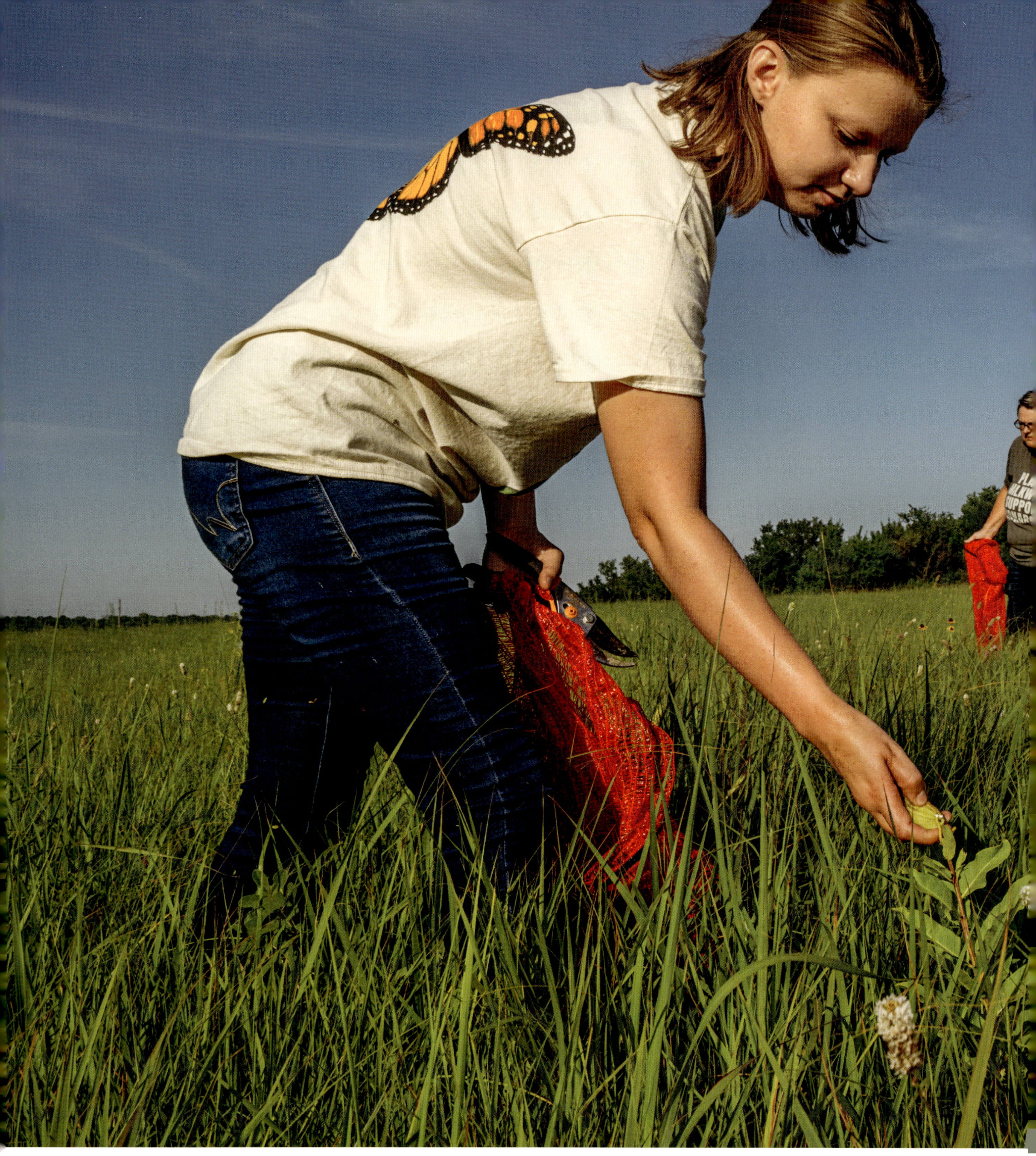

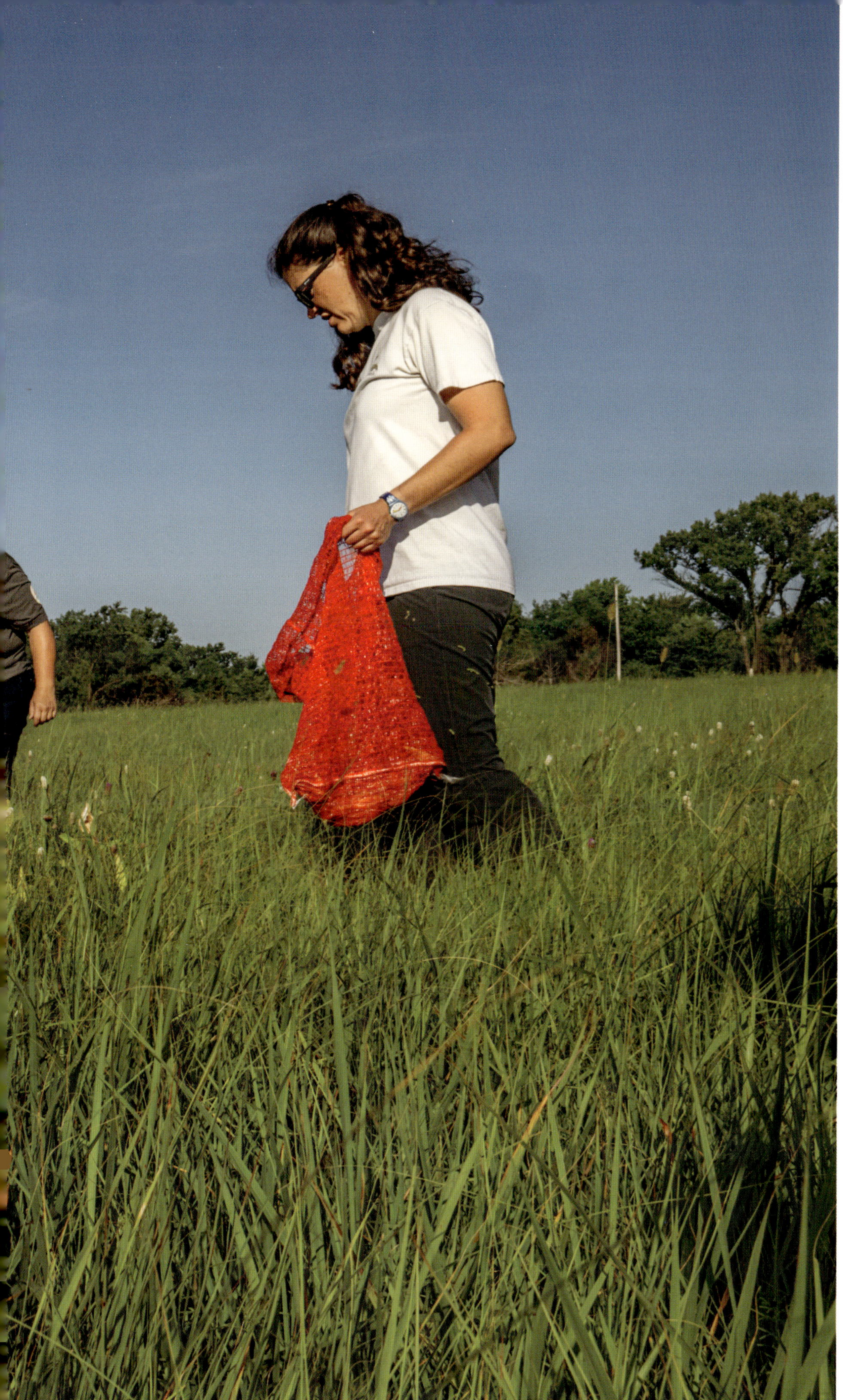

Hayley Allison (left) and fellow Monarch Watch workers collect milkweed seeds in a restored prairie near Lawrence, Kansas.

PAGES 72–73: While milkweed supports monarch caterpillars, mature butterflies find nectar on other native plants. Here, a female perches on blooms of goldenrod.

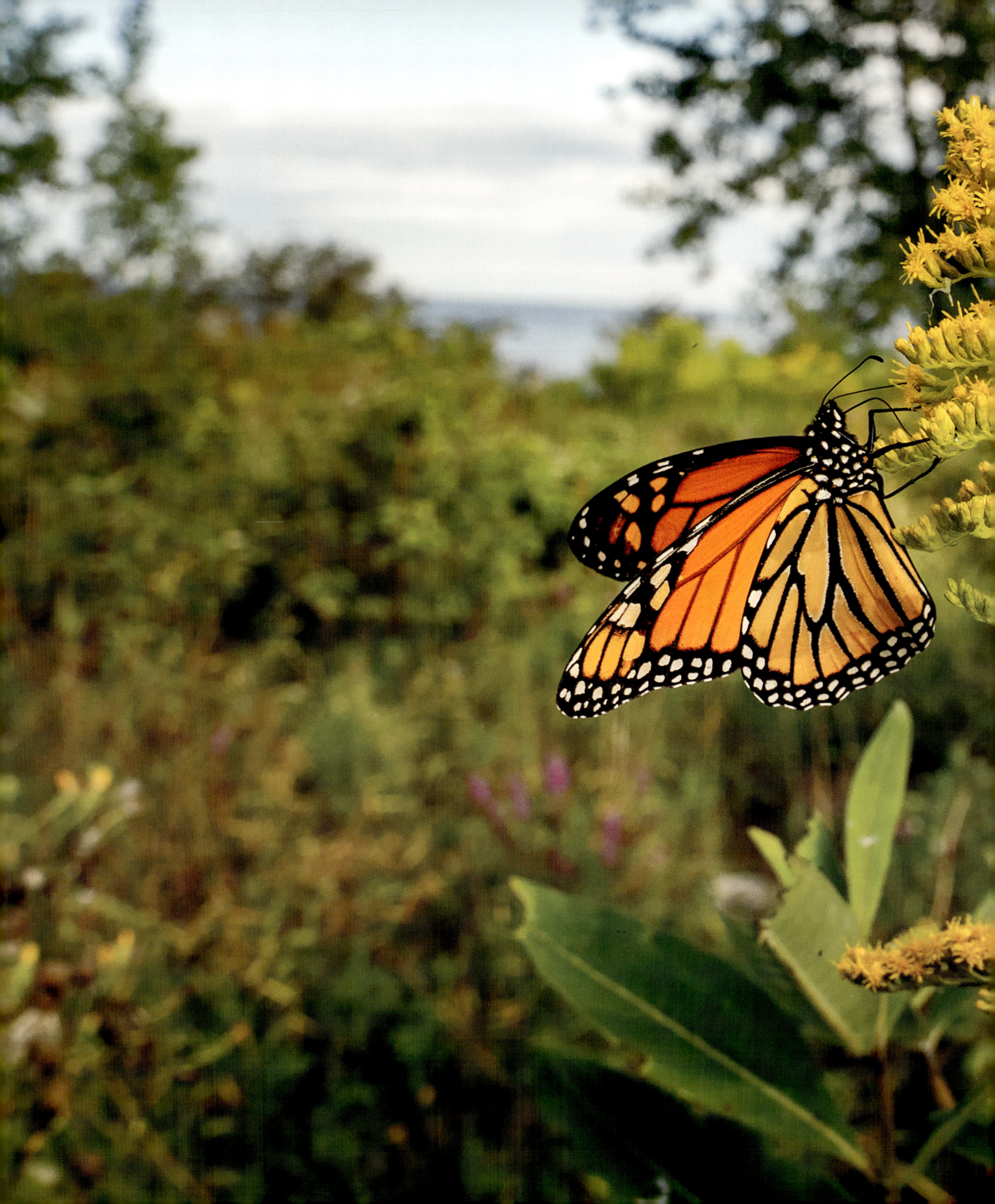

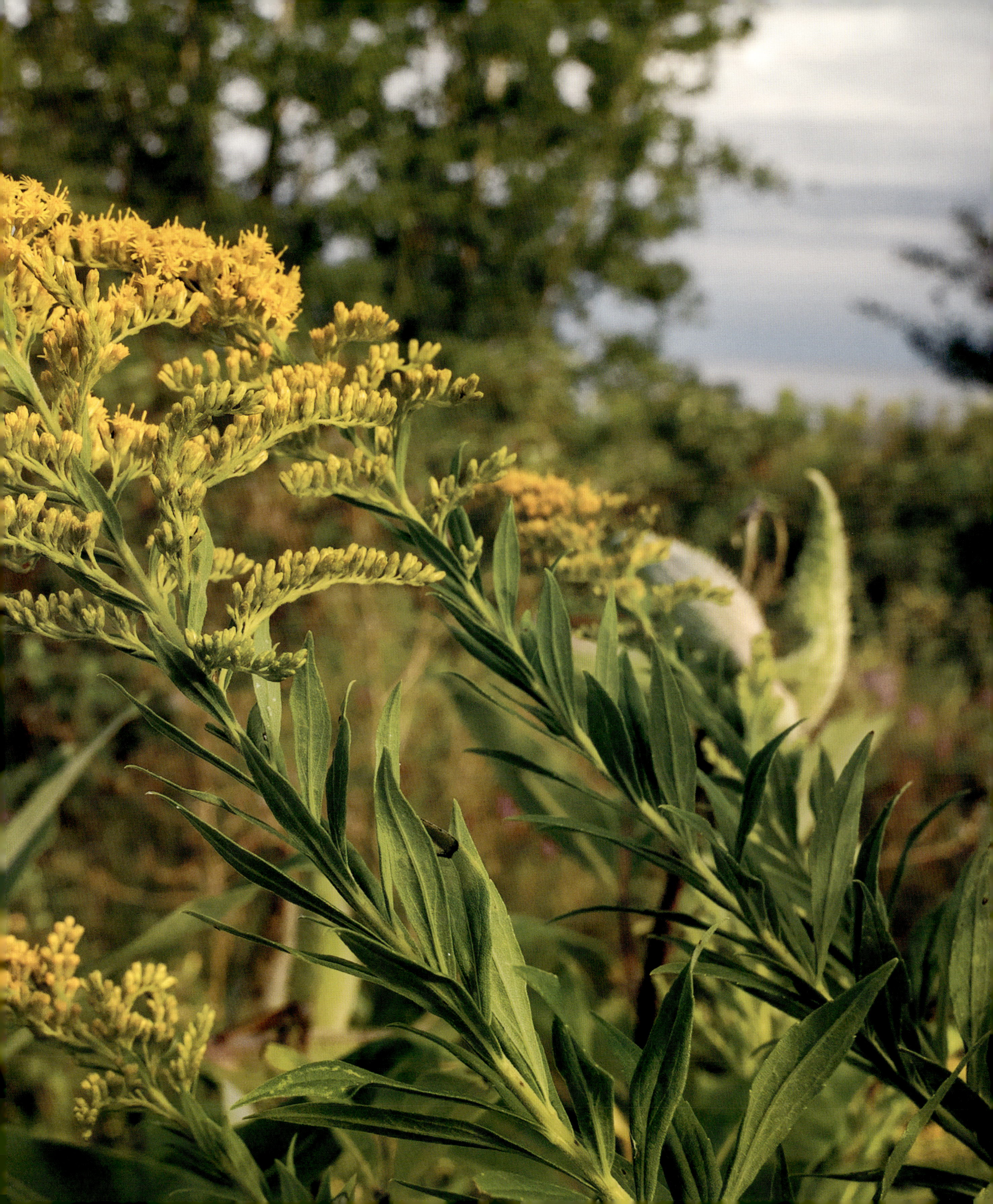

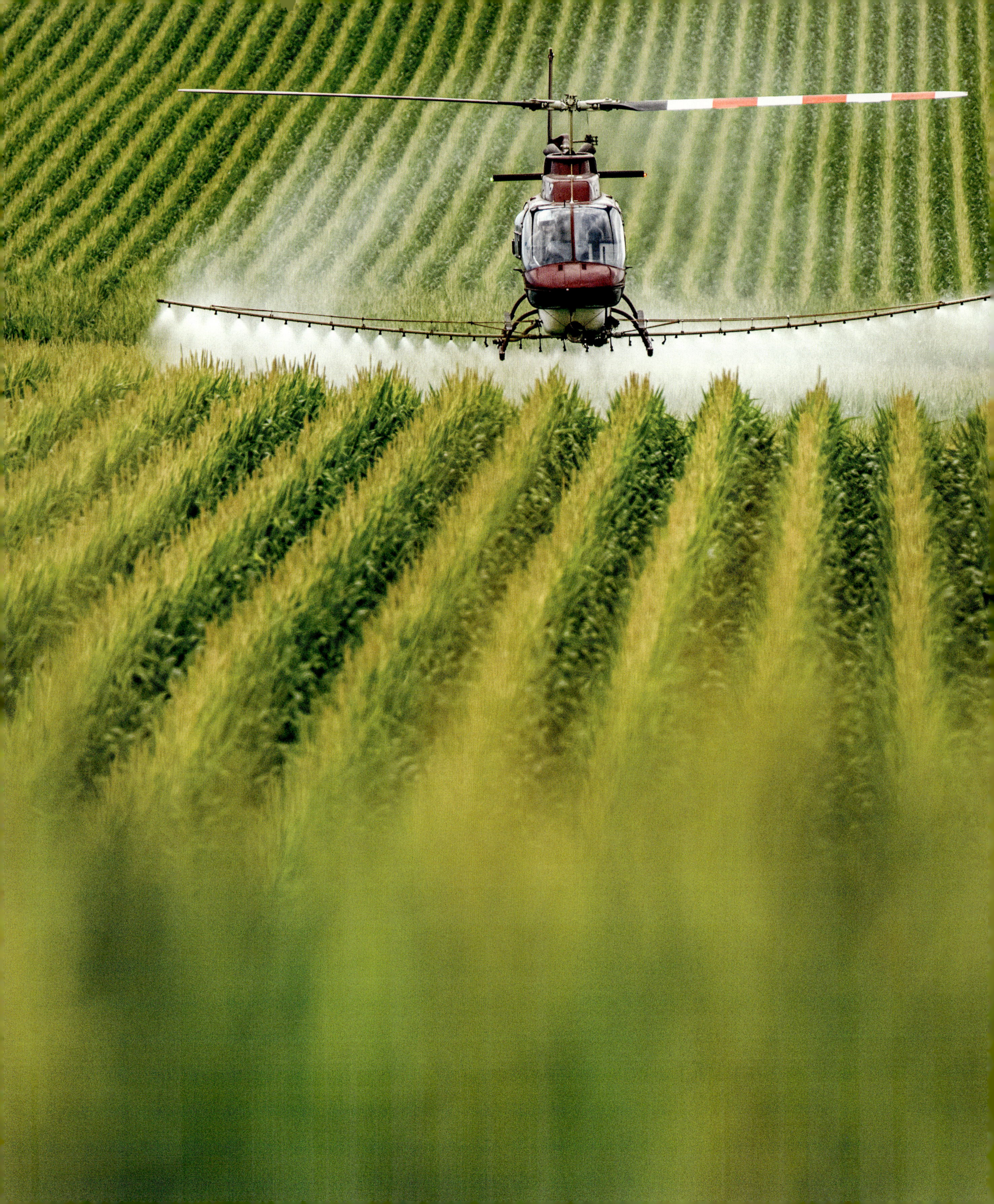

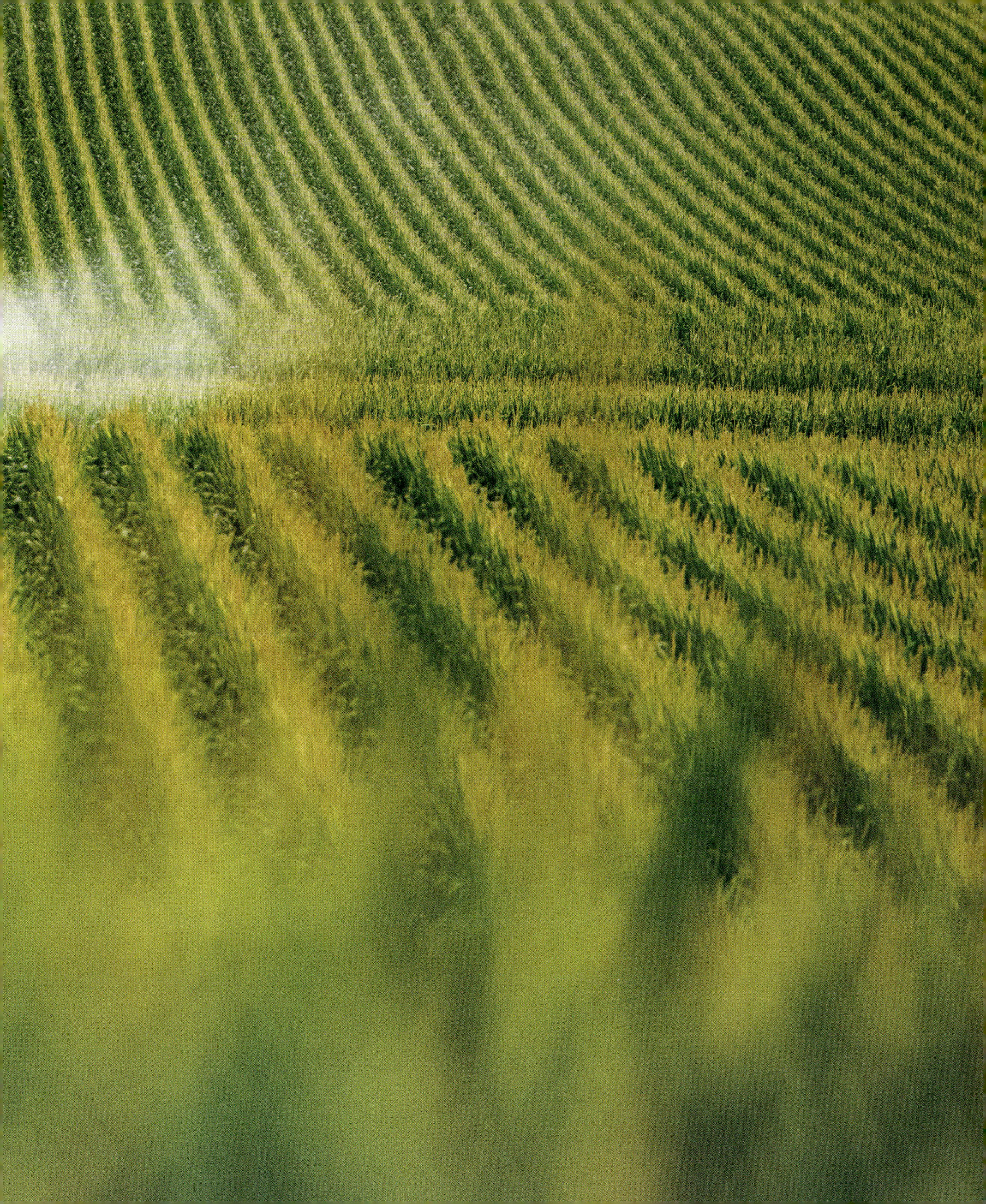

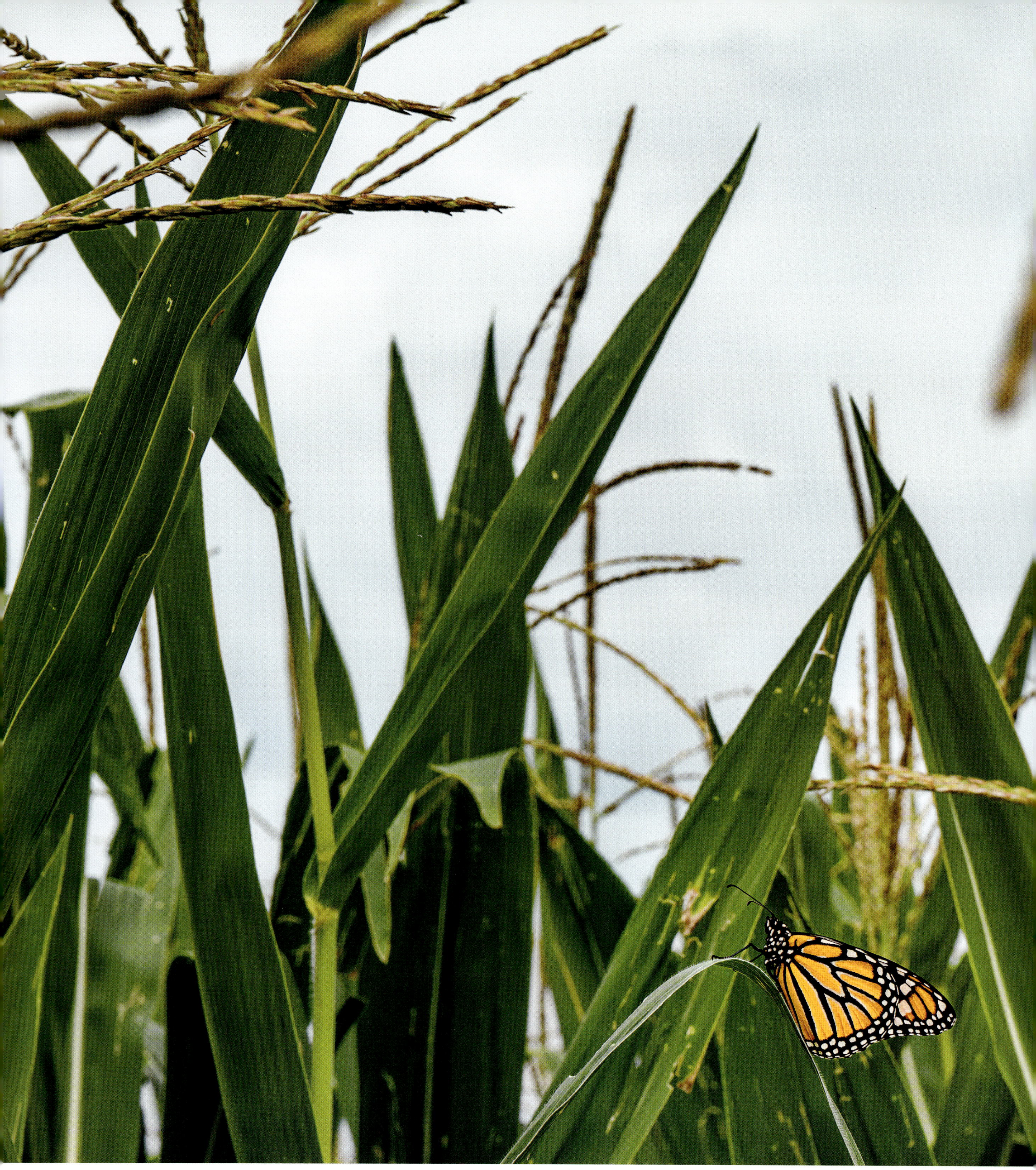

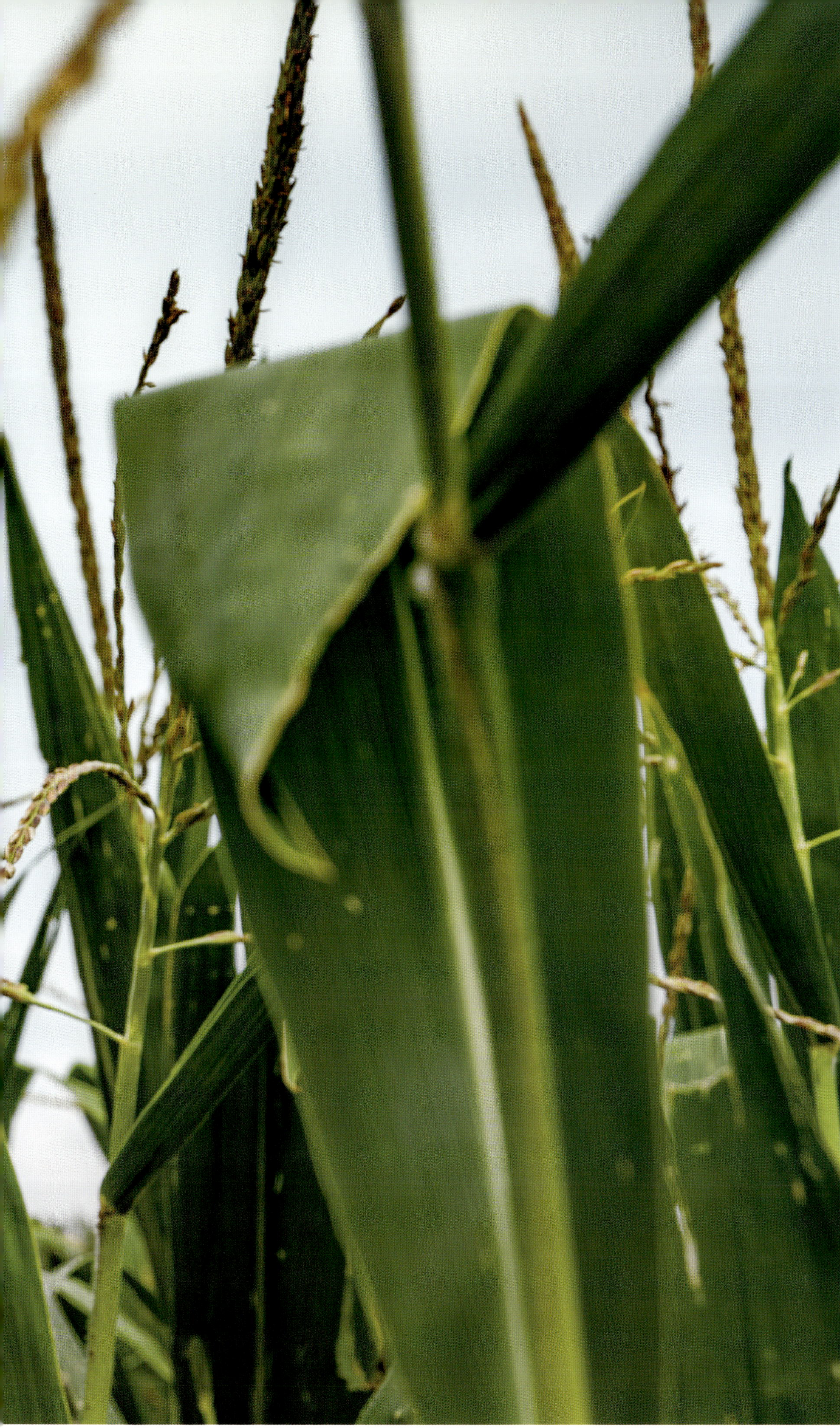

A monarch butterfly perches on corn growing near Henderson, Minnesota. Corn and soybean fields have replaced native prairies, and the widespread use of herbicides has drastically reduced milkweed growing across the Midwest—a major factor in the decline of monarchs.

PAGES 74–75: A helicopter sprays chemicals over cornfields near Osage, Iowa, a common practice of intensive row-crop farming across the Midwest.

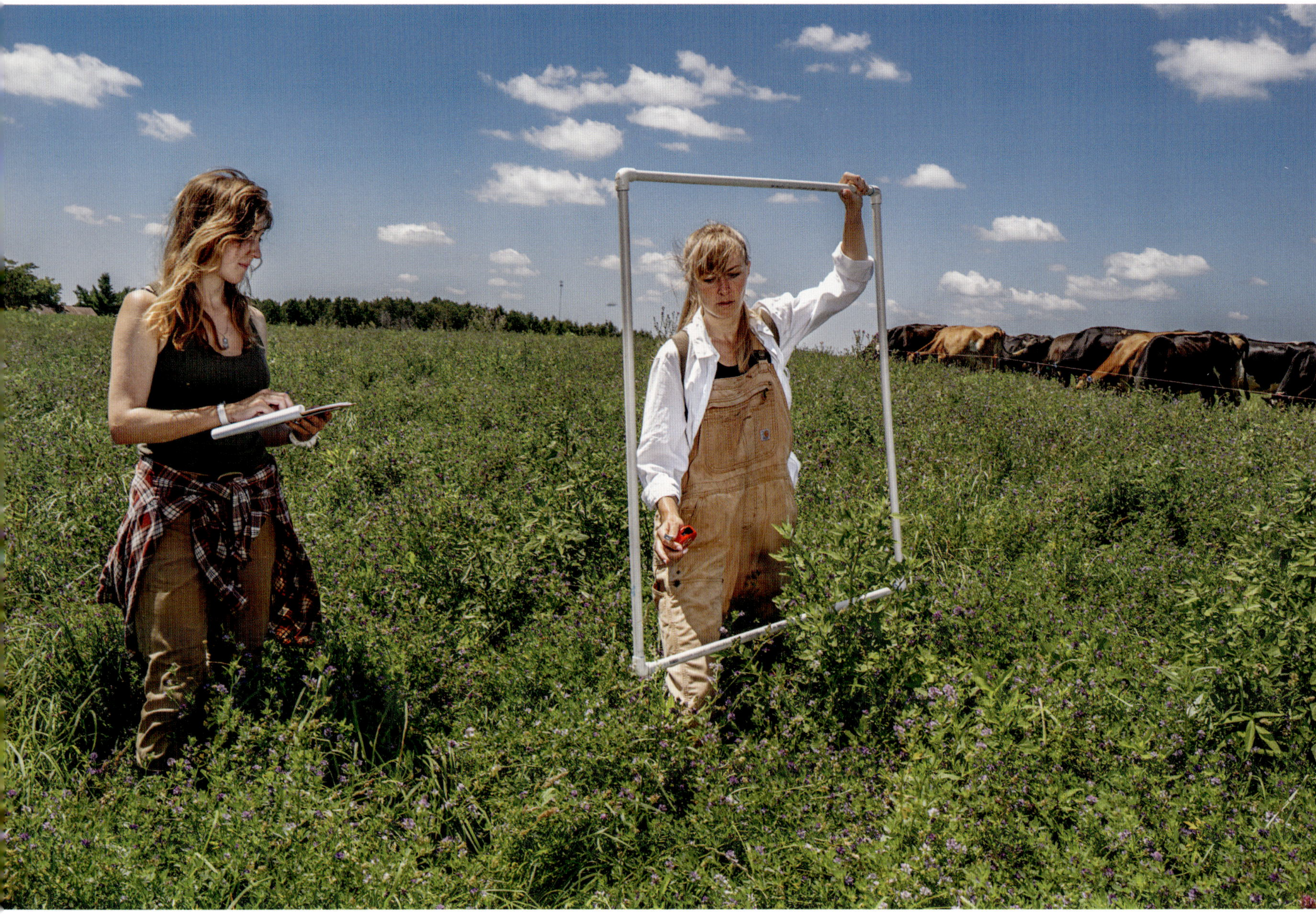

Using a one-square-meter plot frame, entomologist Skye Bruce (right) and field assistant Isabelle Labus measure grass and flower diversity at the Uplands Cheese farm in Dodgeville, Wisconsin, contributing to research on grazing, ranching, and butterfly interactions.

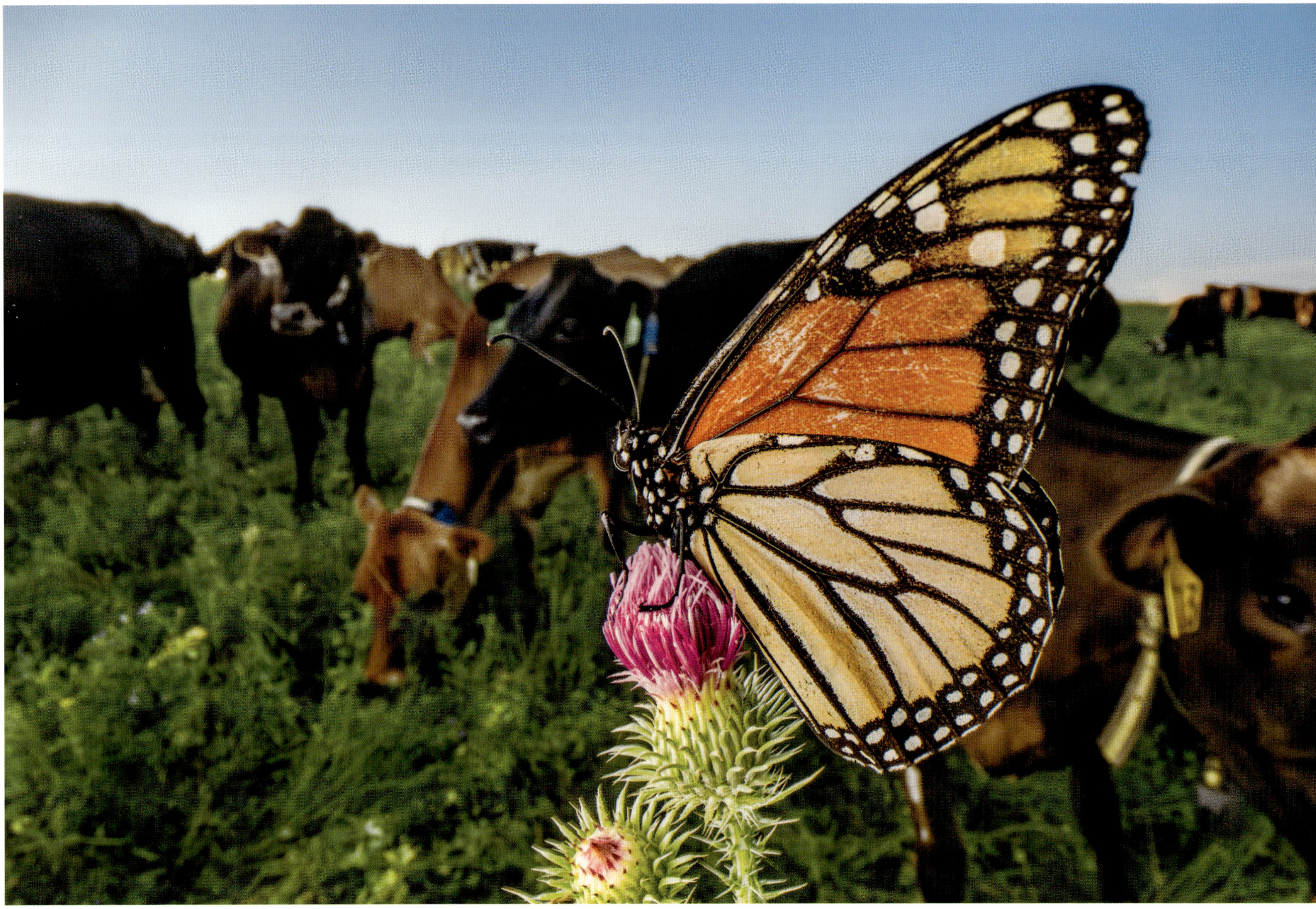

A monarch butterfly rests on a broad-winged thistle flower amid cattle grazing at the Uplands Cheese farm. As part of Bruce's research, butterflies such as this one are briefly captured to train students in accurate identification and then released.

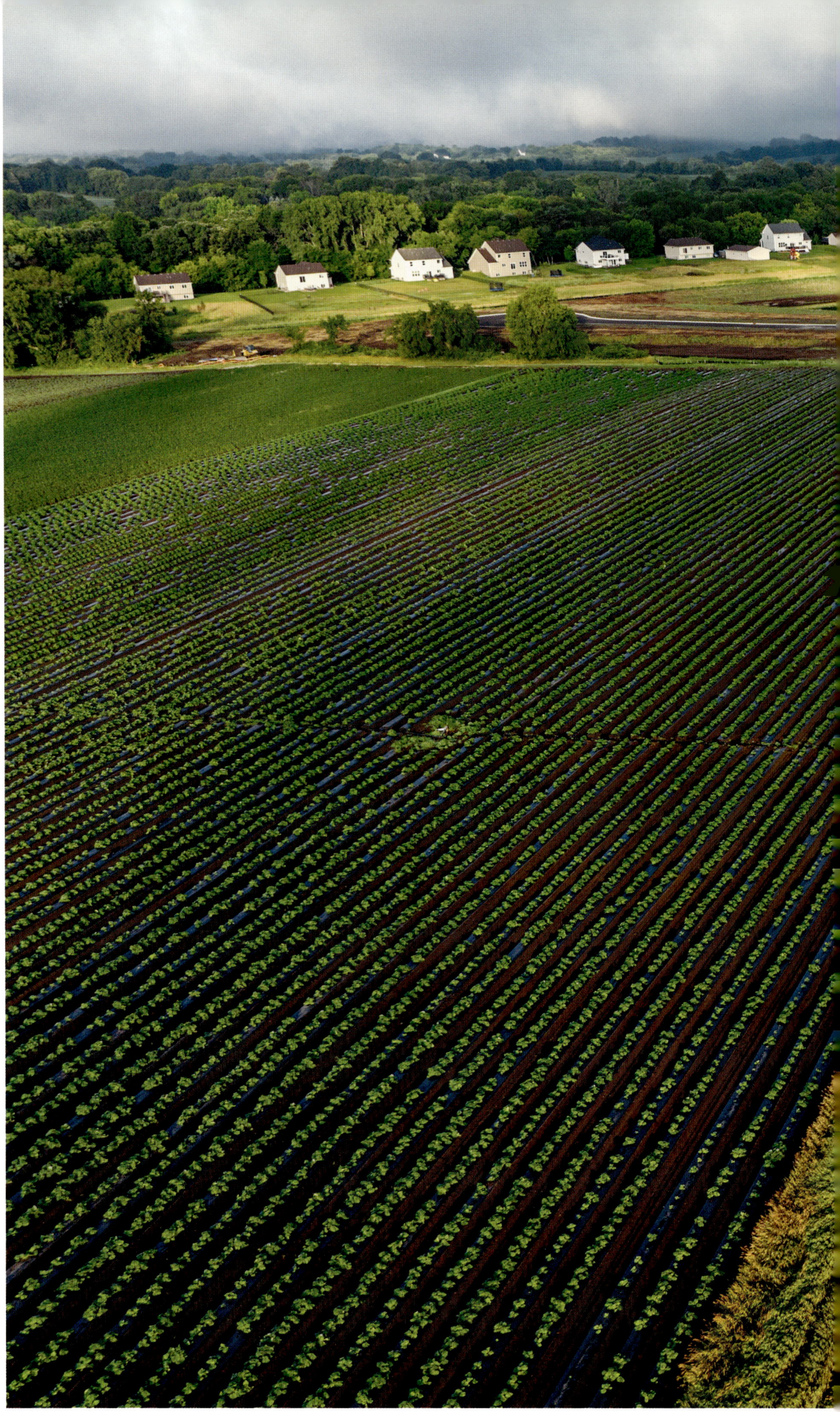

A solar installation operated by ENGIE, a renewable energy corporation, sits by fields and houses near Anoka, Minnesota. Pollinator-friendly vegetation planted beneath and around the panels provides habitat for monarchs and native bees amid a rapidly changing landscape of agriculture and urban growth.

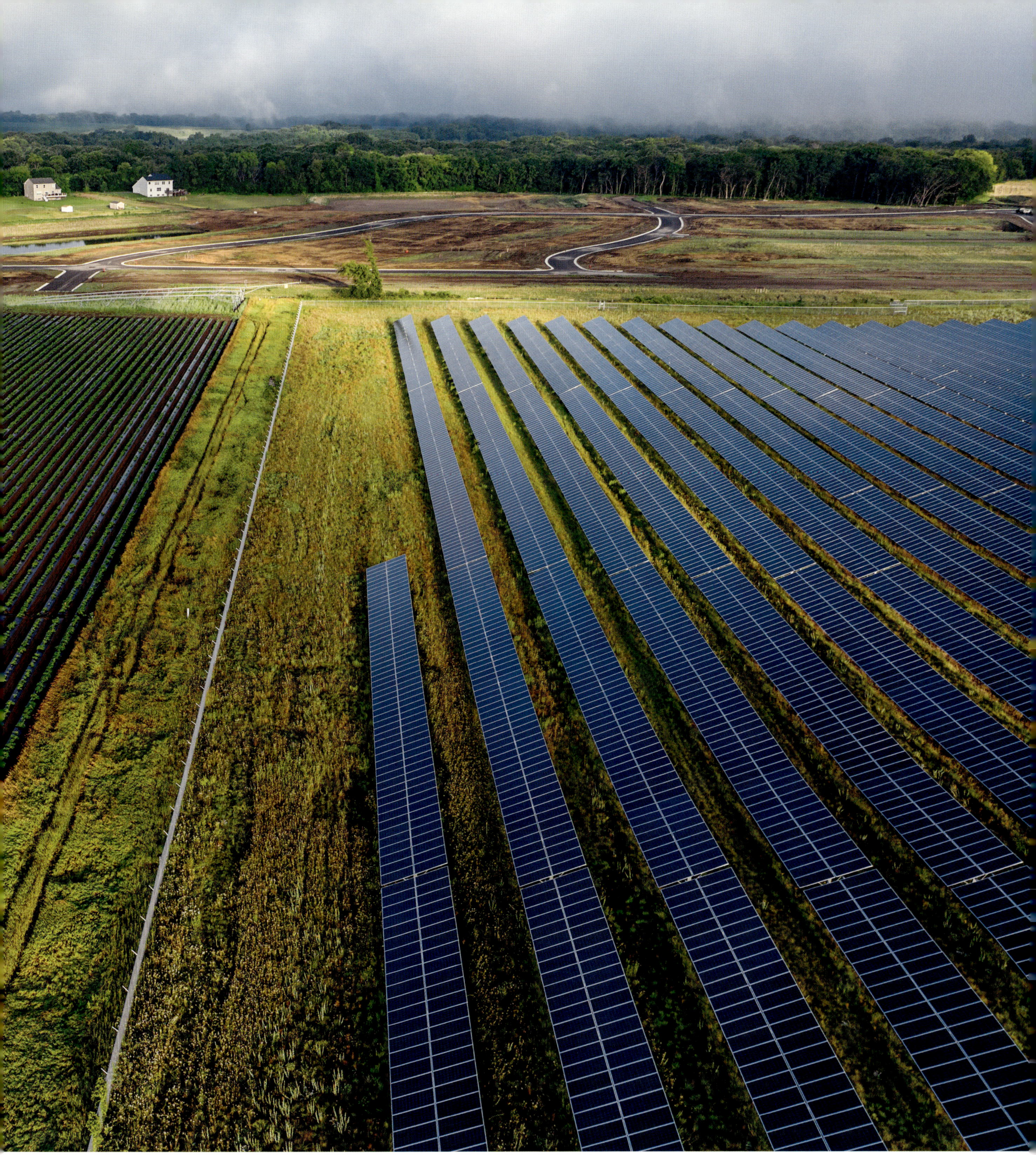

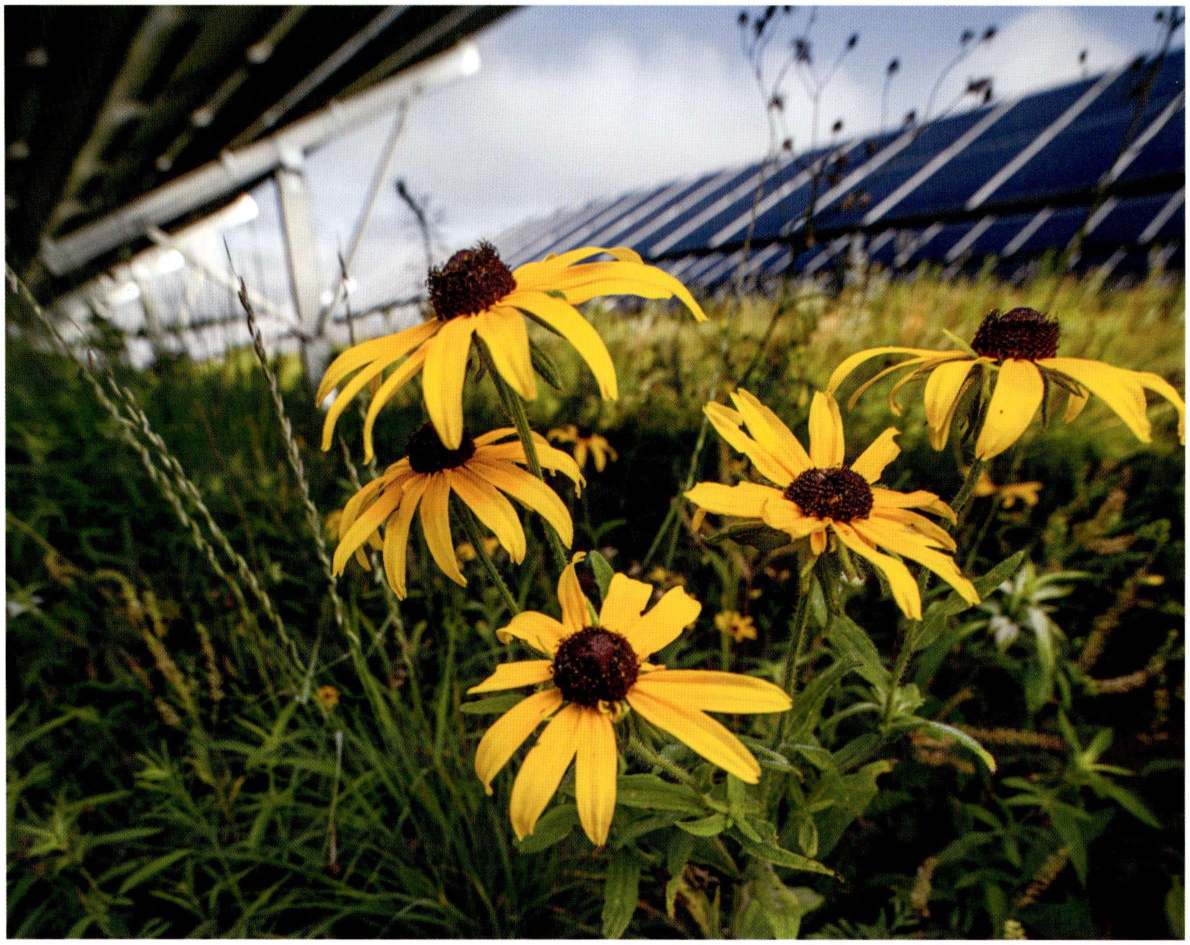

ABOVE AND OPPOSITE: Under and around ENGIE's solar installation, wildflowers bloom along with a mix of native grasses.

PAGES 84-85: Getting very close: A monarch dips into the bloom of a purple coneflower.

PAGES 86-87: Even closer: The scales of a monarch butterfly wing provide the color that signals the toxicity of the species to predators.

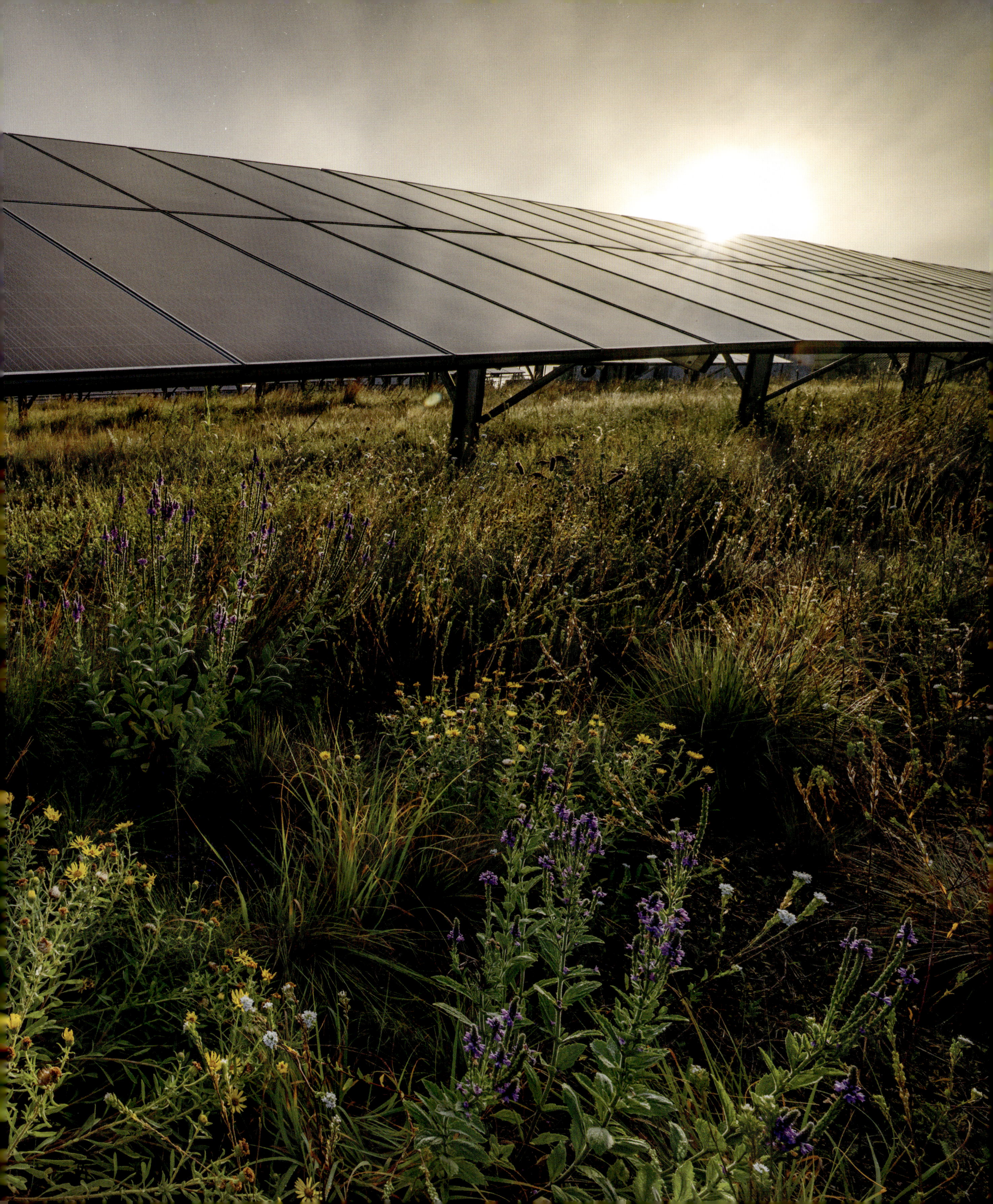

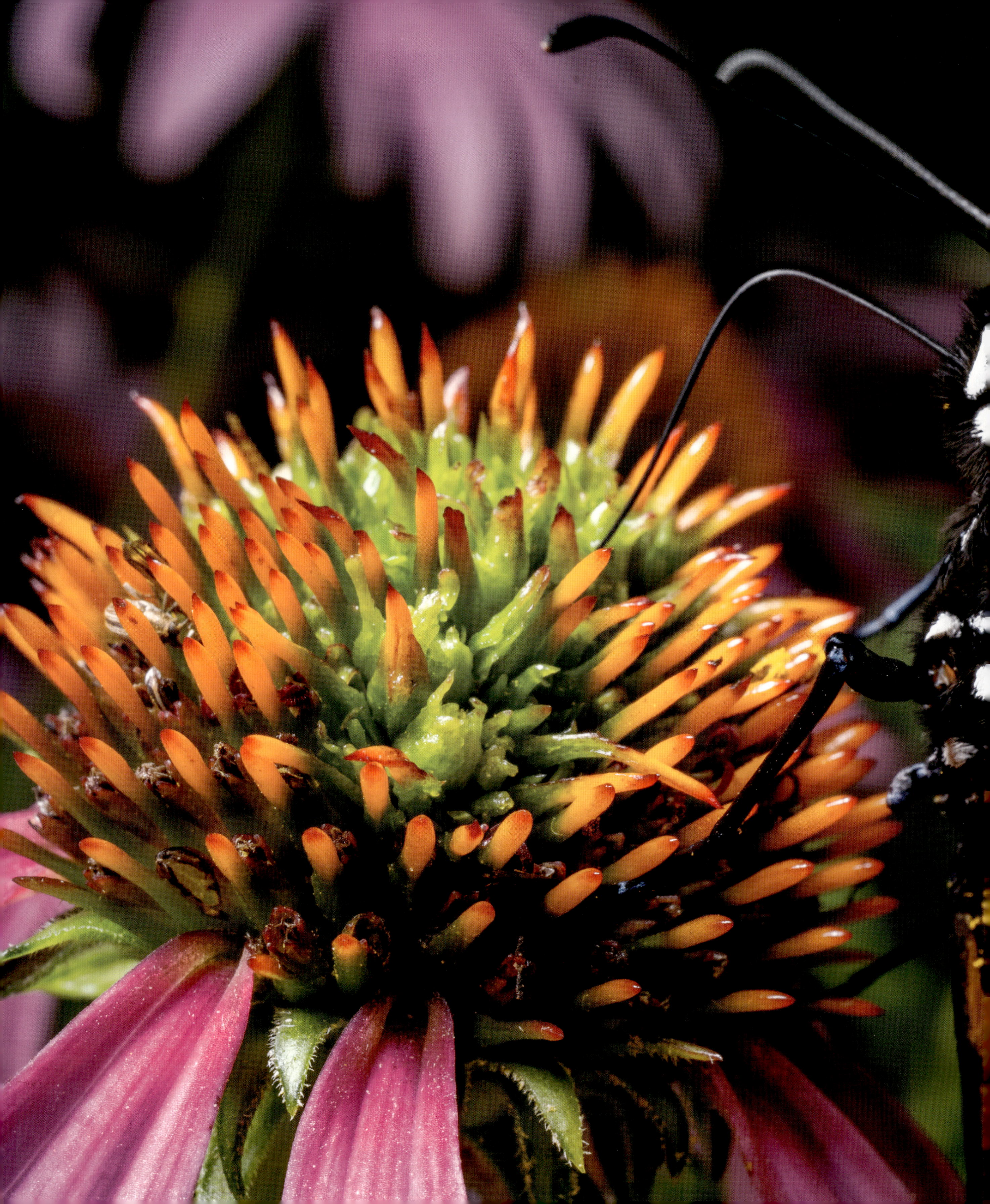

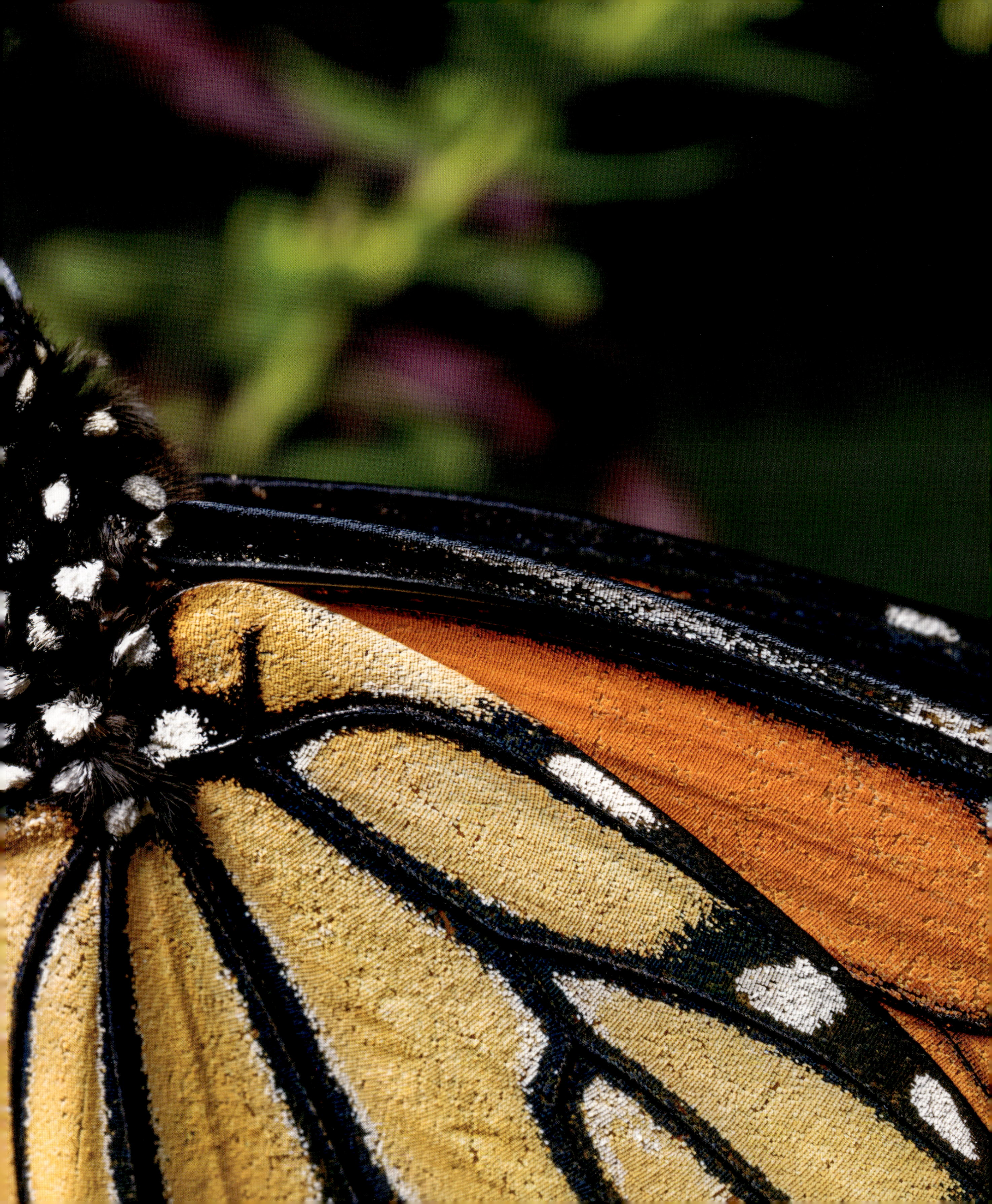

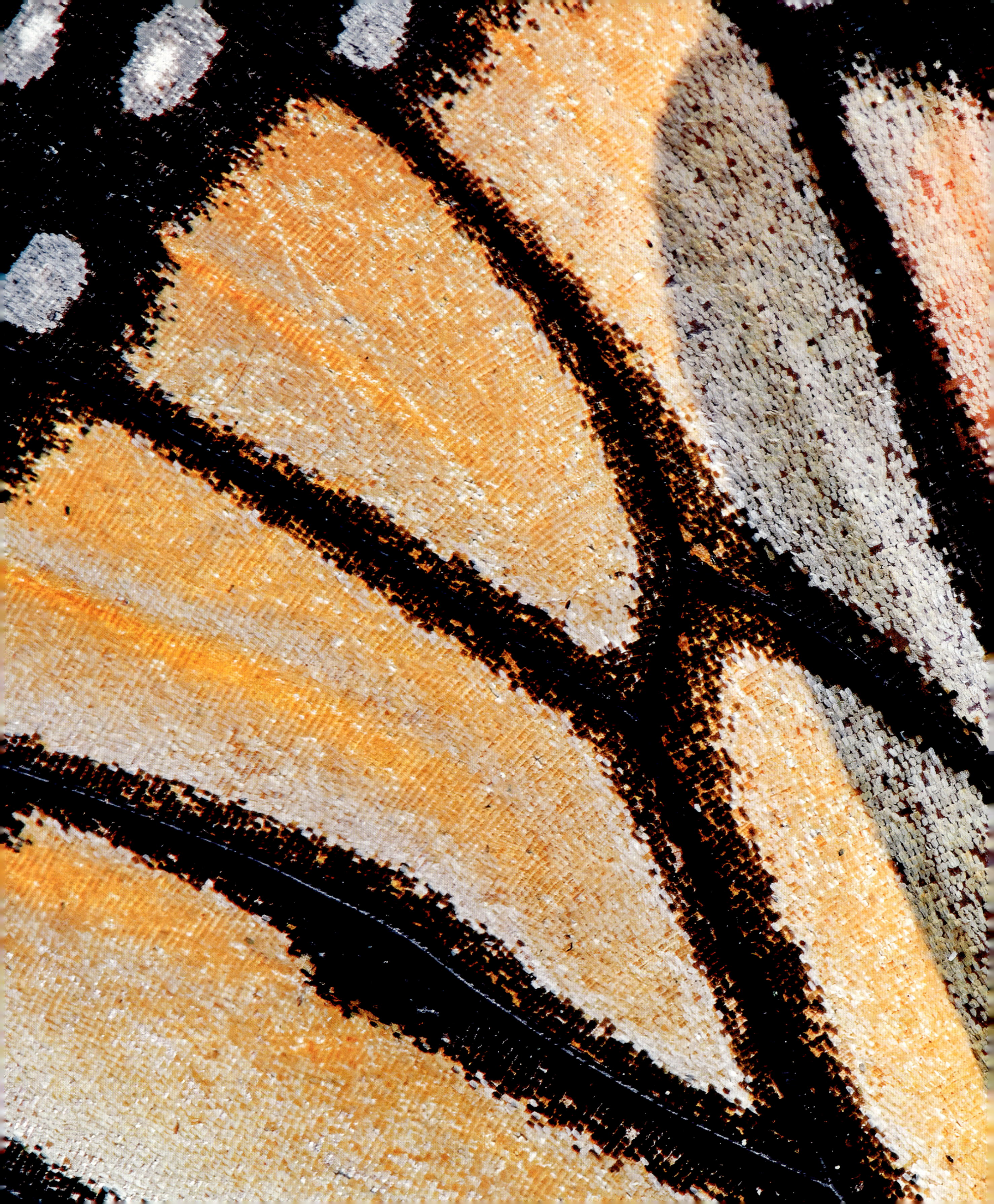

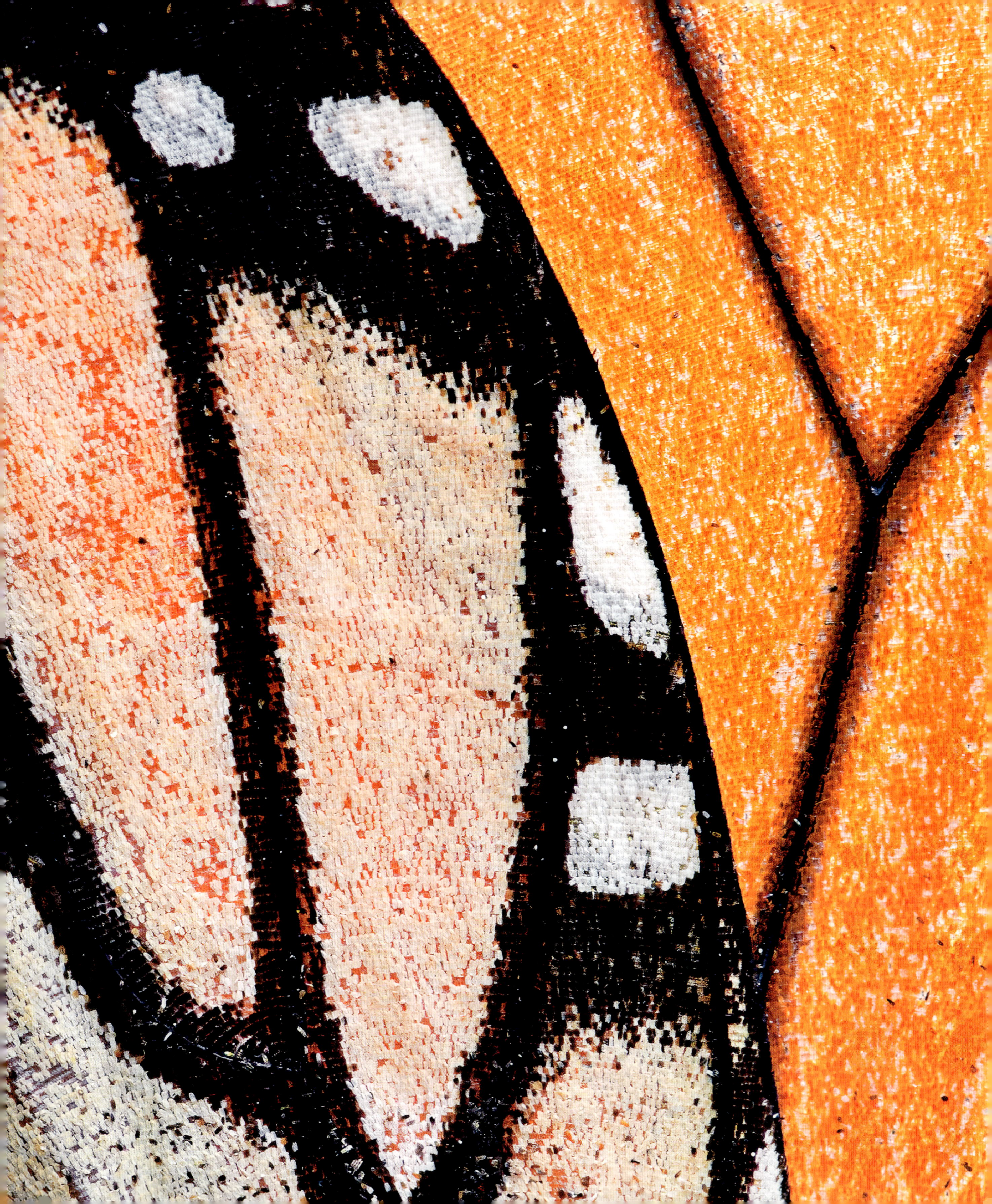

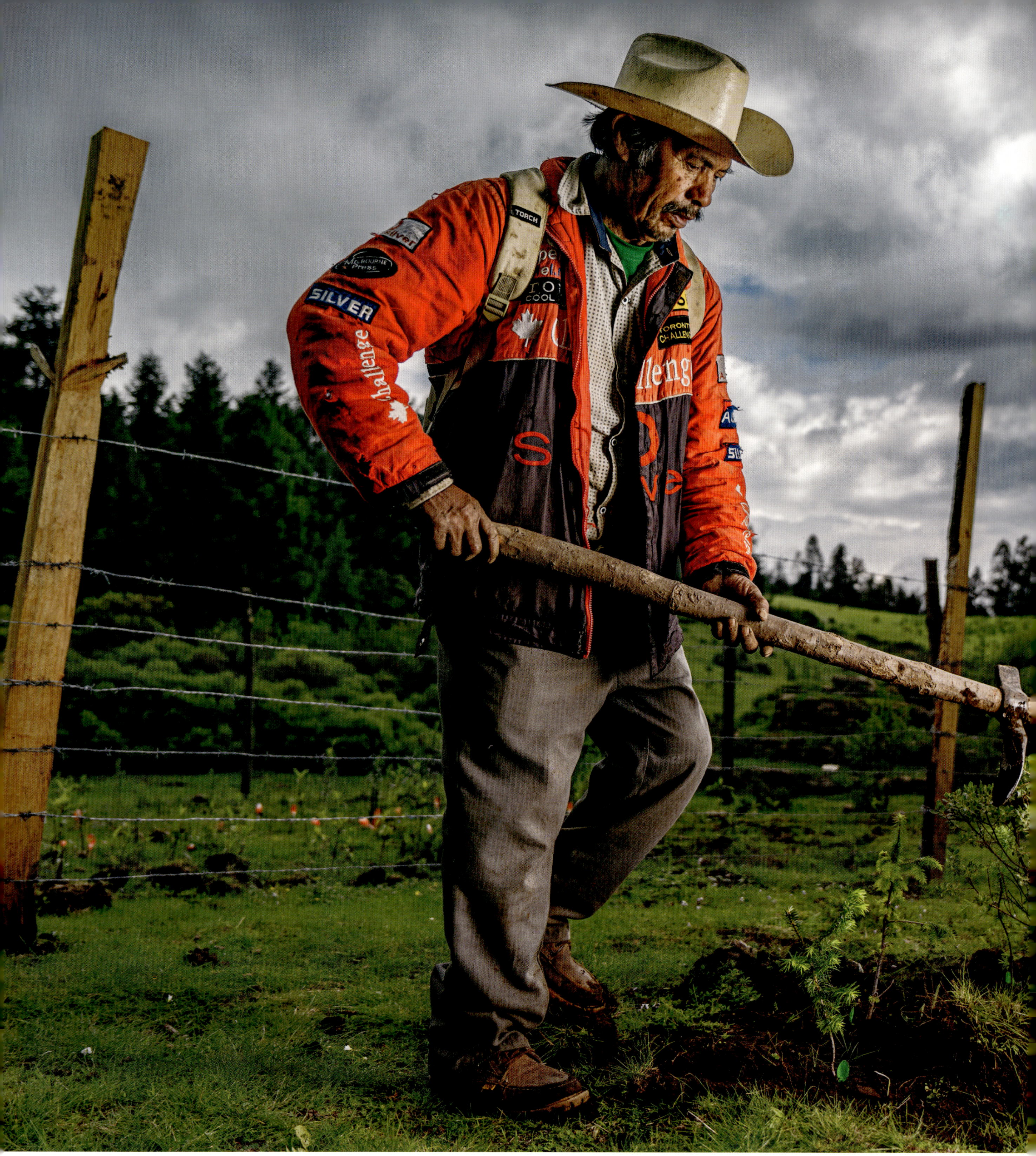
TORCH
SILVER
Challenge

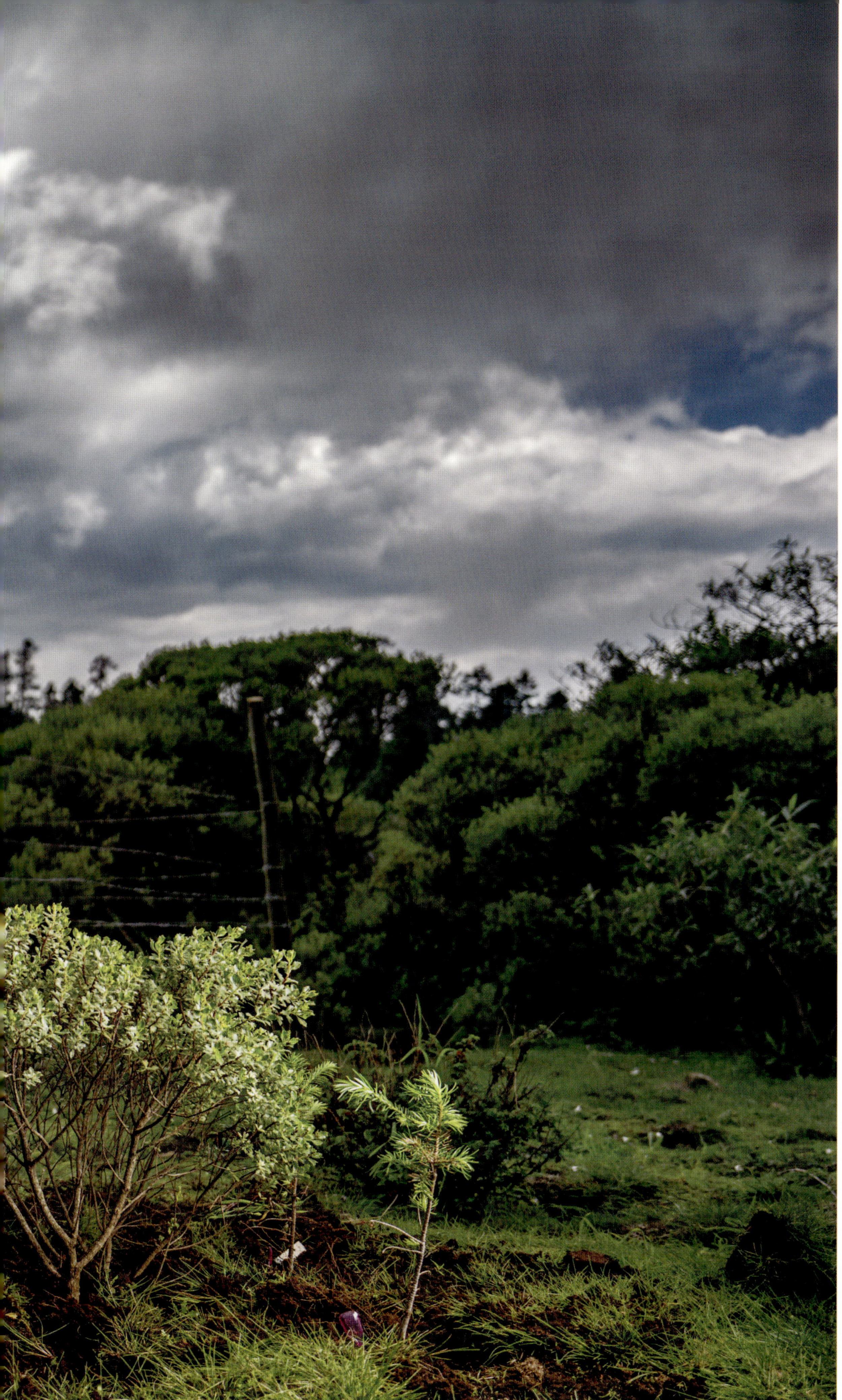

José Humberto García Miranda, from Ejido El Rosario—the community that is home to the largest sanctuary in the Monarch Butterfly Biosphere Reserve—plants oyamel saplings using a reforestation technique that employs *Baccharis* shrubs as nurse plants, providing shade and improving survival for the young fir trees.

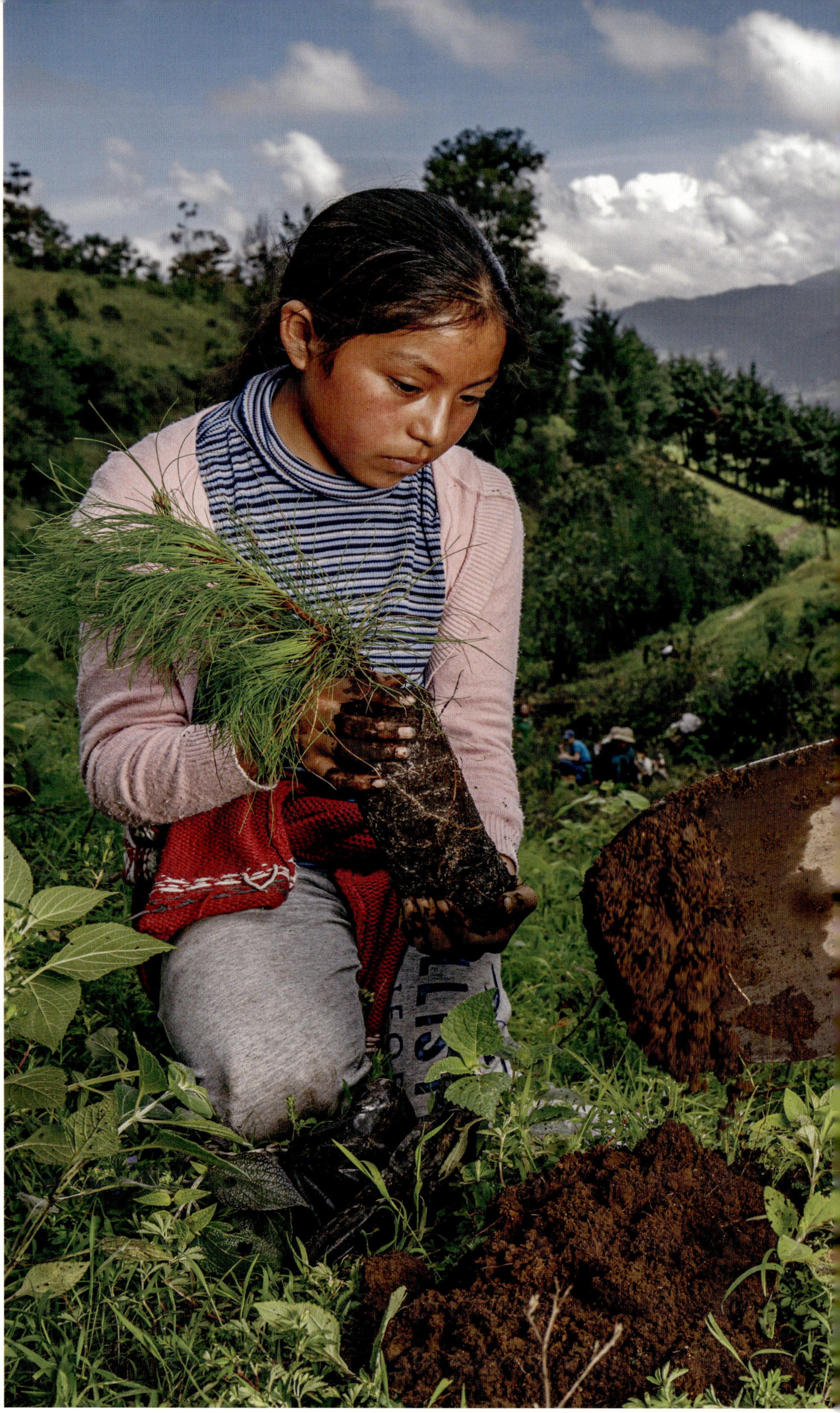

Paula Gómez de Jesús (right) and Brisa Alejandra Mercado Estrada plant oyamel fir saplings during a community reforestation effort. Both belong to the Macho de Agua community, who are neighbors to the Monarch Butterfly Biosphere Reserve in Michoacán.

PAGES 92–93: Mating butterflies, like these in Rowley, Iowa, can stay connected for up to 16 hours, often beginning in the late afternoon or evening and lasting almost until dawn.

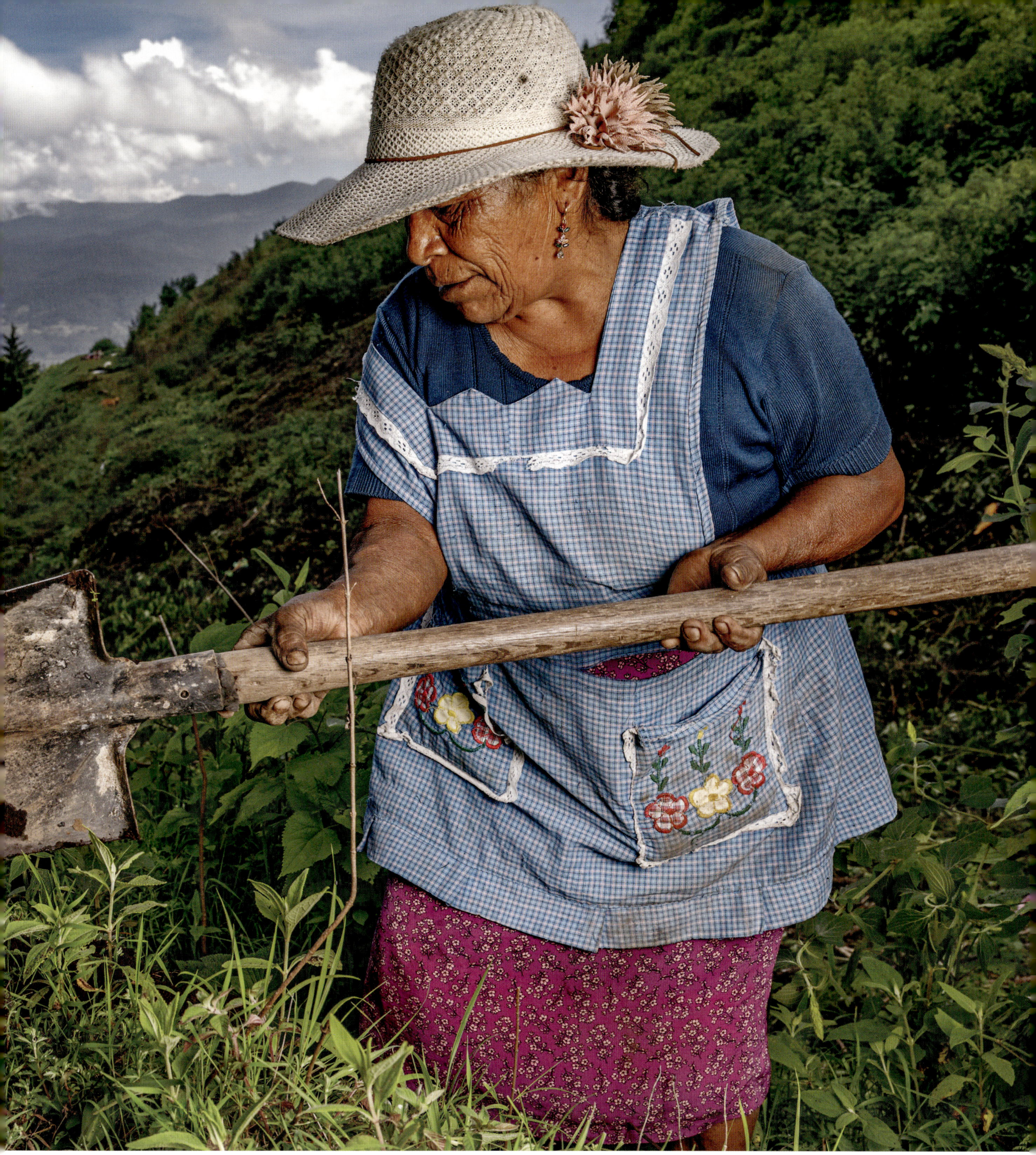

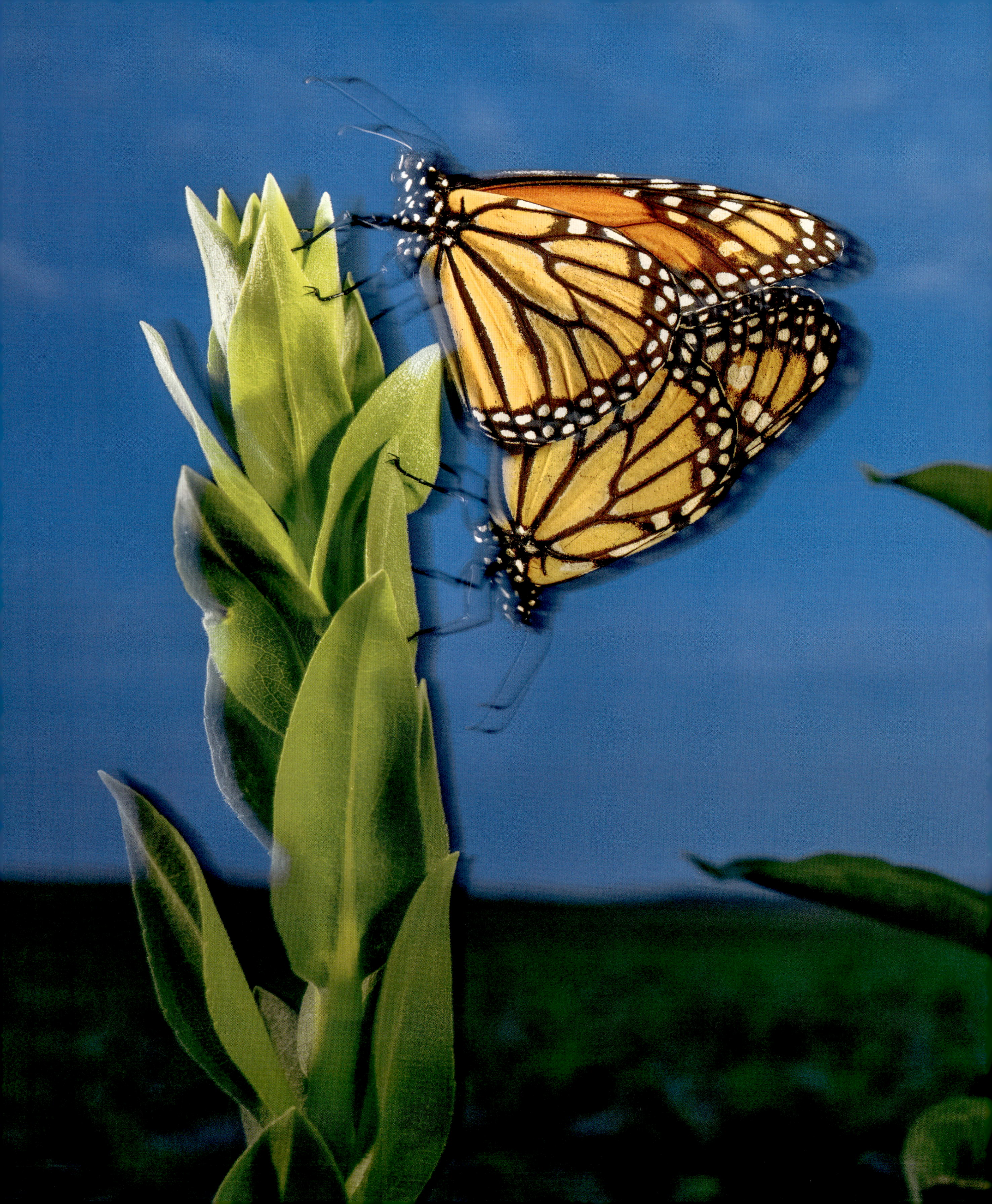

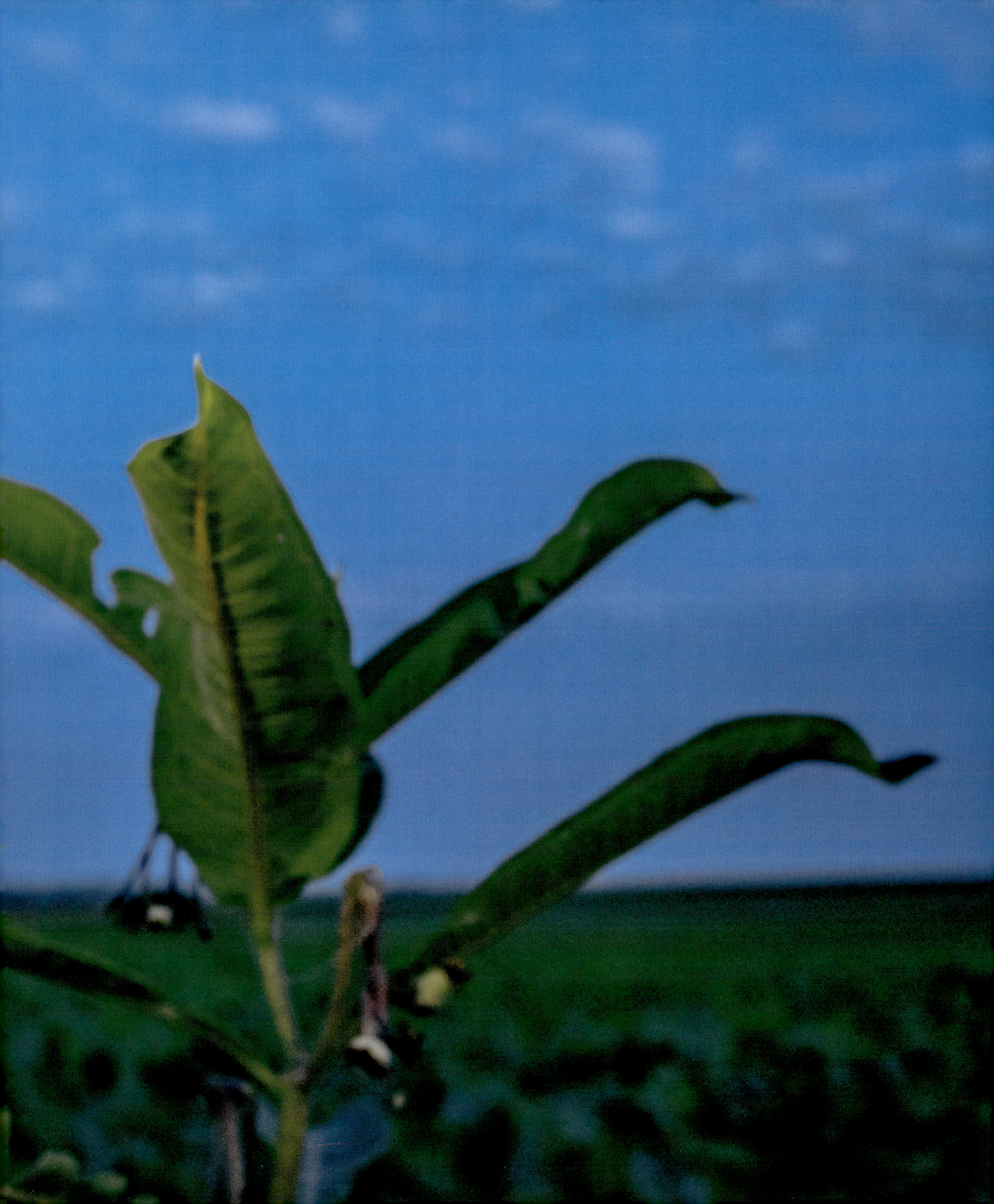

FALL

THE LONG JOURNEY SOUTH

THE MONARCH STOOD STILL on the goldenrod for a few seconds, then stretched its wings slowly—first to the left, then to the right, as if testing the air (pages 98–99). On the back of one hindwing, a tiny round sticker carried a series of numbers and letters, which minutes earlier had been meticulously recorded by Don Davis, the lead volunteer for that event, following the protocols of Monarch Watch, an organization founded in the early 1990s. Monarch Watch is responsible for the success of monarch tagging, now a tremendously popular environmental and educational activity in North America. The butterfly stretched its wings one more time, faced the endless horizon of Lake Ontario, and took off, heading south.

Around us, families and small groups waited their turn with barely contained excitement, all gathered at Ontario's Presqu'ile Provincial Park, northeast of Toronto on the shore of Lake Ontario. Releasing monarch butterflies does not leave anyone indifferent. At the center of it all was Carol Pasternak, who introduced herself as the "Monarch Crusader." Dressed head to toe in butterfly-themed gear—shirt, socks, and, of course, wings—her enthusiasm was infectious. She greeted everyone, gave instructions, and made sure every child left with a smile.

Meanwhile, Don worked quietly in the shade, carefully tagging and recording each butterfly before handing it over for release, just as he had done more than 50 years ago when he took his first steps as a research assistant with Fred and Norah Urquhart, the pioneering Canadian scientists who first traced the monarchs to their wintering grounds in Mexico. Now a retired social worker, Don is probably the longest-running monarch citizen scientist in the world. I can confidently guess he has tagged more monarchs than anyone else on this planet.

The Presqu'ile Provincial Park gathering was one of the most productive monarch-tagging events I have ever attended. The lead gray sky blended with the horizon of the lake, its surface glittering under the clean, bright sun of late summer. There was a nice, refreshing breeze. The mood was festive, almost like that at a fairground, with the crowd cheering as butterflies lifted into the sky (opposite). One large family released their tagged monarchs all at once and then applauded. The light couldn't have been better, and my camera was bursting with equal enthusiasm, while Isa Betancourt—an entomologist and National Geographic Explorer who was doubling as my photo assistant—tried to keep up with adjusting the lights and getting model releases from all the people in the photographs.

It was early September, and these butterflies were ready to travel to Mexico. Ahead of them lay the path they have been flying for thousands of years: an obstacle course that still amazes the scientific community—a path we are only just beginning to understand.

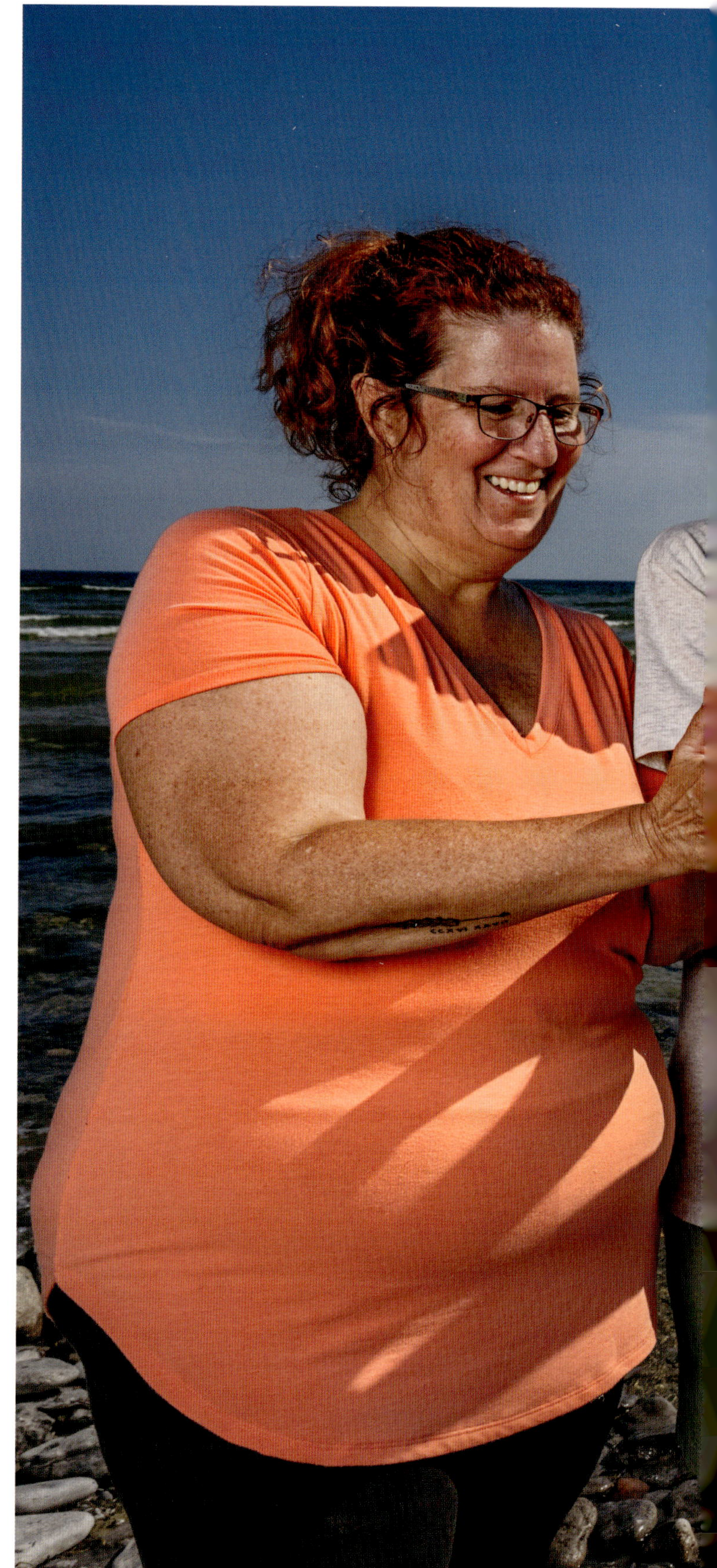

Laurel Merriam, Don Davis, and Carol Pasternak (left to right) release tagged monarch butterflies during a public tagging event at Presqu'ile Provincial Park, Ontario—part of a long-lasting citizen science effort to track the species' migration across North America.

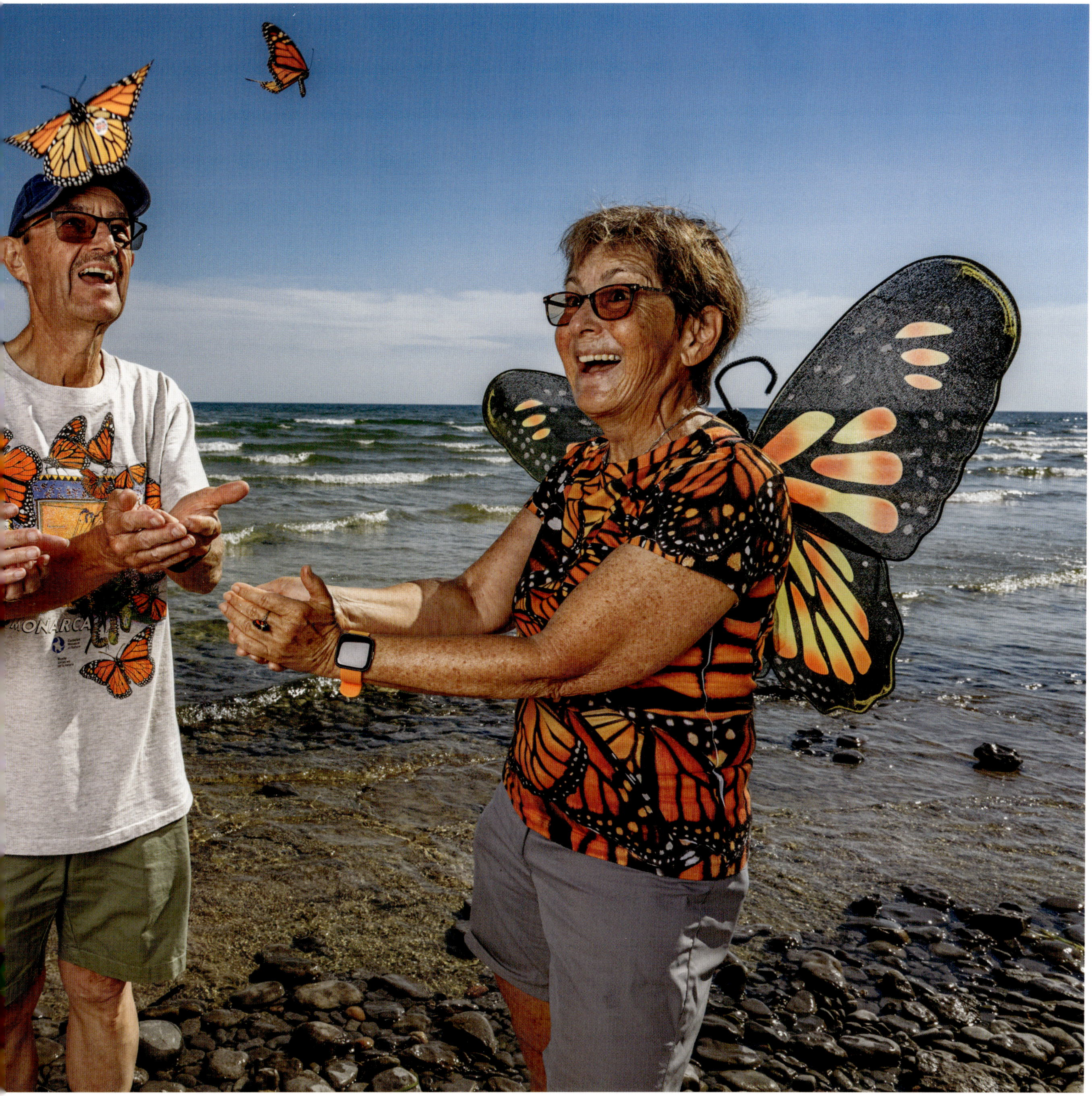

EMPTY PELEE

Of all the monarchs born each year, those from the final generation, born in late summer, are the most fascinating. These butterflies, which we call the super generation, are from the third and fourth generations of the year, and sometimes the fifth. The whole species owes its reputation as an epic migrant to them. The truth is, most monarchs live quiet, unremarkable lives: They eat endlessly as caterpillars, then mate and lay eggs as adults. Yes, predators are always there, and dangers abound, but overall, it doesn't sound like such a bad life.

The monarchs born in late summer and early fall, though, are different from all the others. Their bodies transform in ways unlike those of any other generation. Triggered by shortening days and cooler nights, they enter reproductive diapause: Their mating is postponed, they will live many times longer than the generations born earlier in the year, and their behavior changes. Suddenly, instead of living fleeting, solitary lives, they gather in massive flocks, driven by a single instinct—migration. While their parents and grandparents lived short lives of indulgence, fluttering from flower to flower, these monarchs are born with a sense of purpose that will carry them thousands of miles south.

I was determined to photograph them, to capture that moment when the migration begins. For months I researched maps from Journey North, a citizen science program that keeps track of the migration; its website is both useful and addictive. I read all the available literature and reached out to experts by email and social media. Slowly, a map took shape, dotted with names that sounded almost mythical—Lake Erie, Presqu'ile, Point Pelee. Among monarch enthusiasts, Point Pelee is a pilgrimage site. If you're into monarchs (or warblers), you eventually make your way there.

The reason is geography. Point Pelee is the southernmost tip of mainland Canada, a narrow finger of land that juts into Lake Erie. It acts like a funnel, pointing toward the shortest stretch of water between the two shores—the safest crossing. At different times of the year, migrating birds and monarch butterflies gather here before daring to cross, stacking up in great numbers when the weather is right. For monarchs, it can be one of the most spectacular stopovers of the whole journey south. Those who follow monarchs at Pelee probably also followed the detailed reports of Darlene Burgess on Facebook. A Monarch Watch steward and one of Journey North's most prolific contributors, she logged more than 700 monarch butterfly observations over the years from her lookout at Point Pelee.

So when I finally went to Pelee, I reached out to Darlene. She welcomed me warmly and showed me around her home turf. The place was striking. Rough waters pounded the sand spit, the horizon blurred under thick gray skies, and the wind carried the smell of the lake across the marshes and Carolinian hardwoods. But once again, I ran straight into reality: Monarch populations had been fluctuating dramatically from one season to the next, and as luck would have it, the year I visited was one of the bad ones. All summer I had heard the same refrain at every stop: *I'm sorry, this is a poor year. In other years, this grove would have been covered.* Point Pelee was no exception. The great clusters were absent, and my visit became another chapter in the frustrating chase for vanishing monarchs.

Still, I loved spending time with Darlene. She was funny, dedicated, and inspiring. Later, I learned she had passed away, and the news saddened me. I had promised her I would come back one day, in a better year, at a better time, to see Point Pelee in its full monarch glory, the way she'd always described it. I still owe you that one, Darlene.

TRACKING MONARCHS

Very close to Pelee, in Ann Arbor, Michigan, monarchs were gathering too, starting to roost before the great journey. It was a busy time for D. André Green II, a scientist from the University of Michigan and a fellow National Geographic Explorer, who is leading an interesting experiment studying the detailed movements of monarchs during migration. He and his team are building a sensor small and light enough to be attached to a monarch throughout the entire migratory pathway but powerful enough to record its journey day by day. The key is to bypass heavy, battery-demanding GPS technology. Instead, André's sensor records date, time, light, and temperature—basic clues that will provide invaluable information about how the monarchs navigate the landscape.

Tracking the path of a single butterfly from its breeding grounds in the United States and Canada to the wintering sanctuaries in Mexico has always felt like a mythical quest. It is often referred to as the holy grail of monarch science. The problem has always been the same: How do you follow a small, light, fast-flying insect across thousands of miles?

Back in the 1950s, Fred and Norah Urquhart—the founding parents of monarch-tracking science—came up with a deceptively simple solution: tiny paper tags glued to the wings. The idea was that someone, somewhere, would find that same butterfly and send the tag back. They recruited volunteers to help with their tagging program. Sometimes it worked; most times it didn't. But soon, those paper circles began to reveal the monarchs' astonishing continental journey. The Urquharts discovered that once the days grew shorter, the butterflies all pointed themselves in the same direction. Nobody knew where they were going, but that steady southwest bearing was the first real clue to their long-distance migration.

Later, entomologist Lincoln Brower helped explain that these southbound monarchs are not exactly the same as their parents or grandparents. Although genetically identical to previous generations, members of this generation are efficient gliders with broader wings, their metabolism tuned for lipid storage and their bodies built for migration.

The Urquharts kept expanding their tagging program. After years of trial and error, and with the help of a growing network of volunteer "research associates," in 1975 they were finally able to trace the monarchs to their wintering grounds in Mexico, a location long known by the locals in Michoacán but which came as a shocking discovery to the scientific community and the public worldwide. Simple paper tags started it all, but many questions remained unanswered.

Chip Taylor (see Field Notes, page 131) understood that to decipher the mysteries of the migration, he needed to scale up the tagging effort started by the Urquharts. In 1992, he created Monarch Watch, an educational program that recruits citizen scientists and trains them in how to capture, handle, and tag the butterflies. By making tagging more accessible, Monarch Watch exponentially expanded its impact and created a gold mine of data about the migration.

Thirty years ago, in northeastern Mexico, where those tags weren't yet easily available, people like Rocío Treviño in Saltillo, Coahuila, used an even simpler solution: handwritten letters reporting monarch sightings (pages 152 and 153). There was a huge knowledge gap in northern Mexico about the butterflies' migratory routes. Rocío called on enthusiasts in the region to report observations. Fewer than 25 responded, but that small group became the program's foundation. With email still uncommon and no social media, all communication depended on postal mail. They named the program Correo Real—the Royal Mail. Over the years, it grew into the longest-running citizen science program in Mexico, collecting more than 15,000 monarch reports that were key to producing the first migration map in northern Mexico.

Years later, a new generation of scientists, including André, keeps pushing those boundaries. That's how I had ended up in Ames, Iowa, standing in a sunny grassland with Kelsey Fisher and Steve Bradbury, trying to capture the impossible shot of a monarch with a tiny radio transmitter glued to her abdomen. As I described in the first chapter, their data would later show how monarchs perceive the landscape—what distances they can cover, how patches of habitat might connect. Still, all their findings were limited to local movements over a few days, maybe weeks at most. The real mystery was the full migration. Could a butterfly carrying technology ever make it from Canada to Mexico?

With these questions in mind, I visited Christine Merlin in her lab at Texas A&M University, to watch her investigate mysteries the Urquharts could never have imagined. Christine is a physiologist who discovered that monarchs adjust their internal compass through their antennae, which contain circadian clocks that correct for the shifting position of the sun. Her research has since shown that monarchs also sense Earth's magnetic fields via a light-dependent compass that helps them stay oriented when the sun is obscured.

To study these mechanisms, she places butterflies in a Faraday cage—a metal structure that blocks outside electromagnetic interference. Inside, coils generate controlled magnetic fields while sensors record the butterflies' flight direction. The setup looks almost theatrical: a butterfly from her lab colony, wings beating in place, every shift in direction carefully logged. The mystery of the monarchs may be less of a mystery today, but it remains a fascinating, endless well for curious minds like Christine's.

In 2025, as this book was being written, a small company called Cellular Tracking Technologies, in collaboration with Cape May Point Arts and Science Center, reshaped the field of monarch-tracking science with its new BlūMorpho radio transmitters. These groundbreaking microtransmitters are light enough to be carried by a monarch butterfly. They use 2.4 GHz, the same frequency used by Bluetooth devices, and a growing network of receivers, including smartphones, to detect the movements of tagged monarchs across vast landscapes. For the first time, scientists can follow a monarch's journey in near real time, opening an exciting new chapter.

In an unexpected turn of events, and at the very last minute, I joined the Xerces Society for Invertebrate Conservation —a nonprofit that, among other things, runs the most important program for the conservation of the western monarchs—to document the first ever deployment of these new sensors in overwintering monarchs in California (page 139). To say I felt privileged to witness that is an understatement.

From the daunting task of recovering the early paper tags to the hope of picking up a faint signal from André's high-tech sensors,

(Continued on page 100)

THE PROBLEM HAS ALWAYS BEEN THE SAME: HOW DO YOU FOLLOW A SMALL, LIGHT, FAST-FLYING INSECT ACROSS THOUSANDS OF MILES?

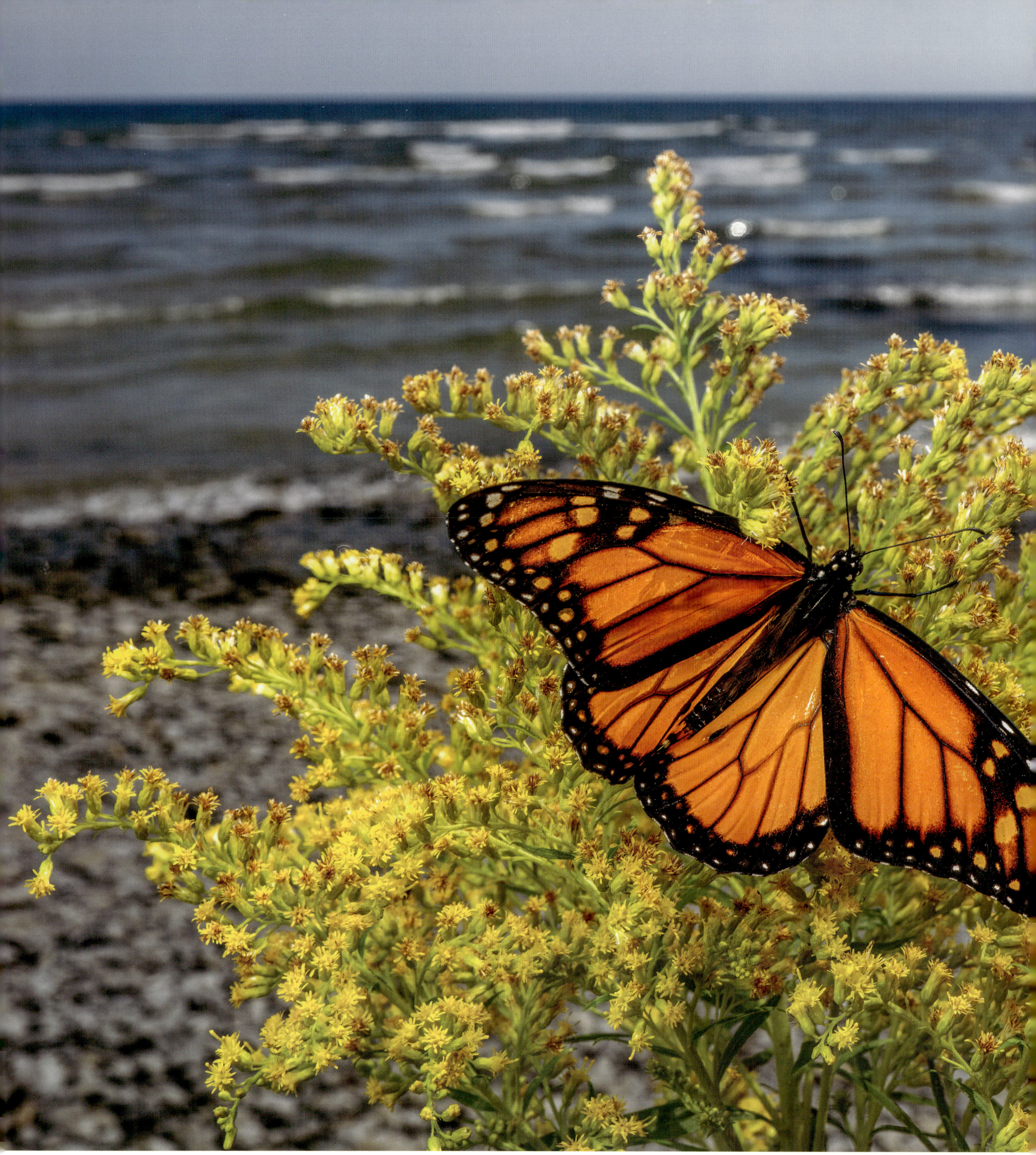

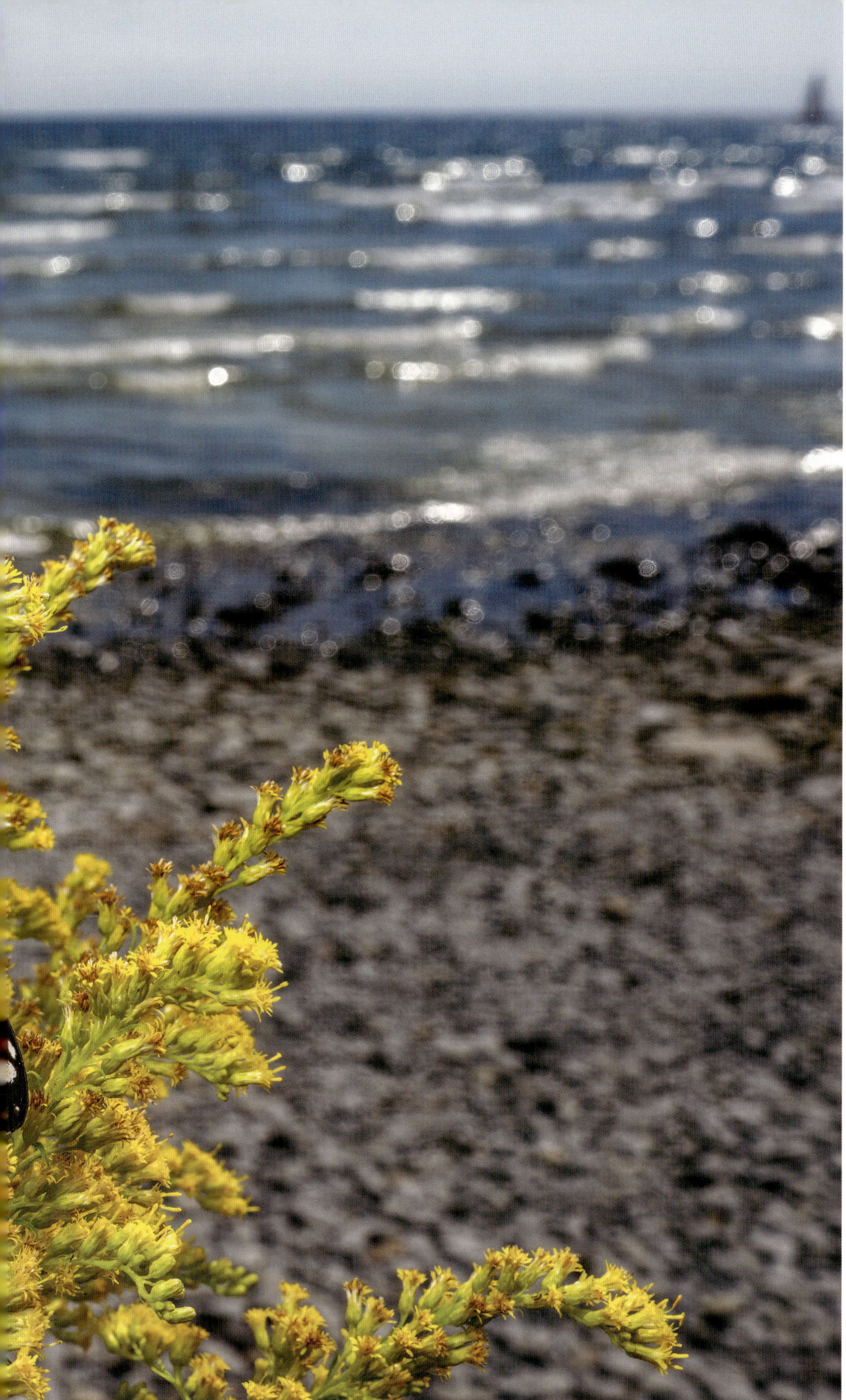

A recently tagged monarch butterfly rests on goldenrod near Lake Ontario at Canada's Presqu'ile Provincial Park. From here, this butterfly will migrate south toward the mountains of central Mexico.

ALL OVER NORTH AMERICA, THOUSANDS OF CITIZENS, ALARMED BY POLLINATOR DECLINE, HAVE STEPPED UP AND DONE THEIR PART.

PAGES 102–103: Urban gardens along the migration route play a major role in monarch survival. Here, a butterfly pauses on the bloom of a butterfly weed, a type of milkweed, in Millennium Park's Lurie Garden, in the heart of Chicago.

(Continued from page 97)
every step is driven by a very human need to make sense of a reality we can't fully observe or explain. Monarchs have always been moving targets that escape our understanding. Maybe that's why we are so obsessed with them.

TAGGING BEYOND SCIENCE

Part of that obsession with monarchs comes from how close they are to our daily lives. We delight when we see them in our gardens, we follow their breeding progress, and we celebrate their successes almost as if it were a household competition. I can't enumerate how many people have bragged to me about having a chrysalis or caterpillar in their backyard. Many now in retirement carry memories of endless summer days when monarchs were everywhere, a staple in backyards along the migratory route. It's no surprise, then, that tagging events—the same ones Monarch Watch helped grow from a few dedicated enthusiasts into popular social gatherings—often feel festive.

There is an almost childish pleasure in recovering a tagged butterfly. It feels like a supernatural achievement, a stroke of luck so random that for a few seconds you feel as if you've been chosen by the universe. Then comes the rush of entering the code on the Monarch Watch website and seeing where the butterfly came from. And then the questions: What happened along the way? Why does it looks so battered—or so fresh? It's like watching people in a train station or airport and trying to guess their stories.

Over the years, I have attended quite a few tagging events, but only once did I go back to the same place in different years. My first visit to Rosetta McClain Gardens came when my life as a photographer was shifting—I was a father of two, we had just moved back to Spain, and I was searching for new work opportunities. The Insectarium of Montréal had invited me to contribute some monarch images for an exhibit. I wanted to use that trip to photograph monarchs in Canada, at the northern edge of their breeding grounds, just as they were preparing to migrate south. As I usually do before an assignment, I read papers, reached out to experts, and asked my contacts about tagging opportunities. A few people pointed me toward Carol Pasternak—the Monarch Crusader, whom I would later join at Presqu'ile—with kind but cautious warnings saying that she was very enthusiastic. What an understatement.

Carol took me under her wing, introduced me to her colleagues, and showed me the best monarch spots around Toronto. The next day she was attending a tagging event at Rosetta McClain Gardens in nearby Scarborough, and I joined her. After a few introductions and explanations of the process, we all got to work. Carol—visor on her head, nails decorated with monarch stickers, sharp stare—was netting butterflies with Don Davis, the legendary citizen scientist, and Betty McCulloch, another regular. Terry Whitman and Kathy Haughton sat in the shade, tagging at a bench. I enjoyed the scene immediately, and I got my first real images of a tagging event (pages 128 and 129).

In 2022, I returned to Rosetta McClain to get fresh images of a tagging event and met both old and new friends amid the same atmosphere: a vibrant community of monarch supporters whose camaraderie and joy were inspiring. Tagging makes you feel part of a larger mission, and it has helped me understand the commitment, almost obsession, of these monarch enthusiasts and citizen scientists—people with an energy and passion I have rarely seen directed toward any other species in my years of working in conservation.

URBAN MONARCHS

As part of my search for a visual story about the mystery of the monarchs, I also wanted to document solutions to the challenges these butterflies face. There are many initiatives underway, many beautiful, untold stories that are inspiring, and practices that could be replicated elsewhere. One of them, without a doubt, is planting pollinator gardens—and for that I needed to step into urban ecosystems.

I could hear the loudspeakers thumping from my hotel room in downtown Chicago. I packed my gear and followed the music east, crossing street after street under subway rails and rusty metal structures until I emerged onto Michigan Avenue, almost blinded by the bright sun. The music dominated the soundscape, with the hum and honking of cars close behind. It was the World Music Festival in Millennium Park, and even though it was just getting started, the atmosphere was vibrant.

I had come to Chicago with the goal of photographing the quintessential urban pollinator garden in North America: the city hall rooftop. But my appointment to visit it wasn't for a couple of days, so I took an afternoon off to visit the Art Institute of Chicago. And right next door, in the most unlikely place, was my newfound paradise. After losing myself in nostalgia over cloud landscapes by Georgia O'Keeffe—one of my wife's favorites—I stepped outside for a walk and stumbled across the most beautiful and unexpected prairie-inspired garden ... and it was full of monarchs! The Lurie Garden, created in 2004 by American landscape architect Kathryn Gustafson and Dutch garden designer Piet Oudolf, transformed a former rail yard into a living mosaic of grasses and wildflowers. It provides nectar throughout the seasons, making it an oasis for pollinators in the heart of the city.

This was the perfect location for the shot I had in mind: a wide-angle macro of a monarch feeding on nectar with a striking skyline behind. I knew it wasn't going to be easy, but by sheer luck I had managed the hardest part—finding the perfect location with plenty of subjects. Now it was just a matter of patience and persistence. I started early the next morning. There wasn't much activity, but as the sun warmed up, the monarchs started to show. I was using a probe lens—a 15-inch tube-shaped lens that looks remarkably like a rifle. A bad choice for photographing in U.S. public spaces these days. After two security guards came to check me out, visibly worried, I decided to change strategy and switched to a nimble setup with a close-focusing 35mm lens. I spent a delightful morning chasing monarchs, jumping like a happy child from flower to flower until I started to see and understand the patterns and rhythms of the place. By afternoon, rain clouds were forming on the horizon, and the light shifted to that crisp, warm brightness that precedes a storm. I discovered a butterfly weed plant with the perfect background and stuck with it until, eventually, every element fell into place and I got the image I had imagined (pages 102–103).

The following day I headed to city hall to meet my friend Mike Rizo, a program specialist with the U.S. Forest Service's International Programs. Mike engages underserved communities in the conservation of monarchs and other migratory species across the United States and Latin America. He has lived in Chicago his whole life and knows everyone in town, and soon he became my go-to person for everything. He introduced me to the city hall staff, who were helpful and attentive, but as often happens with big institutions, the permits were too restrictive. I was disappointed to find out that I would have only an hour in the rooftop garden, from 1 to 2 p.m., the worst possible time for light. Still, I was determined to make the best use of my time. I found some monarchs there, but they were so active—their metabolism on overdrive from the heat and the abundance of flowers—that it was almost impossible to get a decent shot. Still, I tried my best and enjoyed the visit.

The greening of urban spaces seems like the easiest solution, a win-win for city dwellers and for wildlife, for the climate and for human well-being. But progress is painfully slow. Until recently, urban gardens with native grasses, like the Lurie and Chicago's city hall rooftop, were rare in North American cities. It wasn't until the 2000s, when people like Douglas Tallamy began to speak up, that things started to change. Tallamy's books, including *Bringing Nature Home* and *How Can I Help?,* turned what once seemed utopian into something accessible for the average household: replacing boring, biodiversity-poor, excessively tidy lawns with messy gardens rich in native plants and milkweed. The movement quickly took off. All over North America, thousands of citizens, alarmed by pollinator decline, have stepped up and done their part.

From the beginning I knew I wanted to cover this angle of the story. Since I was traveling to so many locations, I made sure that in every town I visited, from Ontario to Michoacán, I would document at least one pollinator garden. After a shout-out to my monarch network, the response was overwhelming, and it created two new problems. First, I am not great at staged portraits. Second, I knew there was no way all the photos I was taking could ever be published. I turned the first problem into an opportunity: I studied the work of great portrait photographers and learned a lot. The second problem still haunts me at night. So many people opened their doors and shared their beautiful gardens with passion and enthusiasm, but only a handful of the pictures I took of them have ever been published. Still, I have attempted to list everyone in the acknowledgments section of this book as a sincere tribute to their contribution and as big kudos for the important work they are doing.

With only two days left of my stay in Chicago, I wanted to make good use of my time. One of those days I visited Pam Karlson, one of those random connections that emerged during my search for pollinator gardens. It turned out she had a monarch chrysalis in her backyard! Wild chrysalises are hard to find. Even when you follow caterpillars closely, they seem to vanish into thin air when it's time for them to hunker down and transform. But Pam, a professional artist and gardener, had been keeping a careful eye, and when I arrived, she proudly pointed it out. I spent the whole afternoon in her garden trying to get interesting shots of the chrysalis. Pam and her husband, Mike Edsey, were one of the most charming couples I met along the way, and their contribution was essential—it's the only picture of a wild chrysalis in my archive (pages 120–121)! We've kept in touch ever since.

On my last day in Chicago, I had a very special plan. Once again, my friend Mike Rizo came to the rescue, connecting me with El Valor,

(Continued on page 104)

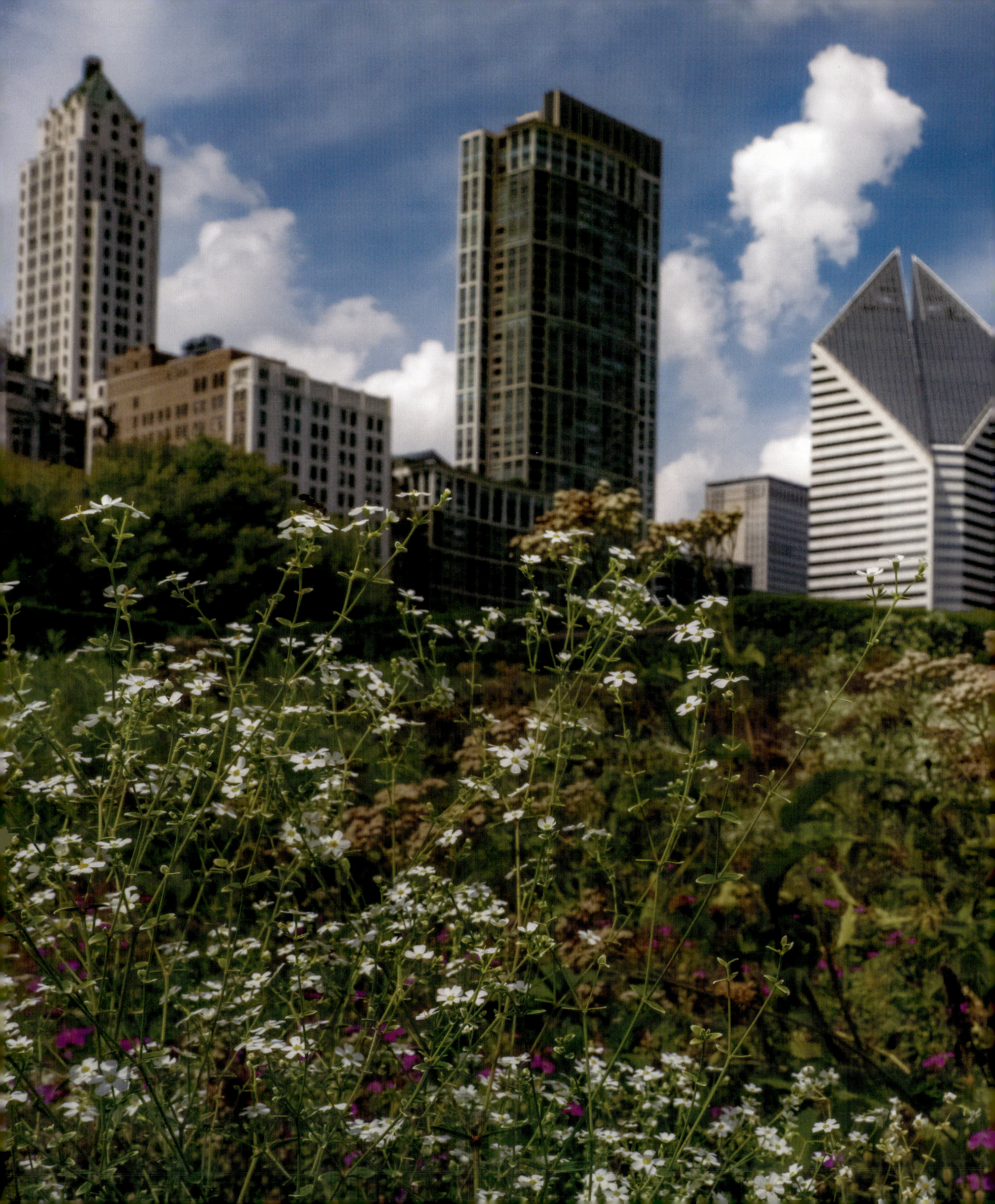

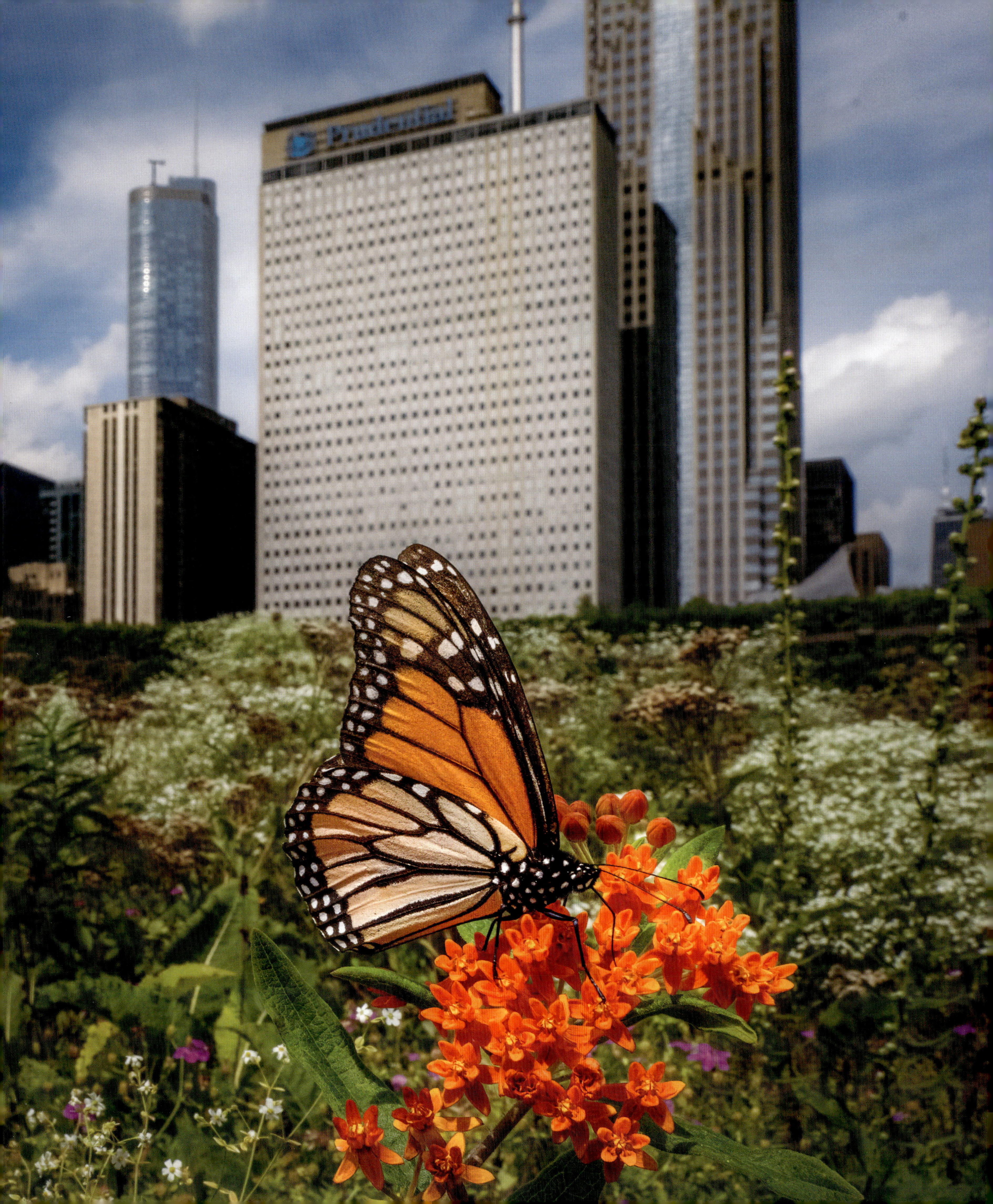
Prudential

(Continued from page 101)
one of his longtime nonprofit partners, which works with children and adults with disabilities and their families, helping build community and belonging in underserved neighborhoods. It also runs the Monarch Butterfly Conservation Program to engage families in science- and nature-based activities. That day, El Valor was holding its annual tag-and-release event, a celebration with a strong presence from the Mexican American community.

There was food, music, dancing, and butterfly props everywhere. The kids tagged and released the butterflies they had raised in class during the spring semester—a practice that is becoming increasingly controversial, since research shows that captive-raised monarchs are not as fit for migration as their wild counterparts. But experts also acknowledge that activities like these strengthen the human connection with monarchs, inspiring new generations and fueling the broader conservation movement. I've read the papers, and I understand the concerns. But that morning, the smiles of the kids at El Valor (pages 146 and 147) made me think otherwise.

TEXAS FUNNEL

After a few weeks of following the fall migration from Presqu'ile to Point Pelee to Chicago and farther south, the butterflies were now approaching what is known as the Texas Funnel. This is where their migration paths from across North America converge, all now heading toward the final route south, and monarchs start to gather in greater densities. Tagging events had been the guiding force of my travels, and the next stop was Fort Hood, Texas, where the U.S. Army runs its Adaptive and Integrative Management (AIM) program, a conservation initiative that carries out an intensive monarch-tagging effort.

It wasn't the kind of place where I expected to find a story about monarchs. Yet here I was, in the middle of one of the largest military installations in the United States, surrounded by soldiers, families, and kids, all chasing butterflies with nets. What started as a pilot program in 2017 has grown into one of the most ambitious tagging efforts anywhere: More than 10,000 monarchs have been tagged on this base since then. For the butterflies, Fort Hood is a vital stopover, a refueling station full of wildflowers that keep them going during a particularly challenging stage of their long flight to Mexico.

I watched Command Sgt. Maj. Calvin Hall, a towering man prepared for the hardest war operations, burst into a sprint with the same energy as seven-year-old Briley Bray, both chasing after a butterfly like it was the most important mission of the day (pages 124–125). The event was coordinated by Christine Luciano, Fort Hood's enthusiastic environmental outreach officer, who had managed to turn a military facility into one of the liveliest monarch-tagging venues I've ever seen.

The day before, I had met Chelsea Plimpton, the lead pollinator biologist with the AIM program, as she tagged in specific locations with her team of seasonal biologists (pages 126–127). They explained that the bulk of the project runs on citizen science. Families, soldiers, kids—they all take part, and they all go home with a deeper connection to the migration. There is something paradoxically beautiful in the fact that a community that usually trains for war also comes together to protect something as fragile as a monarch.

It was a dry October, and I tried hard to capture the role that flowering plants play as pit stops for migrating monarchs in these regions. But it was impossible: I could barely find any blooms, let alone monarchs feeding on them. In recent years, drought has turned the Texas Funnel and northern Mexico into a death trap for many butterflies. Wildflower meadows that once offered reliable nectar are now baked under increasingly frequent heat waves and dry fall conditions. Along highways, roadkill is a quiet but devastating problem. The warm air currents that help the butterflies travel also pull them low over the roads, where cars hit them by the thousands. Some researchers estimate that millions die this way every fall. Drive through Texas or northern Mexico in September or October and you'll see it yourself: flashes of orange squashed on car grills or piled on the shoulders of the highways. And then there is light. Monarchs orient themselves with the sun, their entire migration tied to natural cycles of day and night. But towns and cities along the flyway glow at night—street lamps, billboards, gas stations—and that glow can disrupt their internal clocks, making it harder for them to rest or to take off at the right time.

Retrace, step by step, the most likely path of a migrating monarch, and it feels almost miraculous that any butterfly actually makes it to Mexico. And yet every November they arrive—millions of them filling the oyamel forests again. How does this massive migration continue to succeed? Part of the answer lies in evolutionary biology. Like many r-strategist species (those with short lifespans, small bodies, and high reproductive rates), monarchs are adapted for boom-and-bust population cycles. But another part—one that fuels myths, epic tales, and the wonder we experience in the presence of these creatures—is an inner strength that keeps them flying against all odds.

DAY OF THE DEAD

This certainty of the butterflies always arriving in Mexico—and always more or less at the same time, early November—created the foundation for one of the best known, most beloved, and often misunderstood cultural interpretations of the monarchs.

For the Mazahua and Otomí communities that live around the Monarch Butterfly Biosphere Reserve in the states of Michoacán and Mexico, monarchs are seen as the souls of ancestors returning home. For these communities, the butterfly is not just a migrant insect—it is a messenger, a sacred link between the living and those who came before.

The relationship is also practical and connected to the land. The arrival of the monarchs marks the time for the harvest. These cultures have depended on the forests for generations, using their wood, water, and soils. For centuries, they have been stewards of the very habitat monarchs rely on. Protecting the butterflies is inseparable from supporting the people who live among them. Sustainable forestry, community-based tourism, and cultural projects blend traditional knowledge with livelihoods and conservation. To fully understand the migration, you need to see it as not only ecological but also social, cultural, and spiritual, rooted in the people who share their mountains with the butterflies.

I timed one of my visits to Michoacán with the Day of the Dead. I wanted to document the festive atmosphere of what has become one of Mexico's most popular celebrations, and also a key expression of the local Indigenous communities' deep spiritual connection to the monarchs. On earlier trips I had met Don Sabino Marín, a Mazahua elder in the community of Francisco Serratos, a small Indigenous town on the edge of the reserve. According to some nongovernmental organizations that work in the area, he is one of the last keepers of traditions that are quickly fading.

Don Sabino was a charming man, short, stocky, and energetic. He showed us around his land, leading us up steep slopes until we reached his *milpa,* a traditional farming plot of maize, beans, and squash, surrounded by pines and oaks. We shared freshly made tortillas with his wife, Doña Hilaria, and *hablamos de lo humano y lo divino*—we talked about everything, from the everyday to the sacred. Time runs at a different speed with Don Sabino. Markus Martínez Burman, a National Geographic Explorer who had joined me as part of the National Geographic Society's wonderful Second Assistant Program, was enjoying every minute of it; he could have stayed there placidly chatting forever. I tend to live in a hurry, so that afternoon was a humbling exercise in restraint. Before we left, we made a plan to meet again for the Day of the Dead. Don Sabino invited us to join him early in the morning, when he would clean the graves of his parents in the local cemetery.

We arrived at 6:30. The morning was cold and dark. Don Sabino worked steadily, clearing the ground around the graves, pulling weeds, and tending the site. Around 8:30 he took a break, walked back down the hill to his house, just a few minutes away, and shared a quick breakfast and coffee with us. At 9:30 he returned to the cemetery, this time dressed in a beautiful wool poncho in bone, brown, and black. On his one-wheeled cart he carried flowers, copal (a resin used in many traditional ceremonies in Mexico), and a cup. Step by step, methodically, he decorated the graves, burned the copal, and prayed. The sun was rising, and the last wisps of fog faded away, replaced by the thick, aromatic smoke of the copal (pages 164–165).

Back at home, Doña Hilaria invited us to see their personal altar, or *ofrenda.* It was simple: a mix of fruit (bananas and guavas), herbs, *pan de muerto* (sweet bread), *cempasúchil* (marigolds), and, in the center, a small but powerful figure of a fiery Virgin of Guadalupe, surrounded by lit candles. Like Don Sabino himself, this celebration was quiet and beautiful. No music, no fanfare, no monarch iconography at all. The night before, we had gone to the parade in Zitácuaro, with plenty of *catrinas*—the well-dressed skeleton popularized by Mexican artist José Guadalupe Posada—along with costumes, music, and lots of monarch wings. But what I saw the next morning was completely different. I had anticipated a photo of a cemetery with beautifully decorated graves, rich in monarch iconography, against a blue sky filled with butterflies. But early in my research I realized it was all a fantasy, a false expectation shaped by mainstream culture. And early November is no longer a guarantee of the monarchs' arrival. Climate change is shifting their cycles, making the wintering season shorter: The butterflies arrive later and leave earlier.

Reality is harsh, but also more powerful. I'm not a spiritual person and never have been, but that morning I was moved. The photo I made will always remind me of what the Day of the Dead truly is: a simple celebration of life and death, honoring the memory of loved ones who are no longer with us. Sooner than later, the monarchs would arrive again to these forests, entering the final and most critical stage of their life cycle, and with them would come a social transformation like I have seen nowhere else.

TO FULLY UNDERSTAND THE MIGRATION, YOU NEED TO SEE IT AS NOT ONLY ECOLOGICAL BUT ALSO SOCIAL, CULTURAL, AND SPIRITUAL.

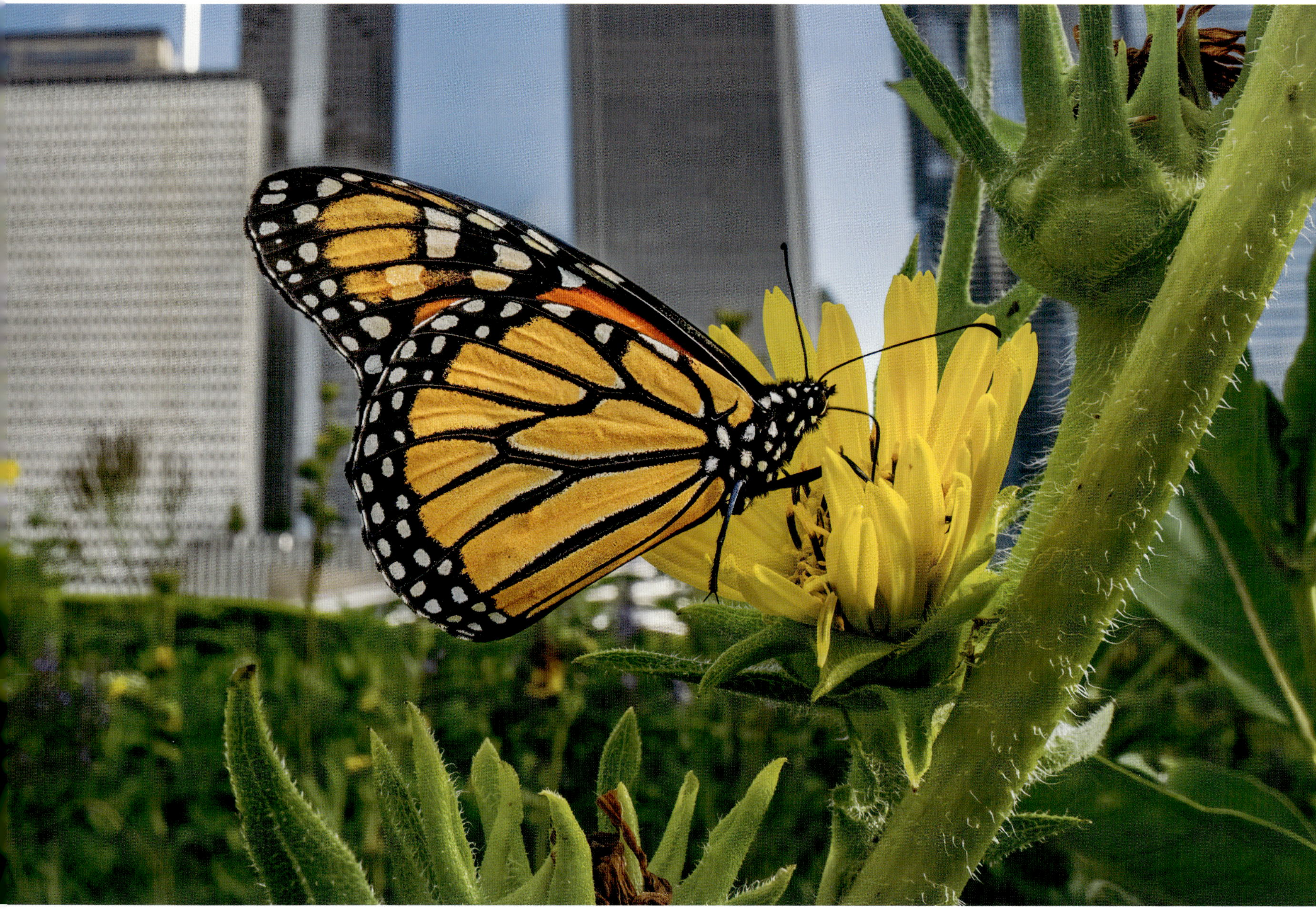

ABOVE AND OPPOSITE: Wild sunflowers planted in Chicago's Lurie Garden help monarchs and many other wildlife species by creating green spaces teeming with food and shelter amid the concrete jungle.

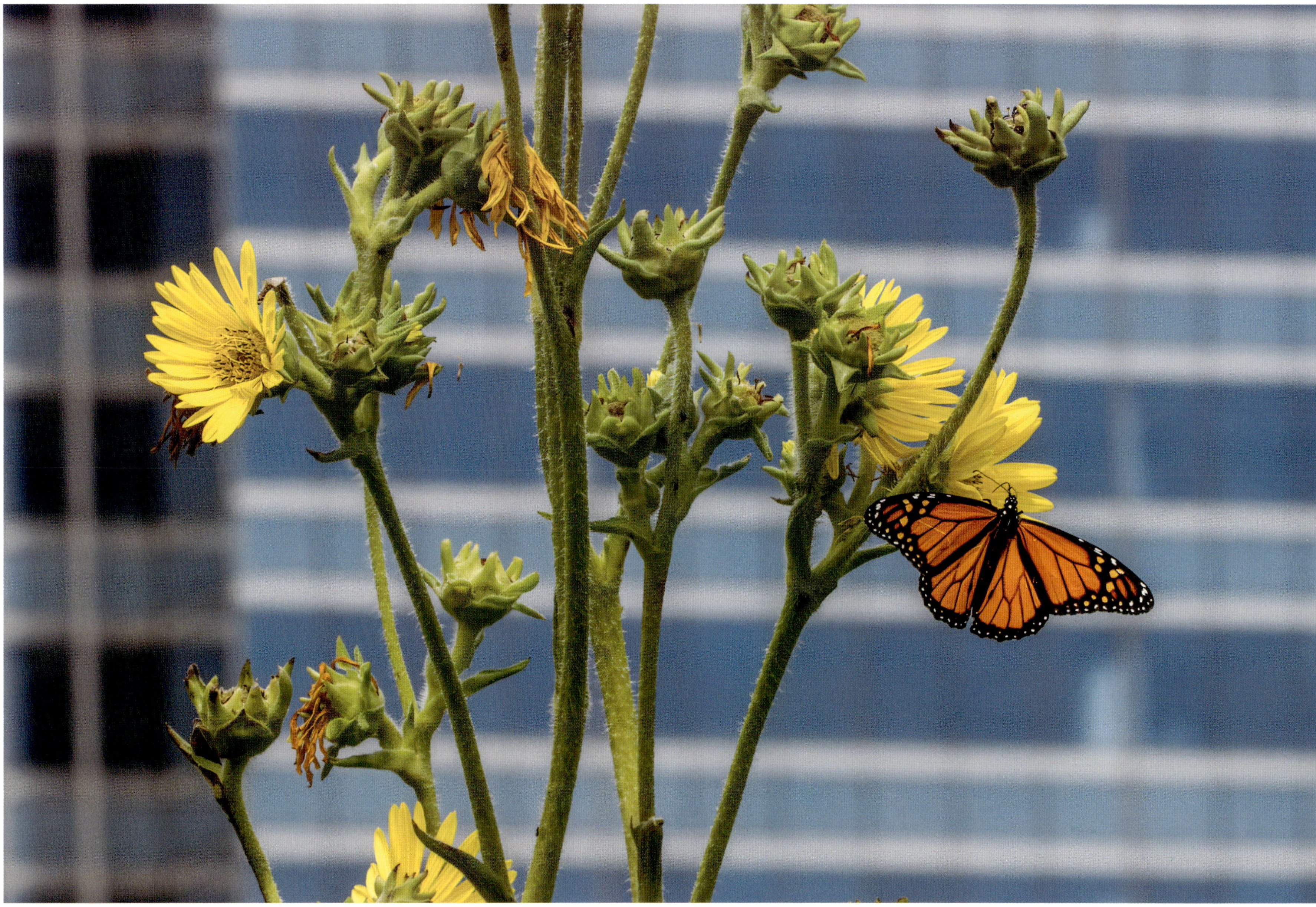

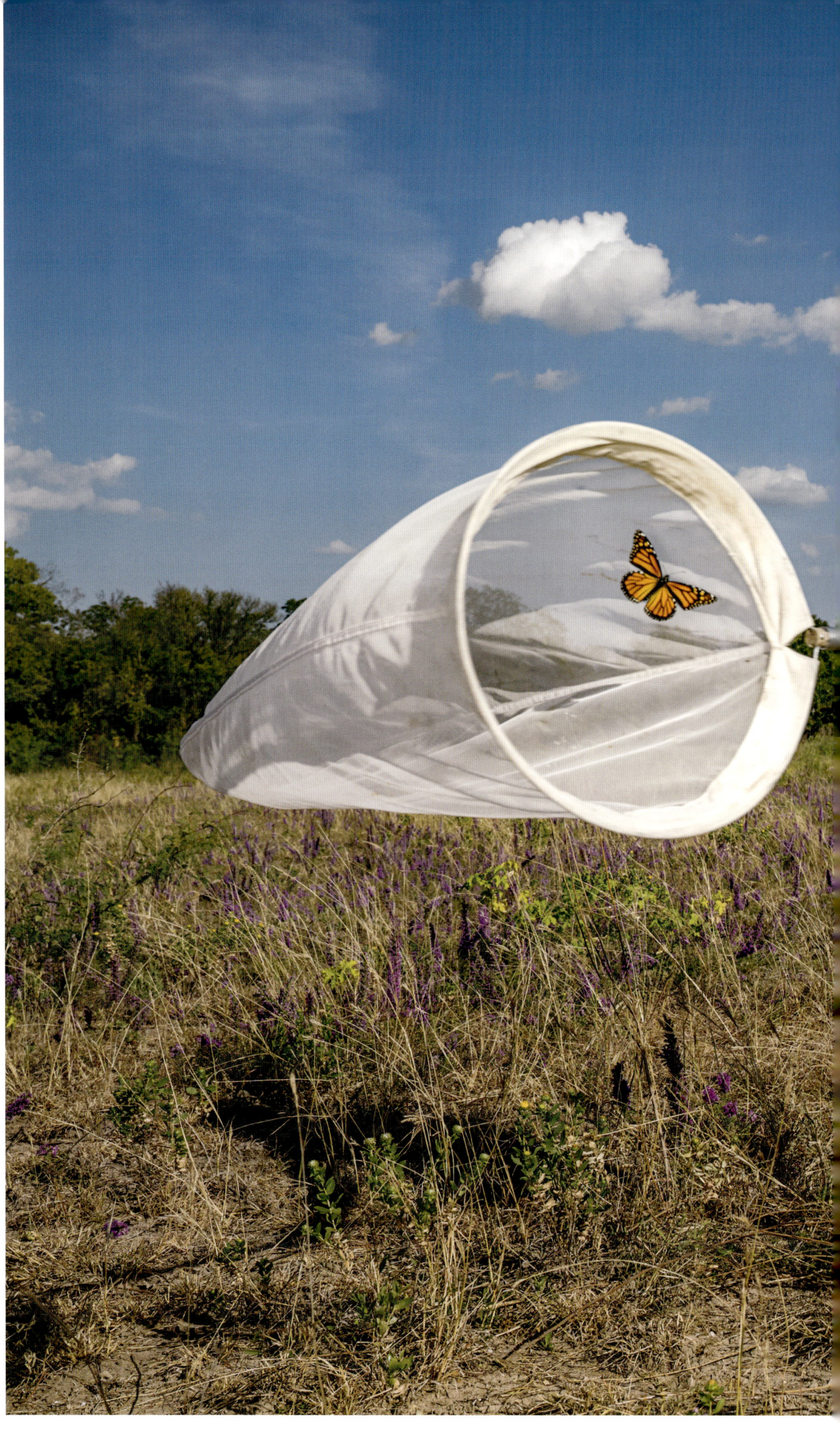

Daniel Knox, a seasonal biologist with Fort Hood's Adaptive and Integrative Management team, captures monarch butterflies for tagging at the Texas military base.

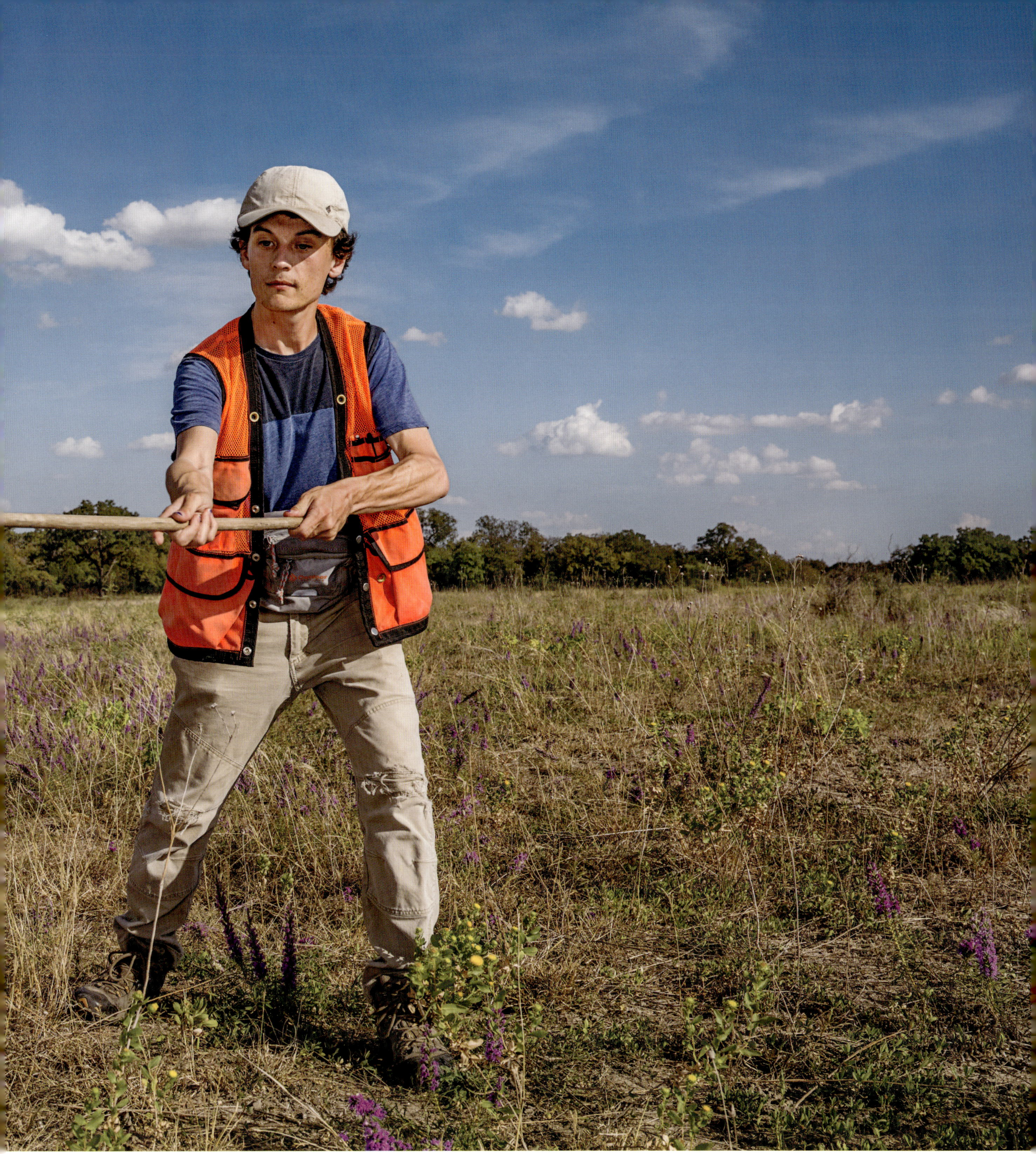

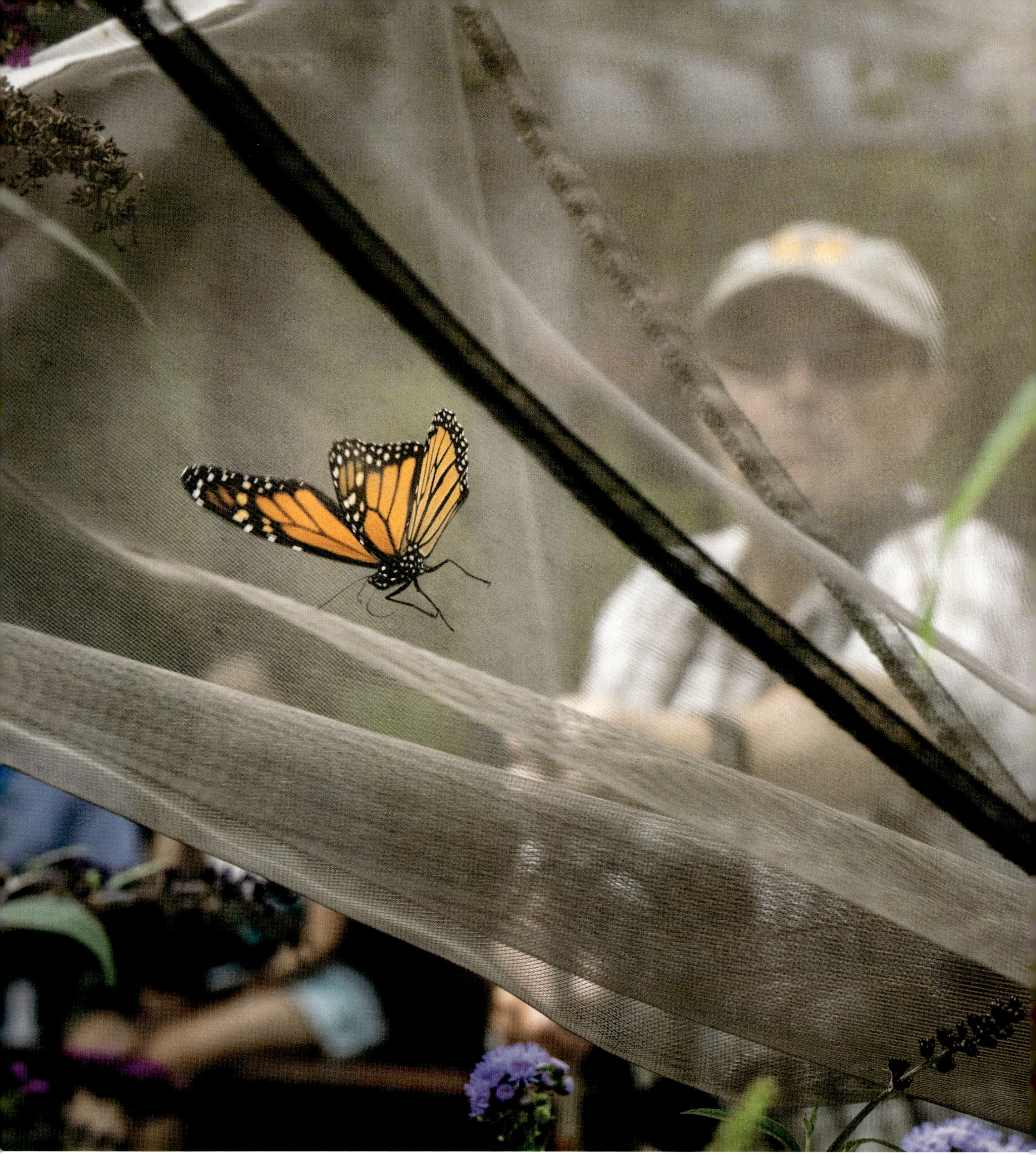

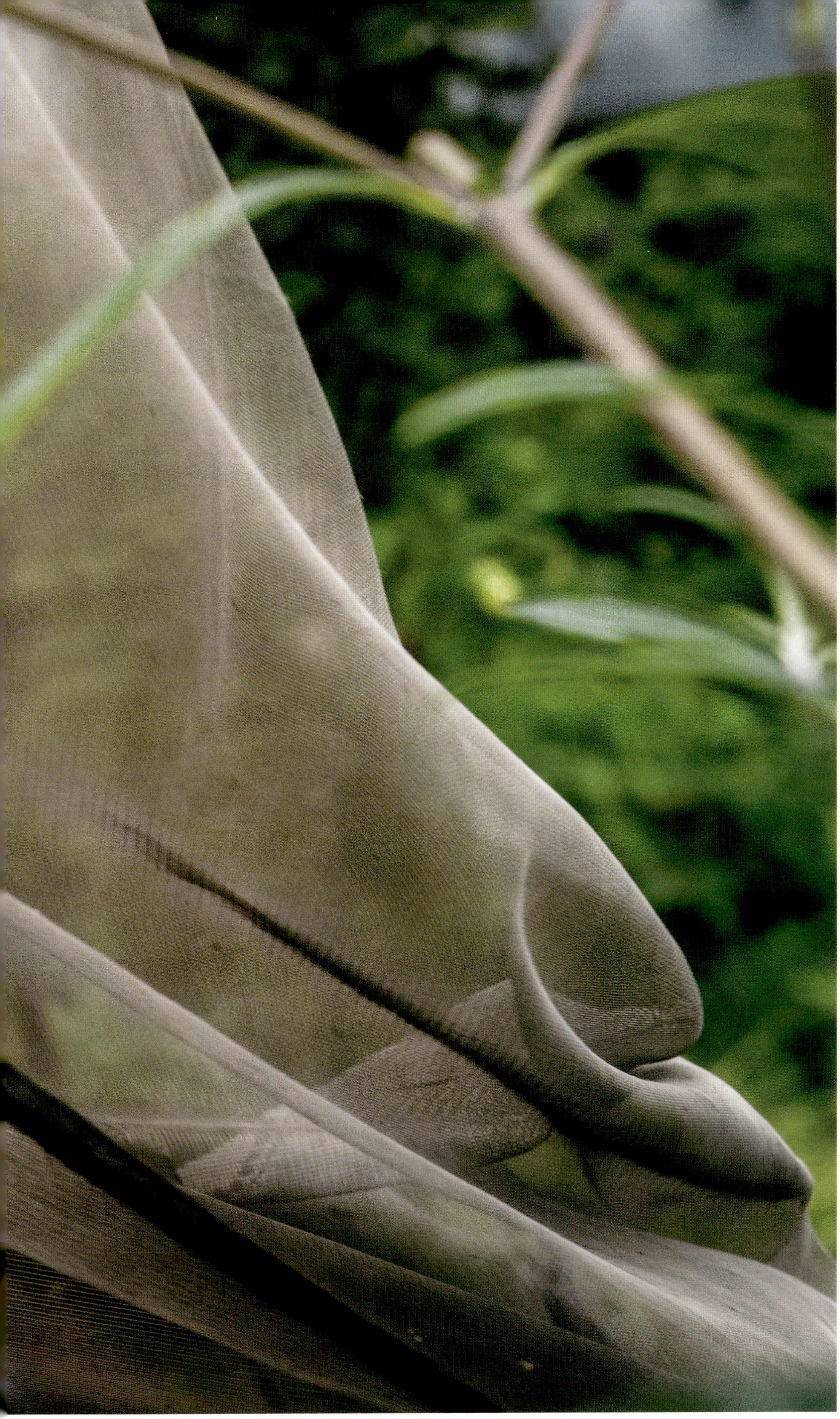

One of the first research assistants to work with monarch science pioneers Fred and Norah Urquhart, Don Davis is among the longest-serving citizen scientists in the monarch community. Here, Davis captures butterflies during a tagging event at Rosetta McClain Gardens in Scarborough, Ontario.

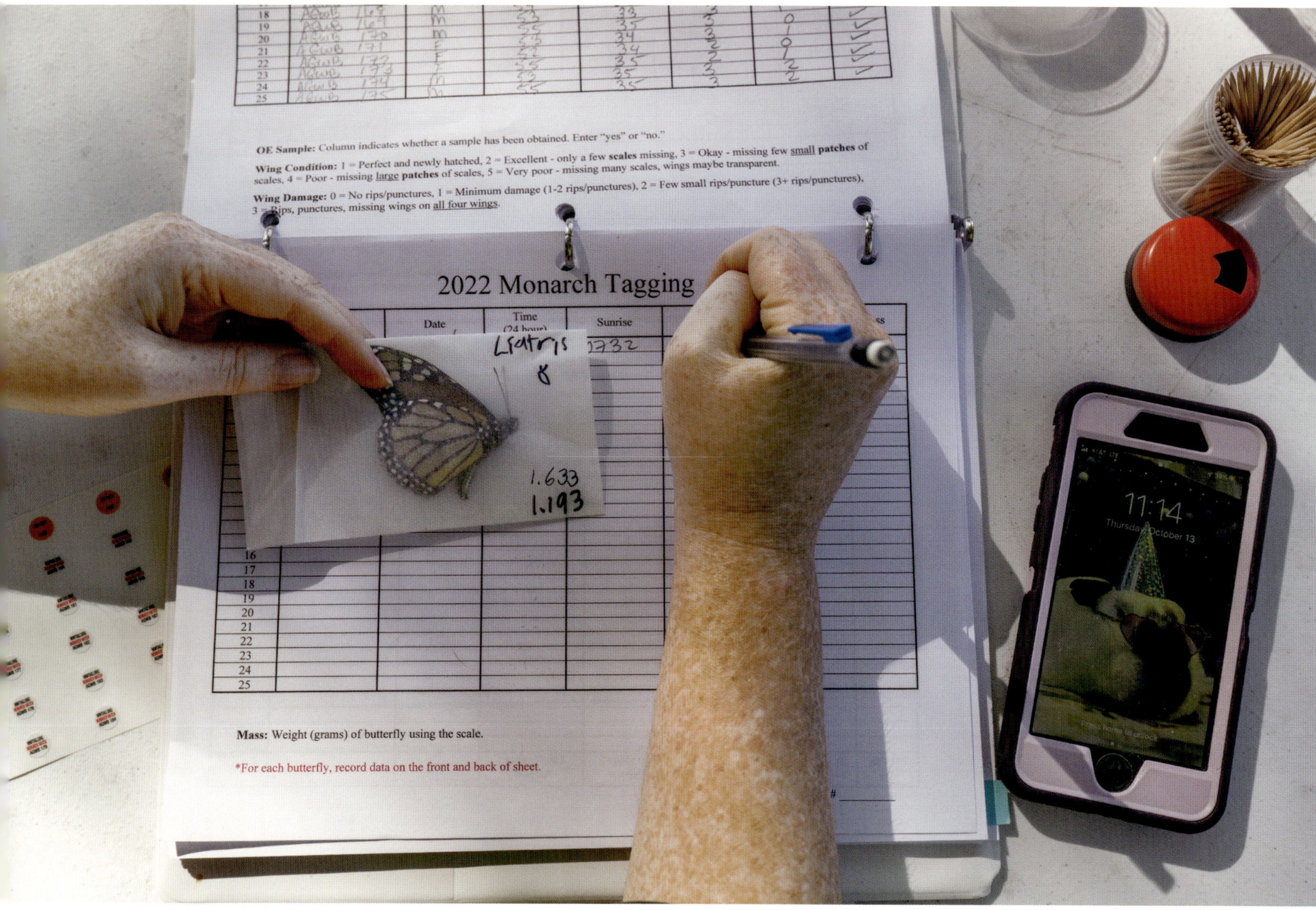

ABOVE AND OPPOSITE: Chelsea Plimpton, lead biologist of Fort Hood's Adaptive and Integrative Management team, gently records and tags a monarch butterfly. She will soon release it, hoping that someone in its ultimate destination in Mexico will find and report it to Monarch Watch.

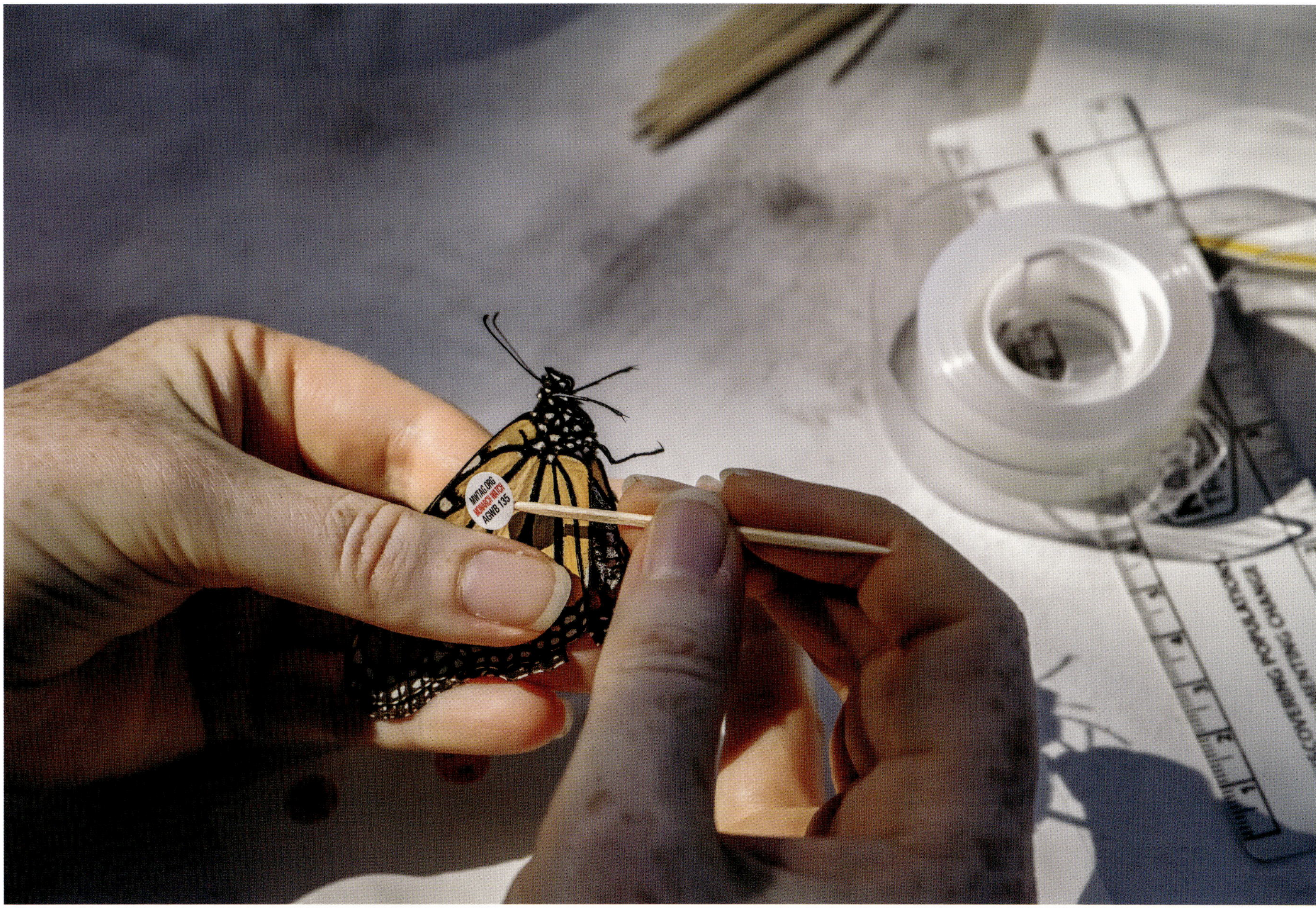

PAGES 114–115: A monarch butterfly feeds on wildflowers called blazing star (*Liatris* sp.) growing on a farm in Foley, Minnesota, that specializes in native plants from the prairies. Because these flowers bloom in late summer, they can provide extra food for monarchs just before they embark on their fall migration.

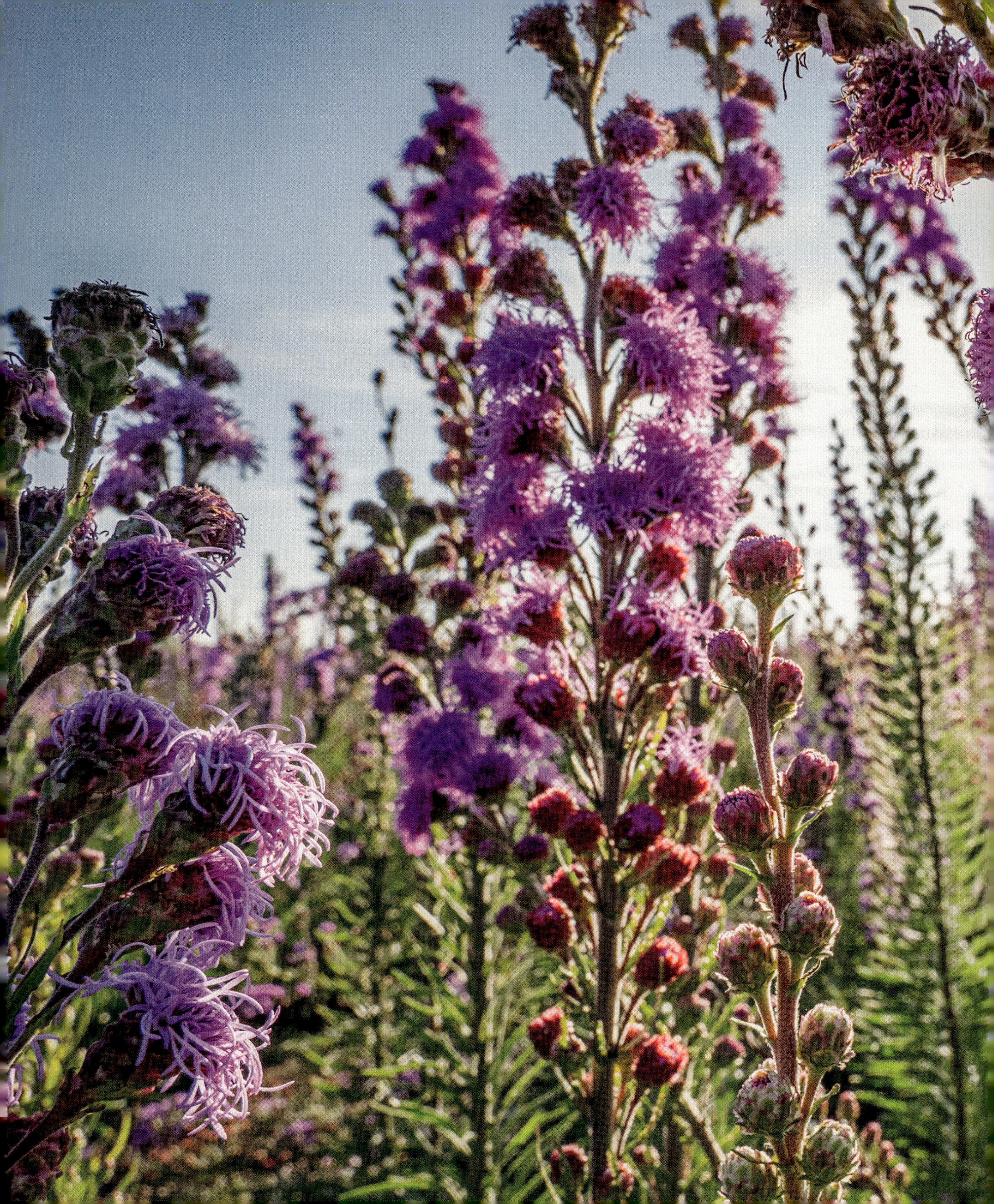

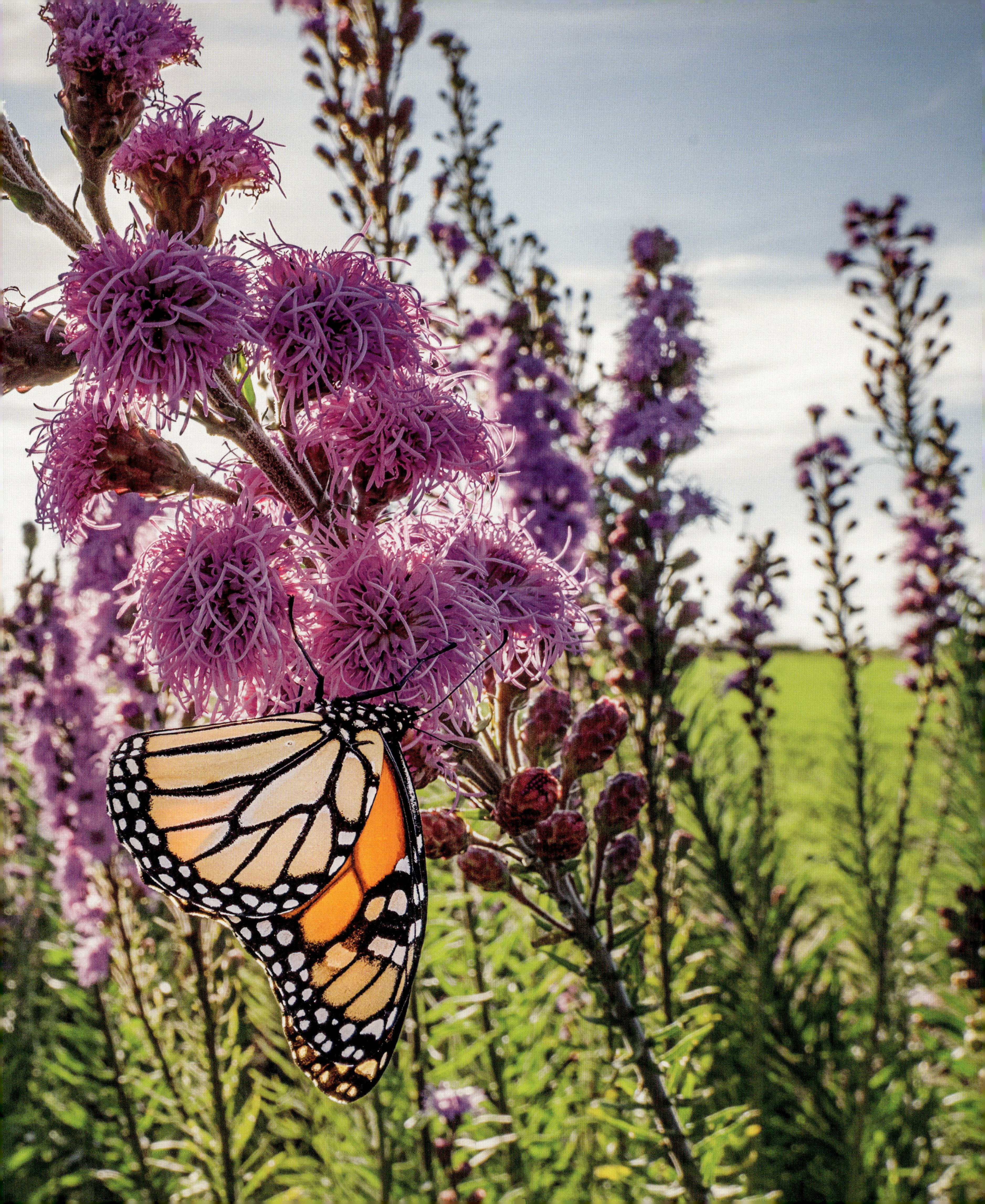

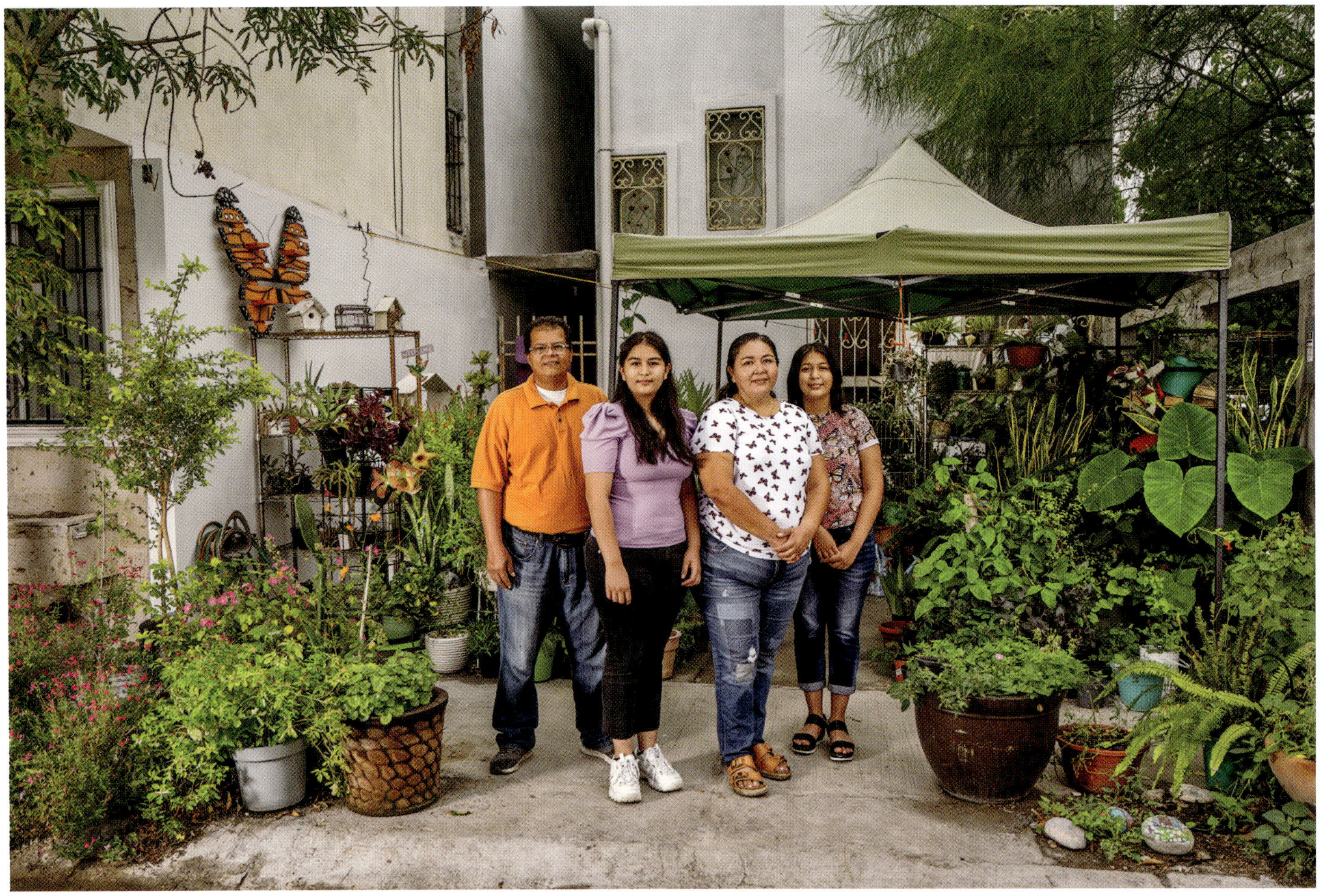

The Martínez Ortiz family in their pollinator garden, Apodaca, Mexico

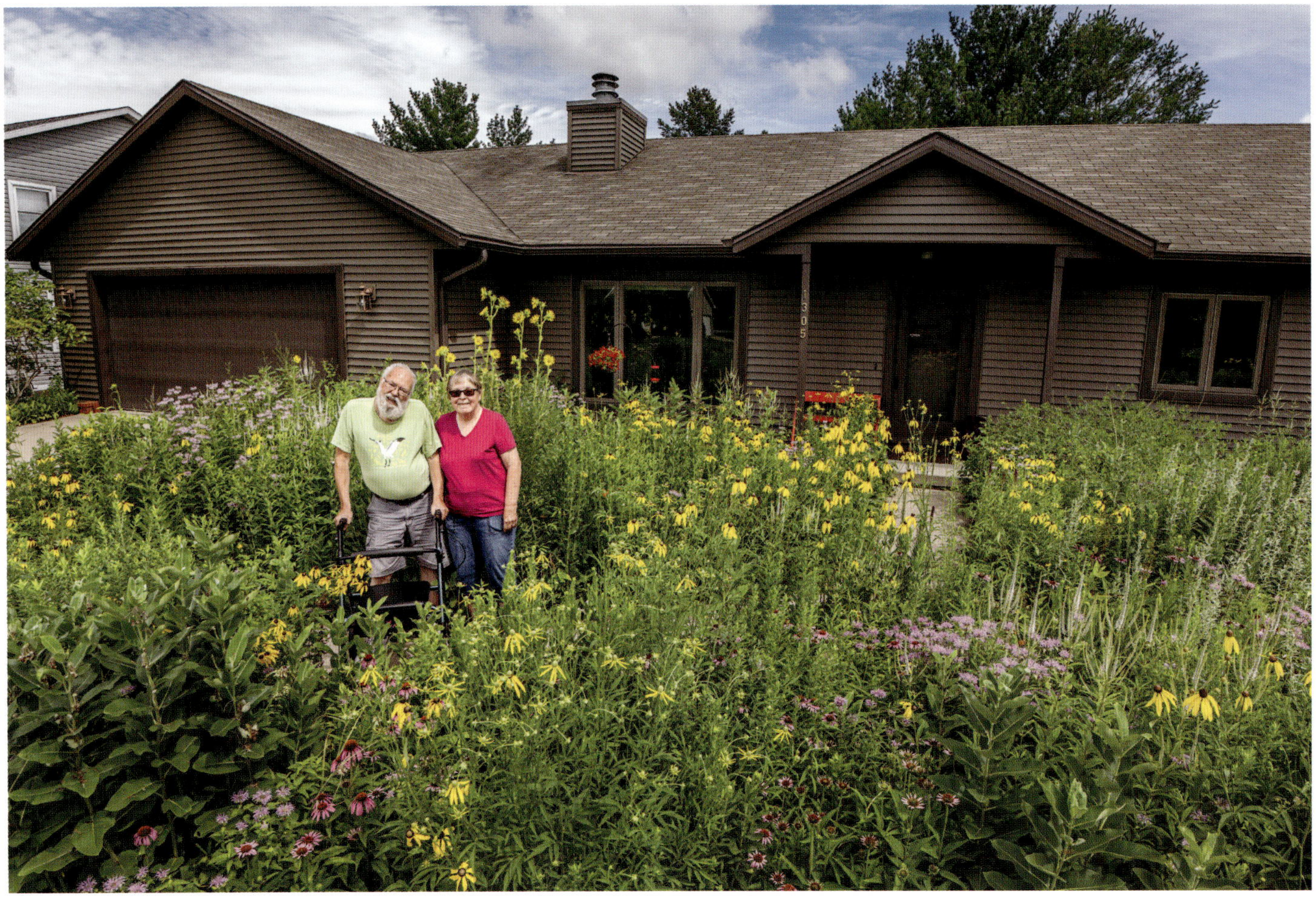

Bill and Ginny Nelson in their pollinator garden, Madison, Wisconsin

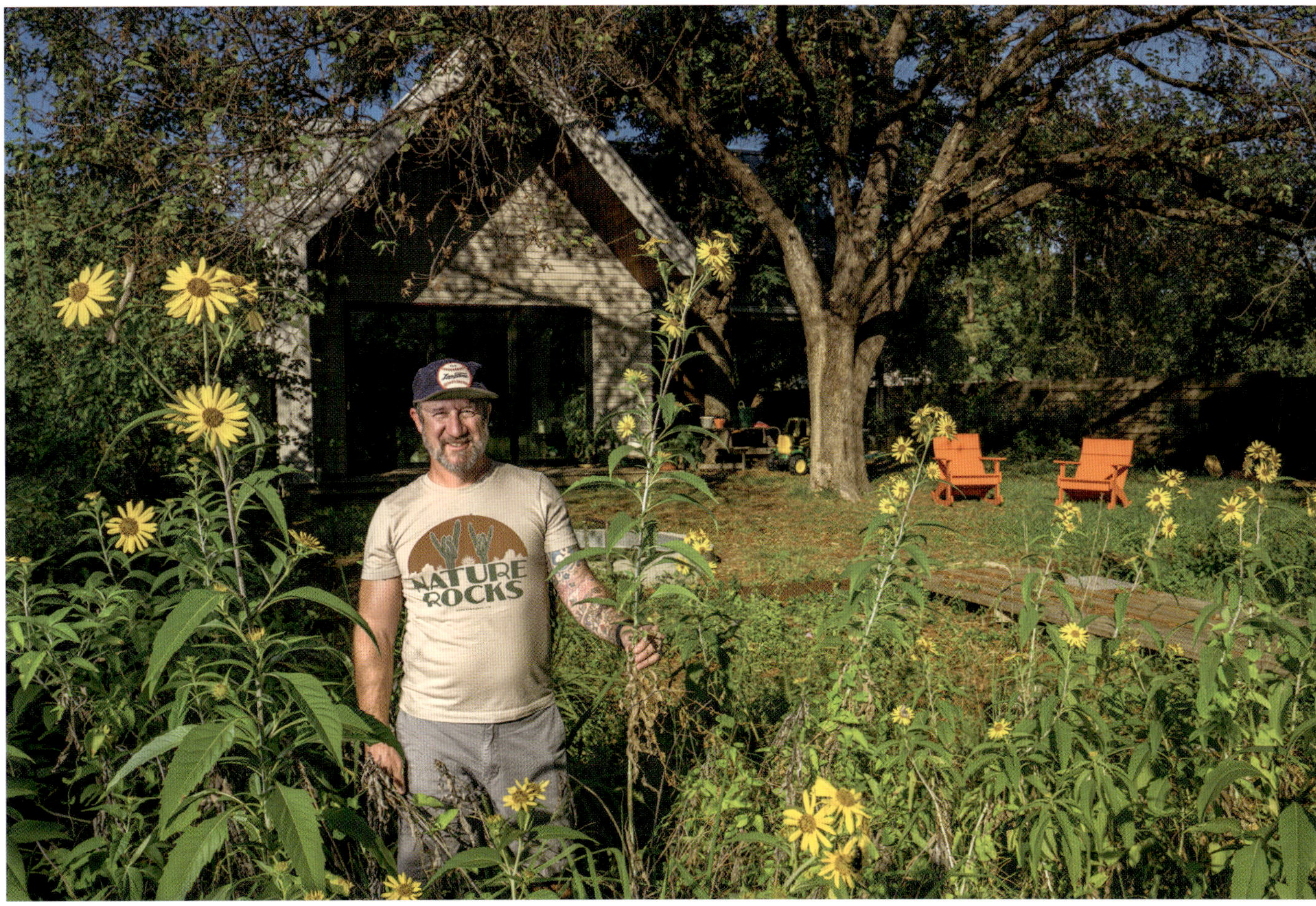

John Hart Asher II in his pollinator garden, Austin, Texas

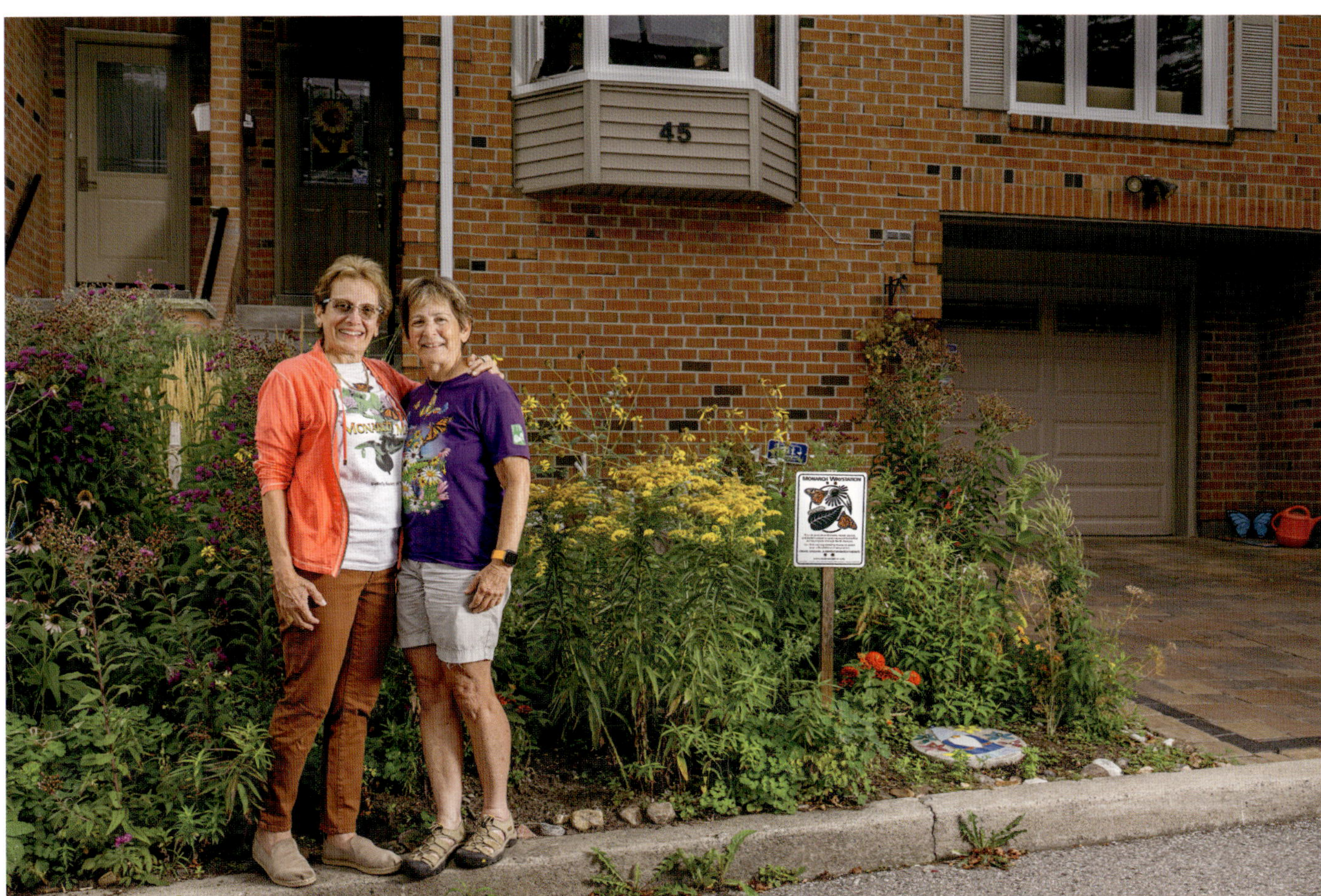

Audrey Kouyoumdjian and Carol Pasternak in their pollinator garden, Toronto, Ontario

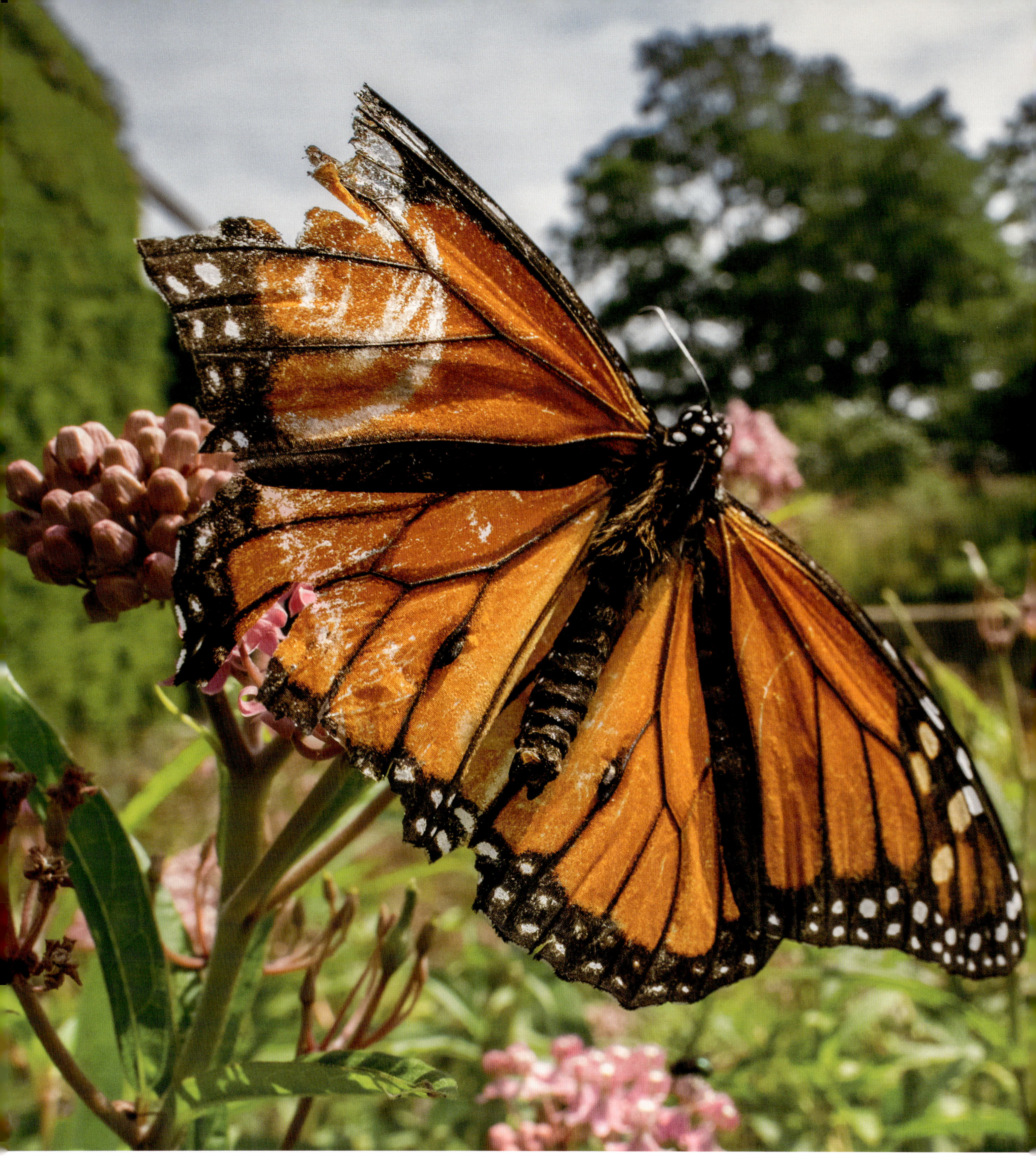

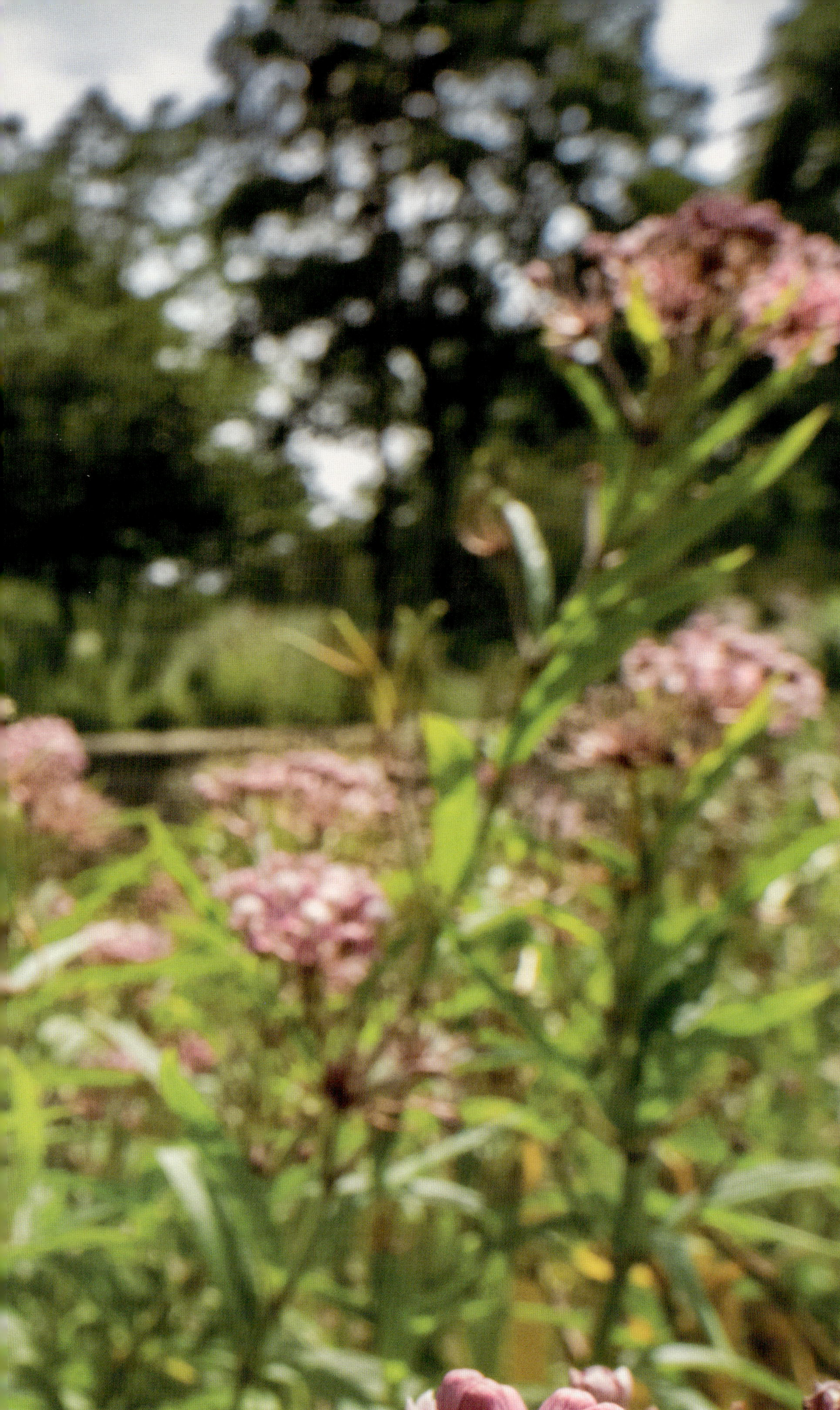

A battered male monarch butterfly in Chicago, on its long southward migration. Its wings show the toll, likely damaged by sun, wind, rain, and encounters with predators along the journey.

PAGES 120–121: A monarch chrysalis hangs from a dwarf gray dogwood *(Cornus racemosa)* in the pollinator-friendly backyard of Pam Karlson and Mike Edsey in Chicago. During this stage, the caterpillar transforms into an adult butterfly inside this jade green casing.

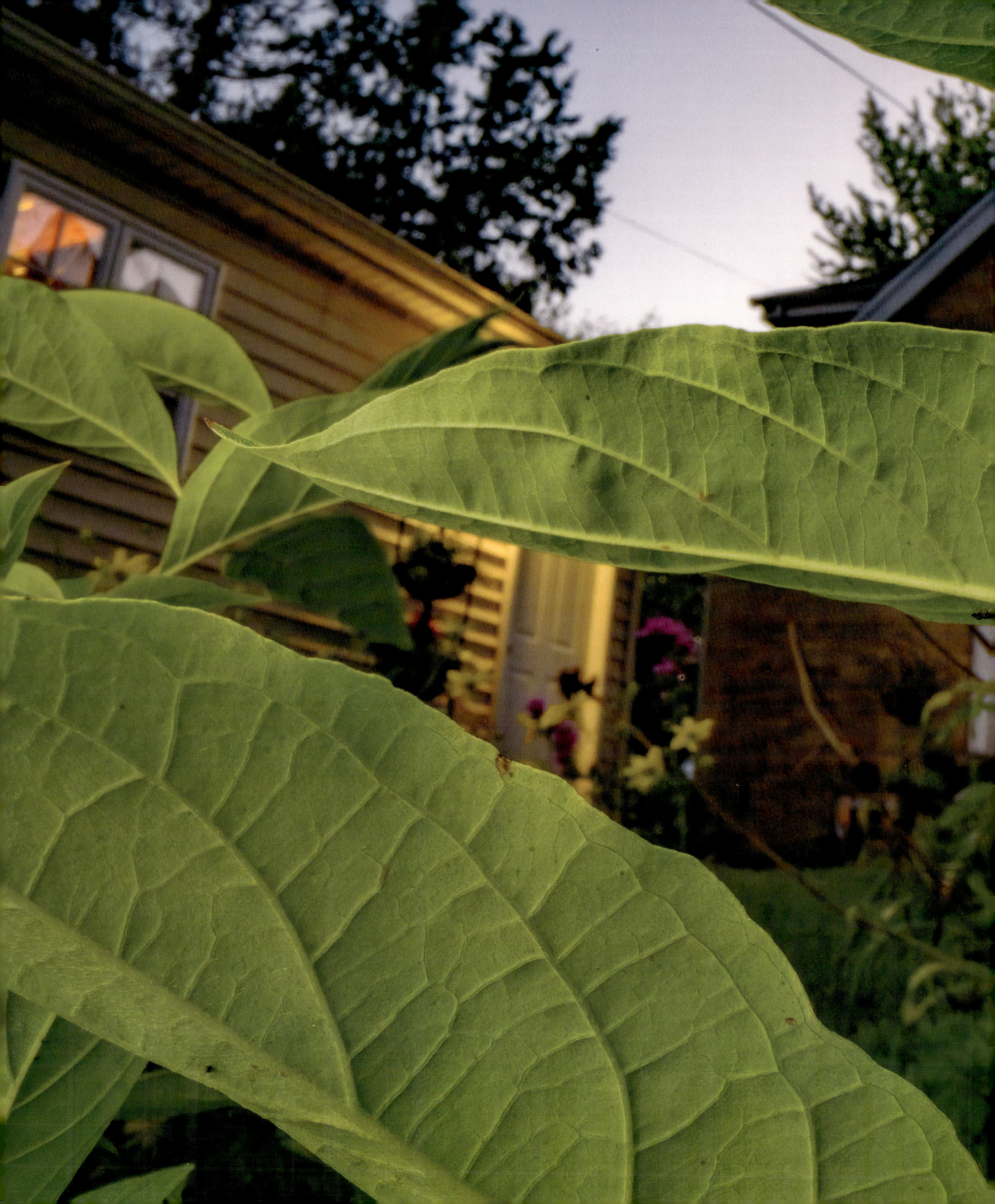

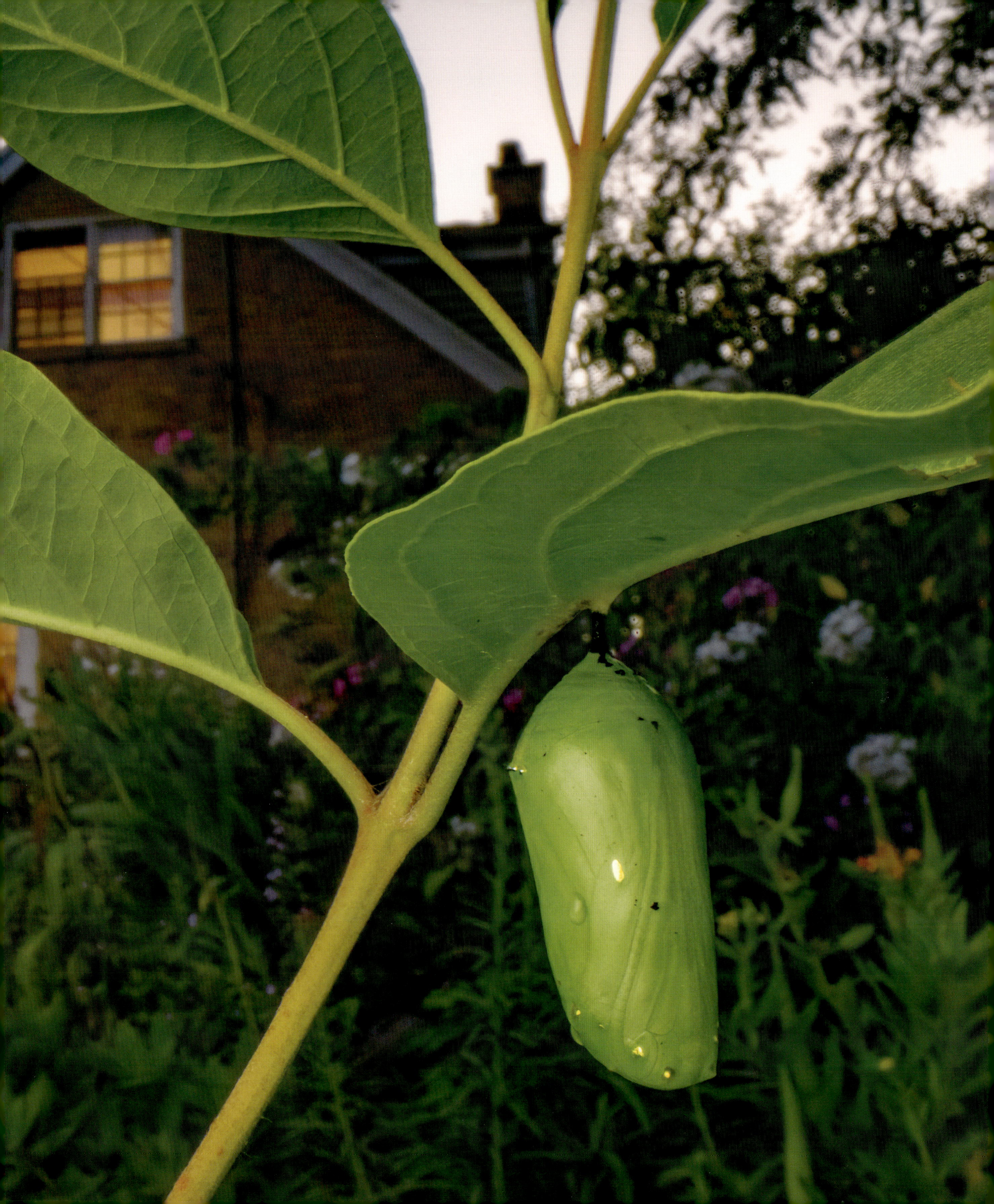

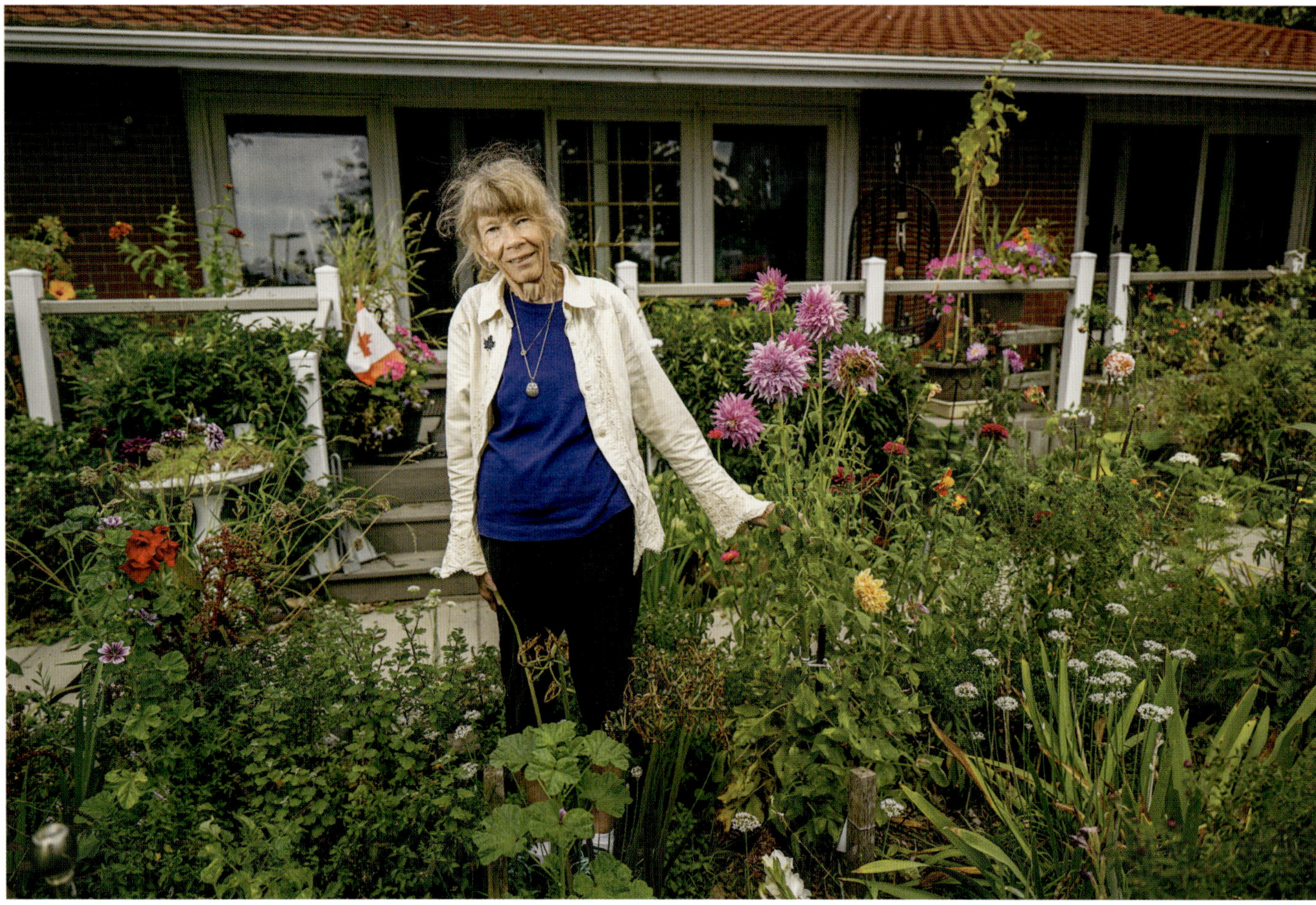

Susan Ross in her pollinator garden, Leamington, Ontario

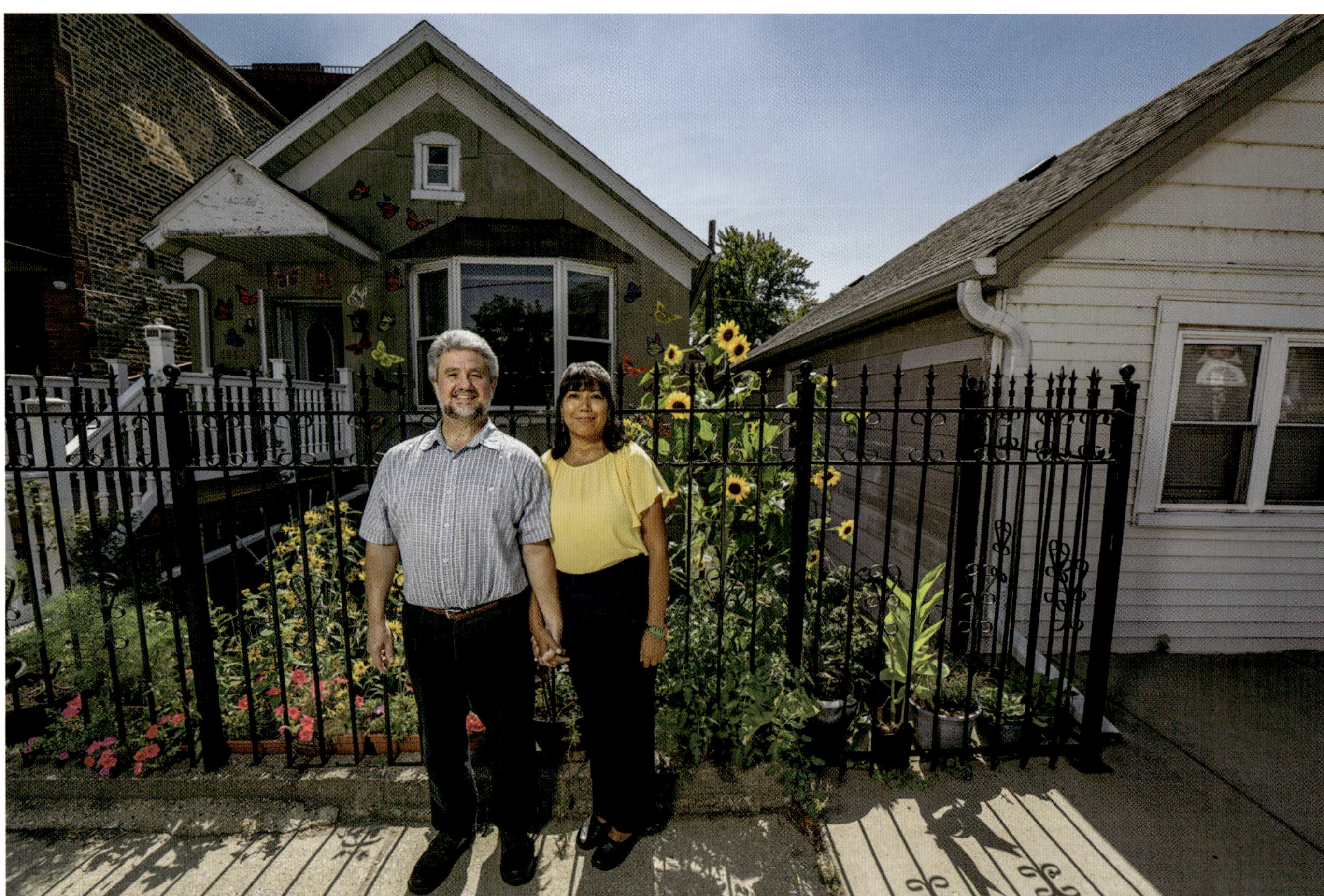

Leone José Bicchieri and Claudia Galeno Sánchez in their pollinator garden, Chicago, Illinois

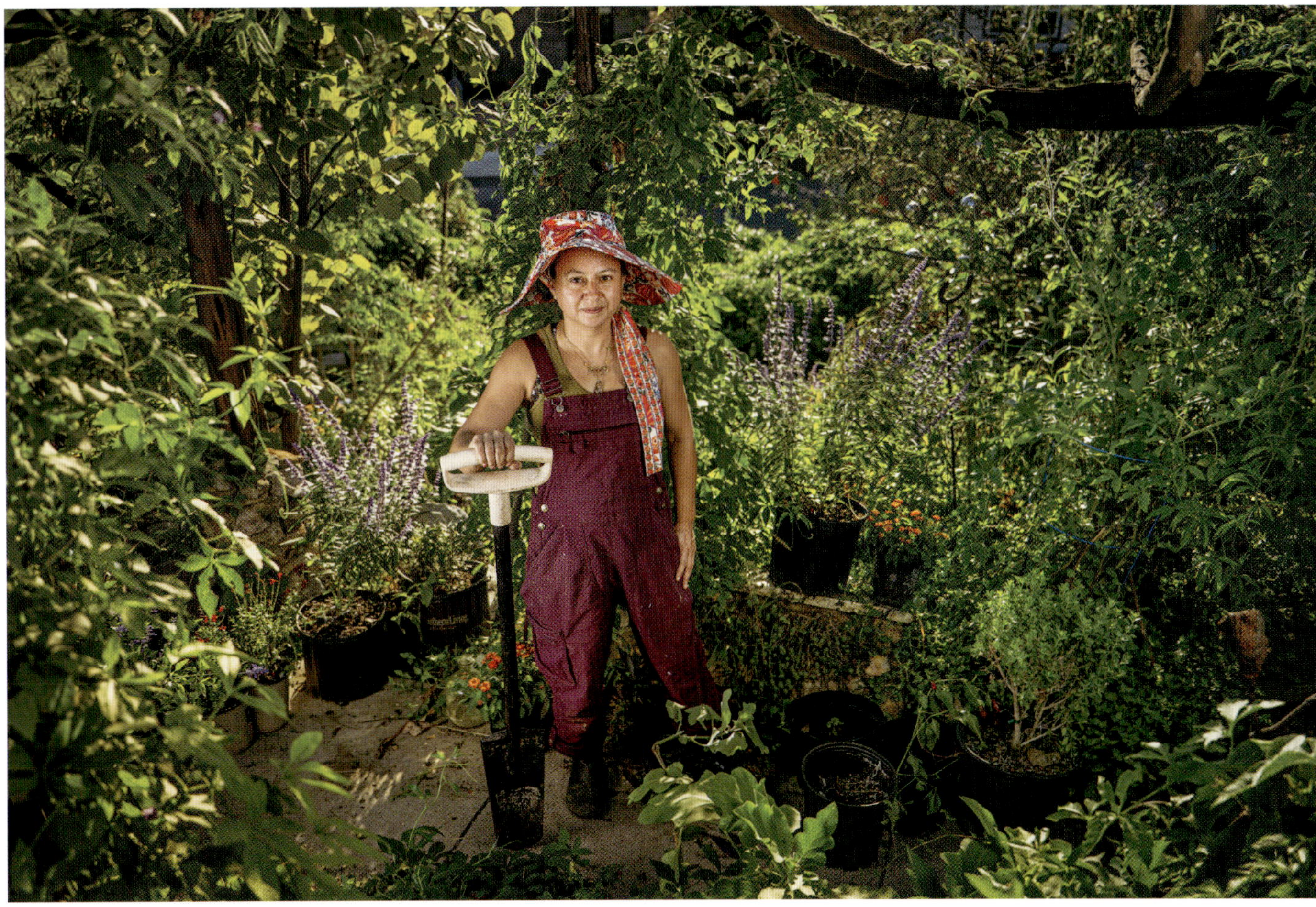

Alexa Villalobos in her pollinator garden, Austin, Texas

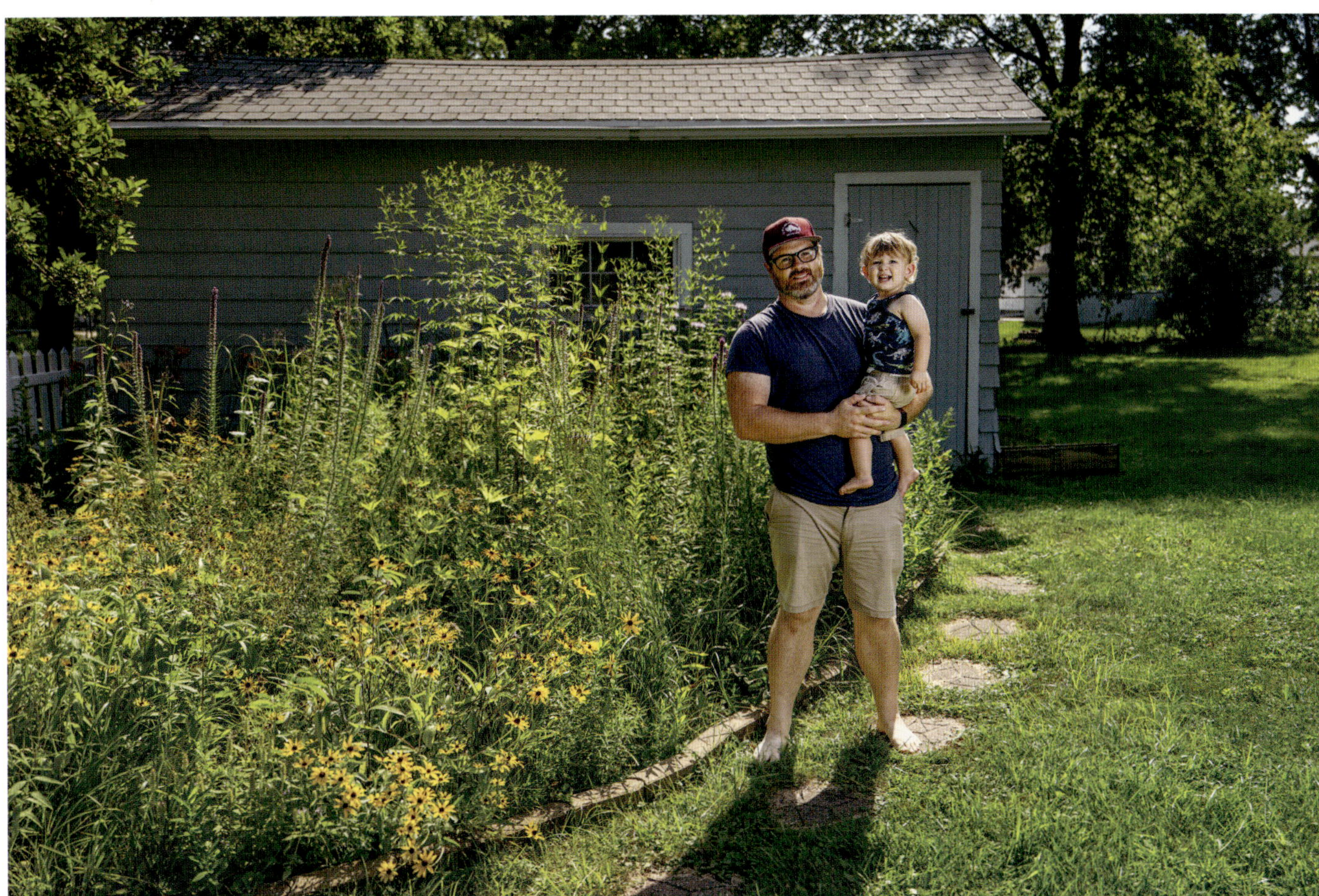

Justin and Owen Meissen in their pollinator garden, Cedar Falls, Iowa

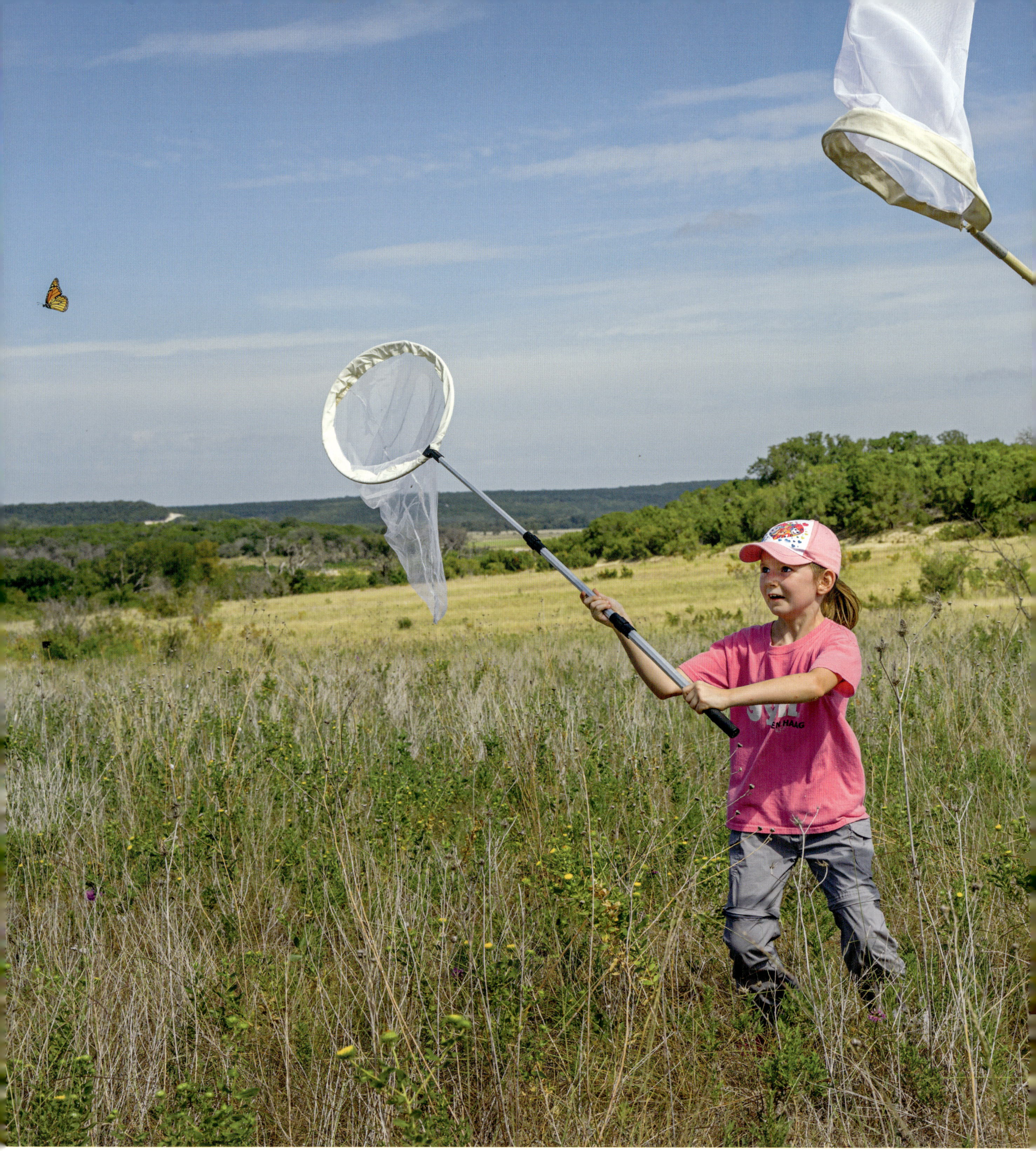

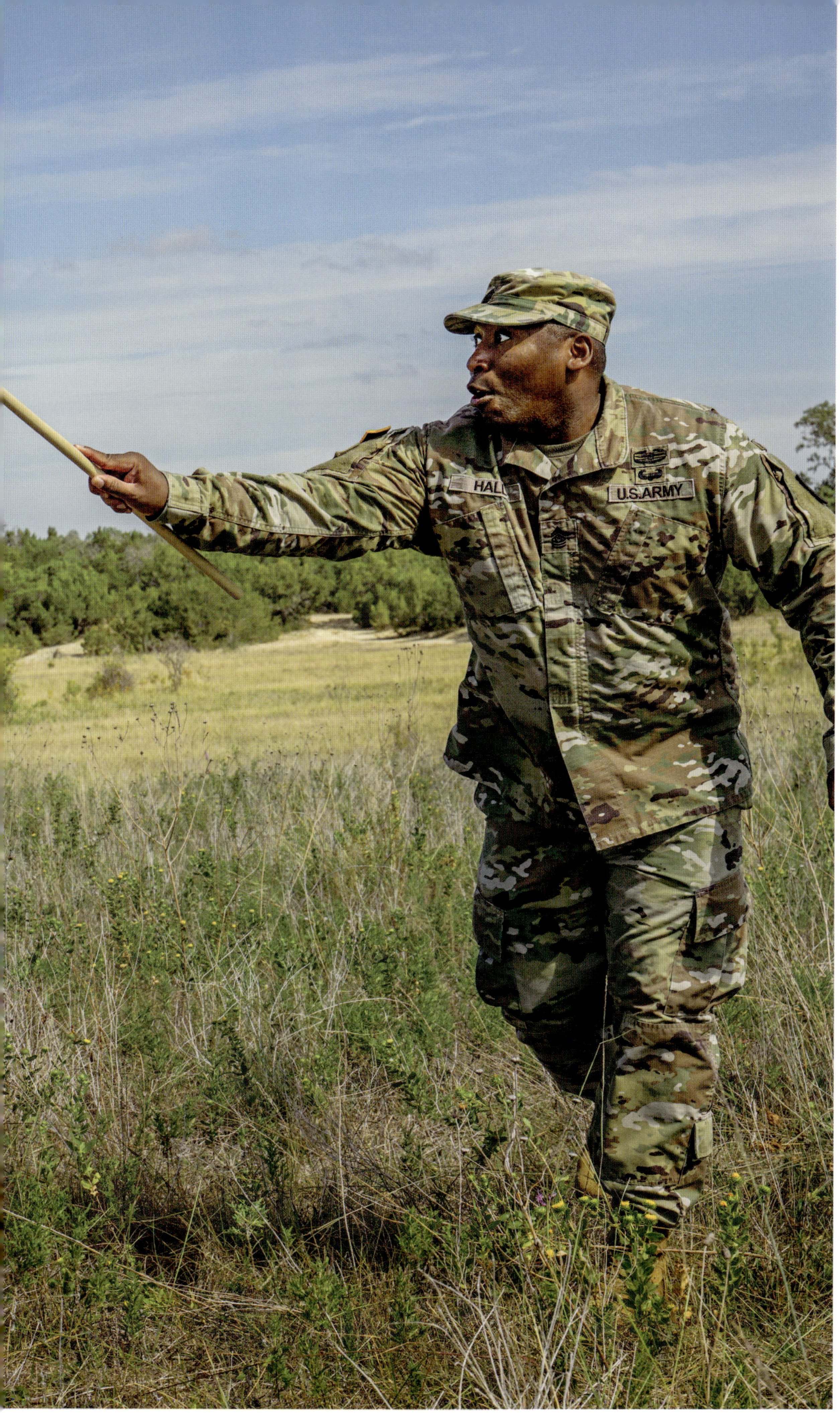

Command Sgt. Maj. Calvin Hall teams up with Briley Bray at a community tagging event organized at Texas's Fort Hood. Monarch tagging has become a popular environmental education and citizen science activity in North America, uniting thousands of people around a common goal.

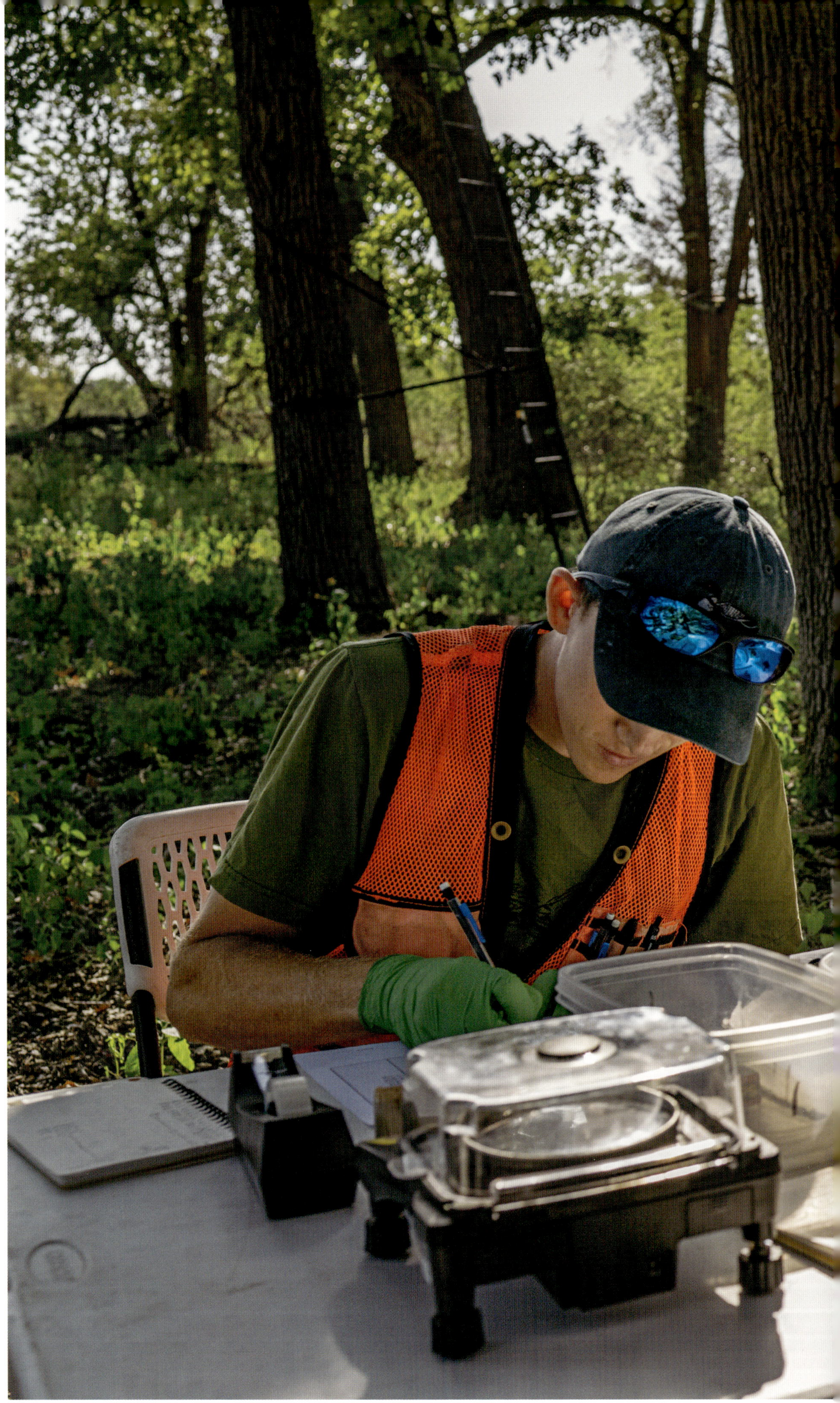

Chelsea Plimpton (right), lead biologist of Fort Hood's Adaptive and Integrated Management team, and her husband, Charlie Plimpton, tag monarch butterflies together.

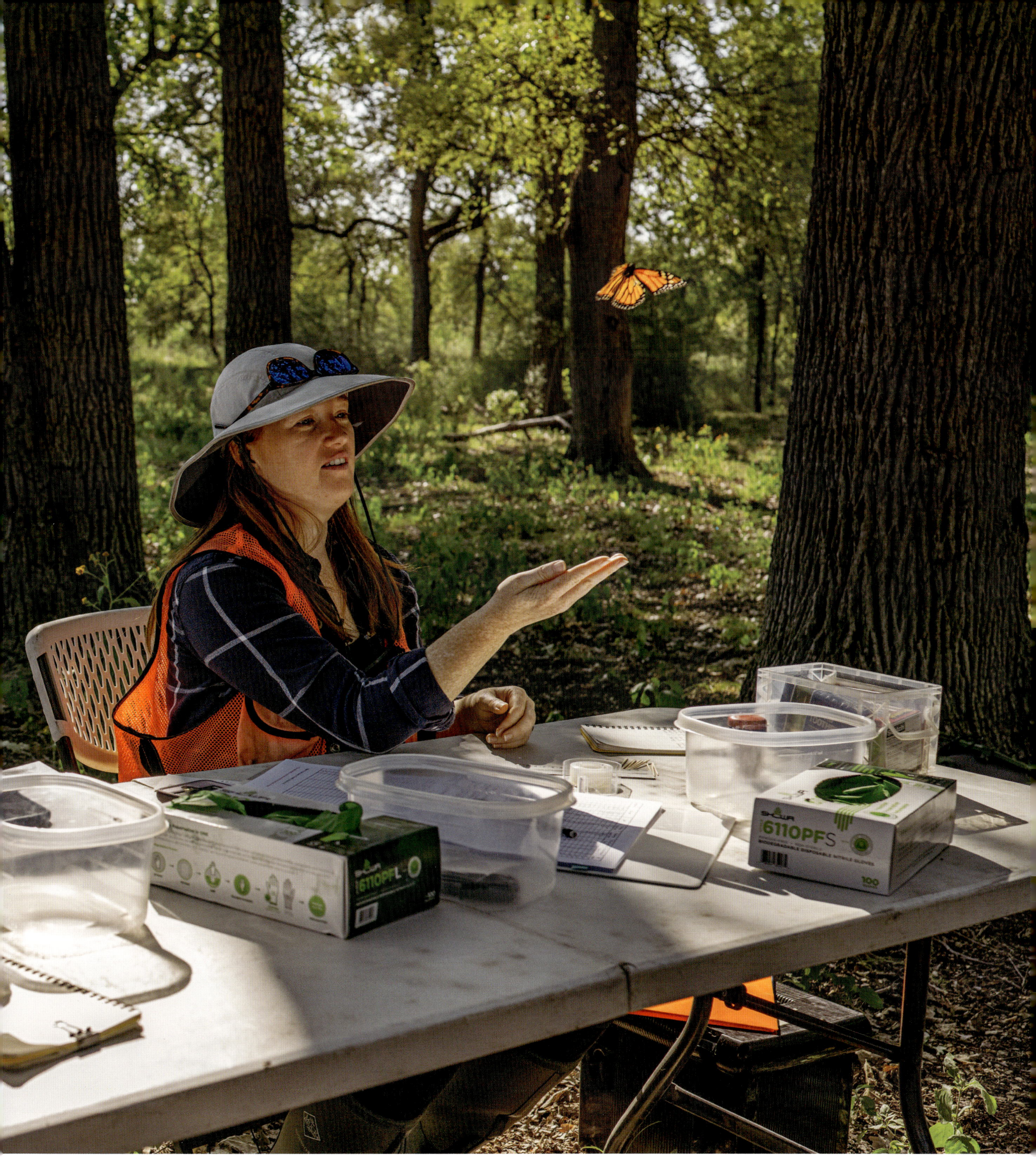
6110PFL
6110PFS
100

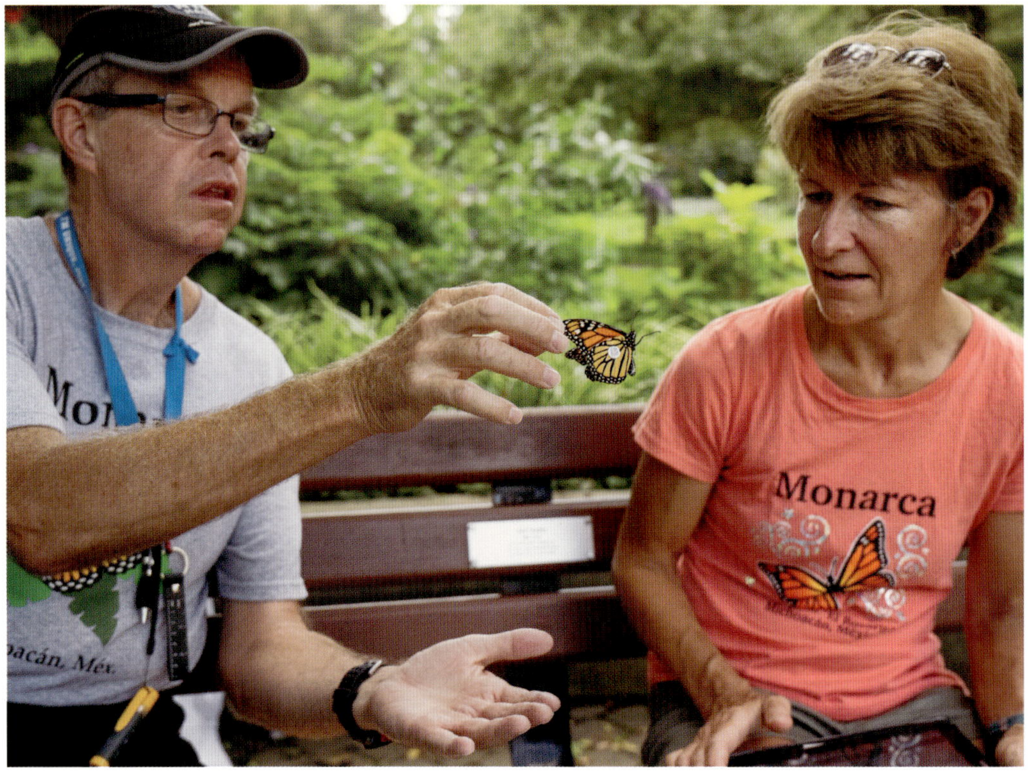

Terry Whitman and Kathy Haughton capture, tag, and release monarch butterflies at Rosetta McClain Gardens in Scarborough, Ontario.

OPPOSITE: A recently tagged monarch butterfly rests on a flower in the gardens before taking off on its southward migration.

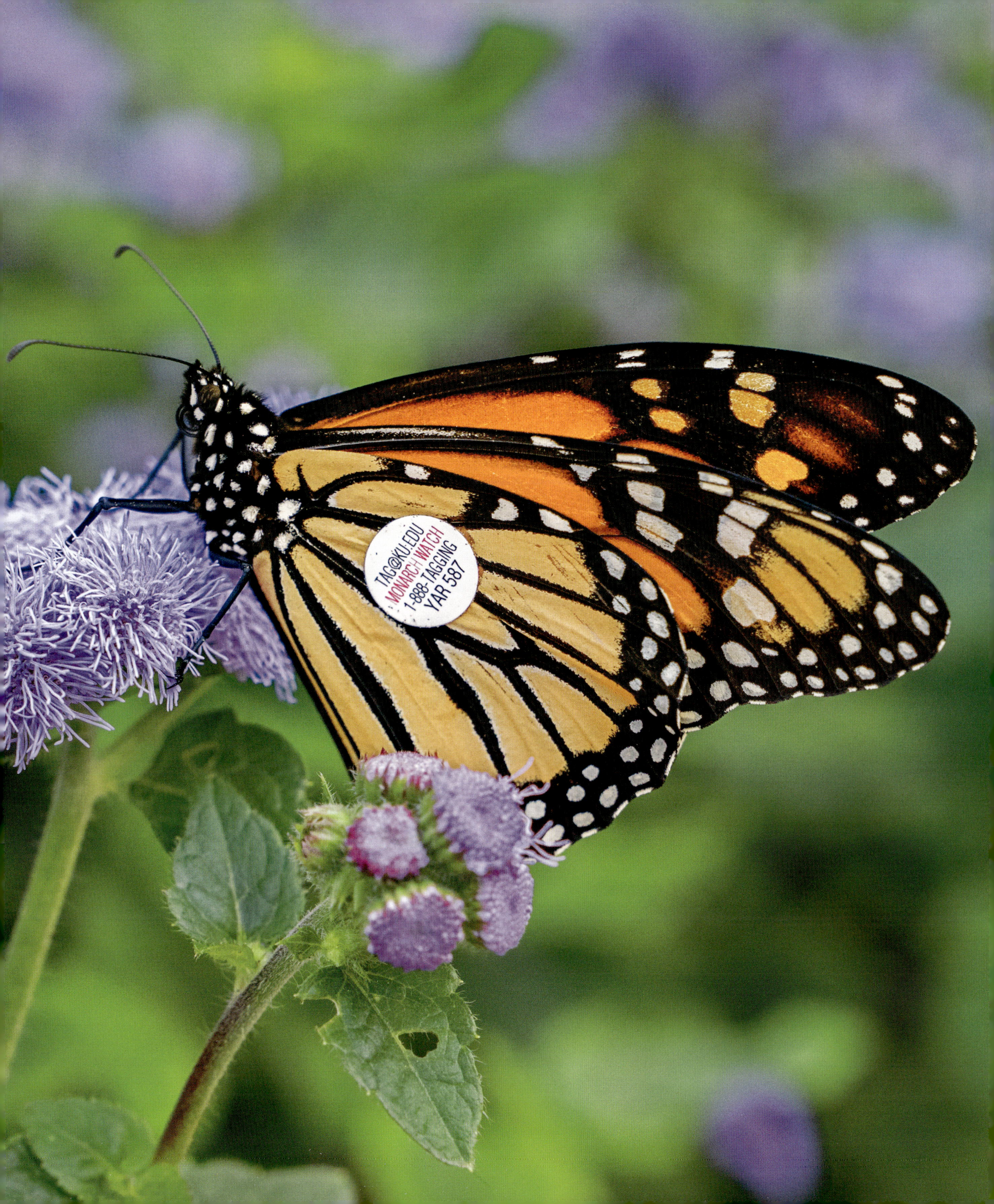
TAG@KU.EDU
MONARCH WATCH
1-888-TAGGING
YAR 587

The founder and longtime leader of Monarch Watch, Chip Taylor, pauses with his butterfly net in a prairie near Lawrence, Kansas.

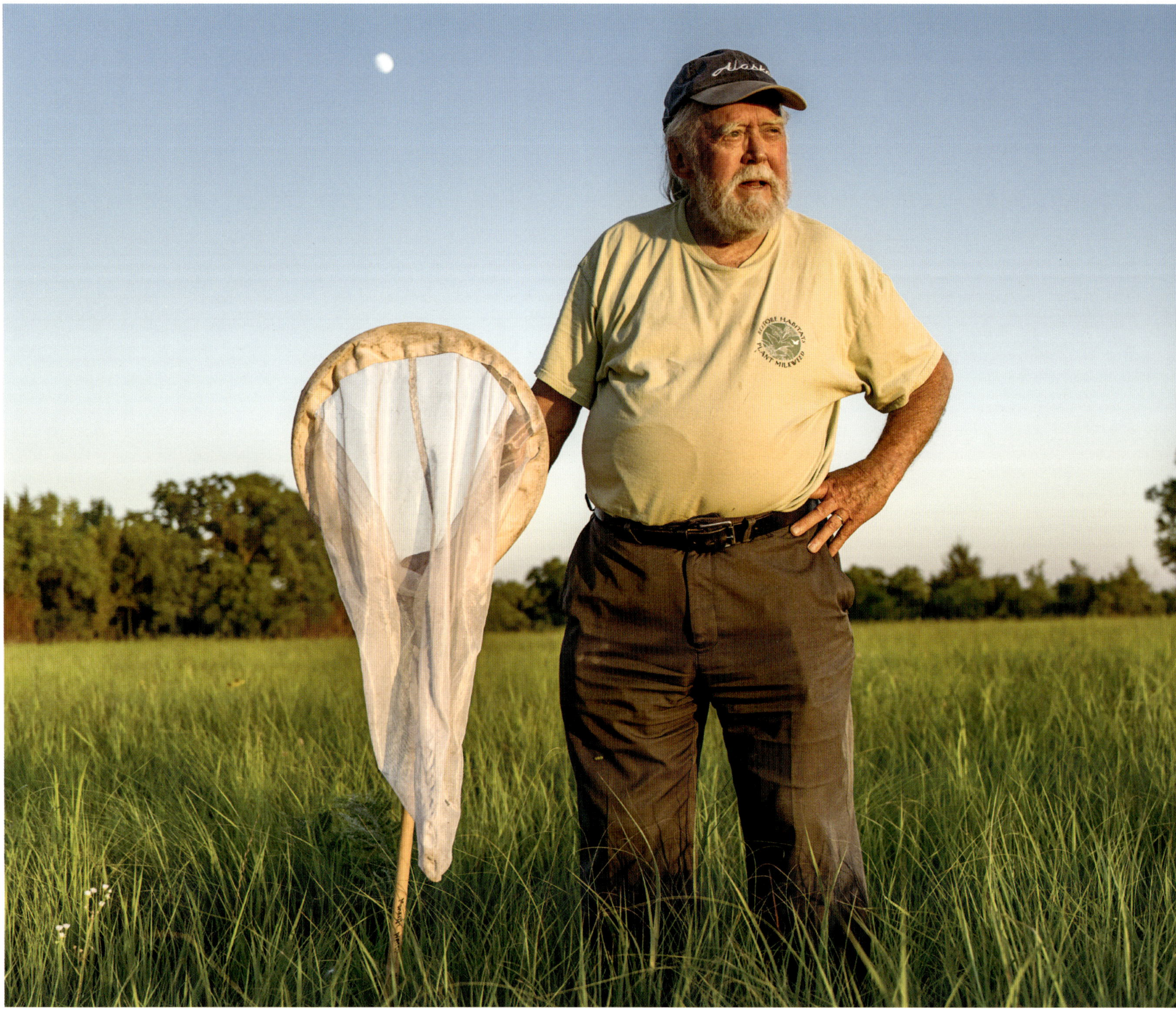

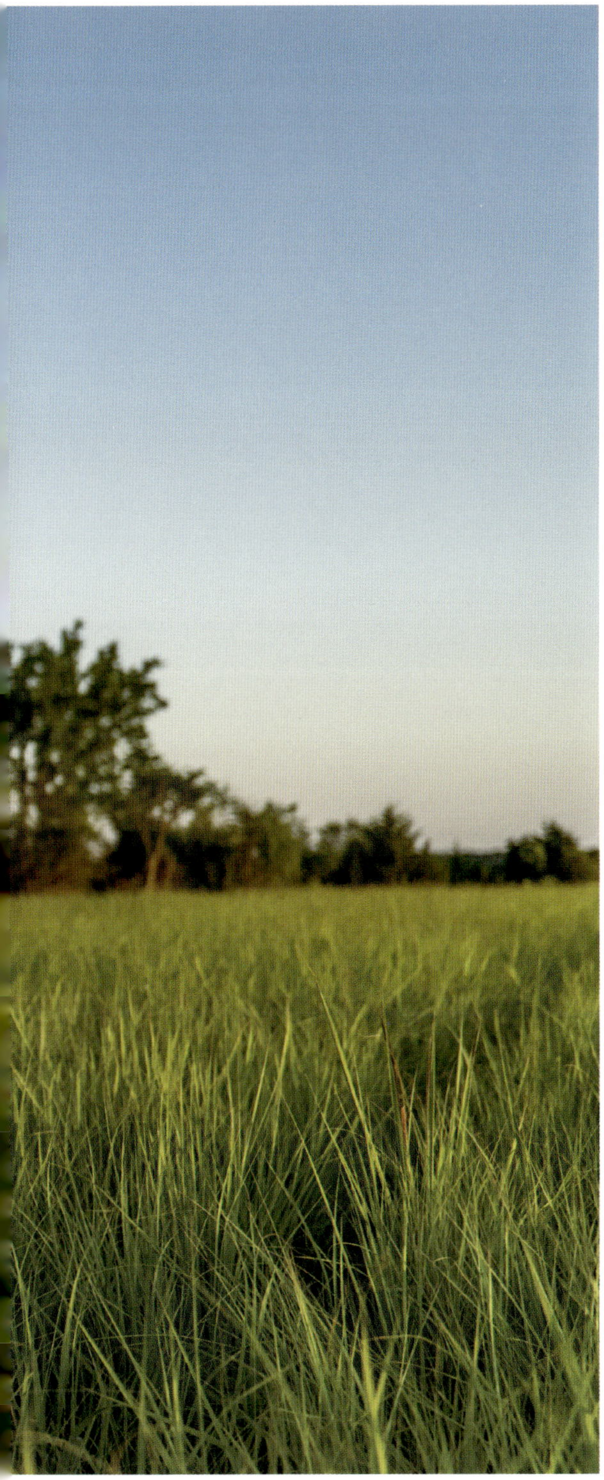

FIELD NOTES FROM

CHIP TAYLOR

I DON'T REMEMBER when I reared my first monarch; I was probably less than 10. At that age, I was already fascinated with the many butterflies and moths I encountered. I even had a small collection by the time I was 12. Honeybees were also of interest, and after much reading, I started my first honeybee colony when I was 14. Both interests have dominated my personal and professional life. I began working with the so-called killer bees in South America in 1974. By 1990 it became clear that funding for those studies would be coming to an end, and I began looking for another biological issue that I might be able to contribute to.

At that time, I was teaching a graduate course in field biology each fall at the University of Kansas, and it was natural to involve students in studies of monarchs migrating through our area each September. These monarch exercises generated many questions, and I began to realize how little we knew about the monarch migration. Those thoughts led to discussions about how we might expand our understanding, and that led to the start of Monarch Watch, a monarch-tagging program, in 1992. Fred and Norah Urquhart of the University of Toronto had developed a tagging program that showed the location of the monarch overwintering colonies in central Mexico in 1975. The Urquharts were interested in learning the geographic origins of monarchs that reached Mexico, and by the early 1990s they had established that those butterflies originated from much of the area east of the Rockies. They ended their program soon after.

I thought we could learn more through a more targeted tagging program. There were still many questions to answer, such as when the migration starts and what changes occur in the fall that cause the butterflies to begin migrating. But to answer those questions, we had to recruit volunteers to tag migrating monarchs from the Rockies in the west to the Canadian Maritime Provinces in the northeast. Boldly, but somewhat tentatively, we fashioned a press release that was distributed to the upper midwestern states in early September, asking for volunteers. We offered to send tags and instructions at no cost if people

I THOUGHT WE COULD LEARN MORE THROUGH A MORE TARGETED TAGGING PROGRAM.

agreed to participate by sending us the dates and locations at which the monarchs were tagged. Our request appeared on the front page of the *Des Moines Register,* and the response was overwhelming. Within days we had signed up over 500 volunteers. A month later, we sent the same message to news outlets in Oklahoma and Texas and quickly added another 500 volunteers. We were off to a quick start, and although the responses from the volunteers were extremely positive—especially from teachers who involved their students—the tagging results were disappointing. To start the program, we had adopted the Urquhart tagging system, which involved placing the tag over the leading edge of a forewing after rubbing off the scales on both sides of the wing so that the adhesive would stick to the membrane. That method didn't work well. Tags fell off and wings were broken.

It was clear we needed a new system. So in the following four years we asked volunteers to use glue to apply rectangular tags to the underside of a hindwing. This method was better, but it was still difficult for the taggers, and the recoveries were too few. We achieved a breakthrough in 1997 by developing a circular tag with a special adhesive that could easily be applied to the discal cell—a large, structurally sound cell in the middle of the underside of the hindwing. This method was a game changer. Recoveries went from one out of every 5,000 tagged to one out of every 100, and even higher from some locations. In Mexico, we started paying guides and residents for recovered tags in 1998 and reporting on the increase in recoveries to our tagging volunteers. The numbers tagged and recovered increased rapidly, as did the geographic area of tagging. With the new tags, it became easier to sponsor events at which we invited the public to participate. These events allowed us, and organizations that partnered with us, to educate the public about conservation issues. And perhaps more important, they became family and classroom affairs that provided parents and teachers the opportunity to experience nature with their children and students. The number of monarchs tagged at these events is certainly less important than the experiences of the children. But the numbers tagged and recovered over all the years are important, too, since they have helped us answer many questions about the migration. Still, others remain.

The numbers are impressive. More than three million monarchs have been tagged since 1992, and more than 21,000 tags have been recovered in Mexico. This is remarkable considering the time (two to three months) and distance (1,000 to 3,000 miles) that it takes monarchs to arrive at the overwintering sites. The hazards on these journeys are many, ranging from natural enemies such as spiders, mantids, and some birds to traffic and weather extremes. Although monarchs are not as fragile as other butterflies, many suffer wing damage that lessens their ability to continue the migration. Overall, the tagging data suggests that under the best of conditions only 15 percent of the monarchs reach the overwintering sites. There they must survive through the four-month winter and then the 800-mile return flight to Texas in March to start over again by laying eggs on milkweed. It's a life-and-death annual cycle that is truly amazing.

Analysis of the dates of tagging shows that there is a migration window: a 30-day interval for each latitude. The window seems to be defined by the decline of the sun in the fall sky. Migrations begin after the sun angle at noon drops below 57 degrees and end after the angle has declined below 46 degrees. Butterflies tagged within the time that the sun declines from 57 to 46 degrees have the best chance of reaching the overwintering locations in Mexico. Those tagged after the sun is lower than 46 degrees have lower prospects of reaching their destination. Failure to reach the overwintering sites can be quite high when the emergence of the last butterflies of the summer is delayed by cold weather or migrations are stalled due to high temperatures that limit flight. Amazingly, the first of the fall monarchs arrive at the overwintering sites in the last few days of October, as the sun angle drops below 57 degrees—just in time for the Day of the Dead celebrations on November 1 and 2, when some local residents believe the monarchs represent the souls of deceased ancestors returning to visit them for the holiday.

Chip Taylor is an emeritus biology professor and the founding director of Monarch Watch, an organization dedicated to monarch education, conservation, and research.

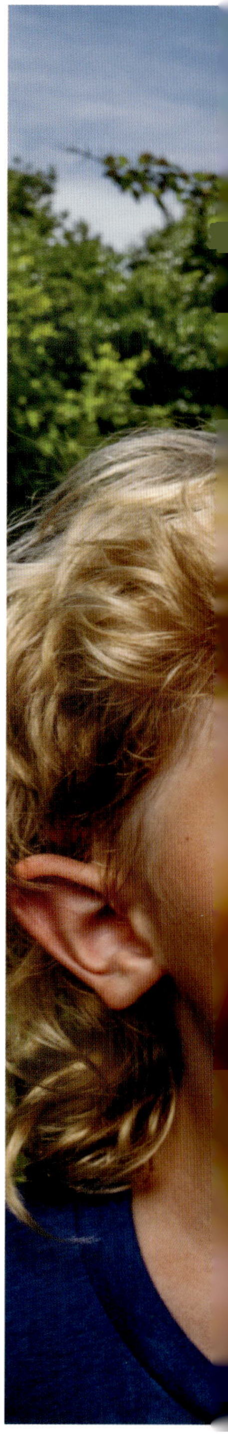

Chip Taylor and young volunteers—Ruby and Rusty Graham in front; Julie, Casey, and Esther Kong behind them—marvel at a monarch caterpillar in the Lawrence, Kansas, Monarch Watch pollinator garden.

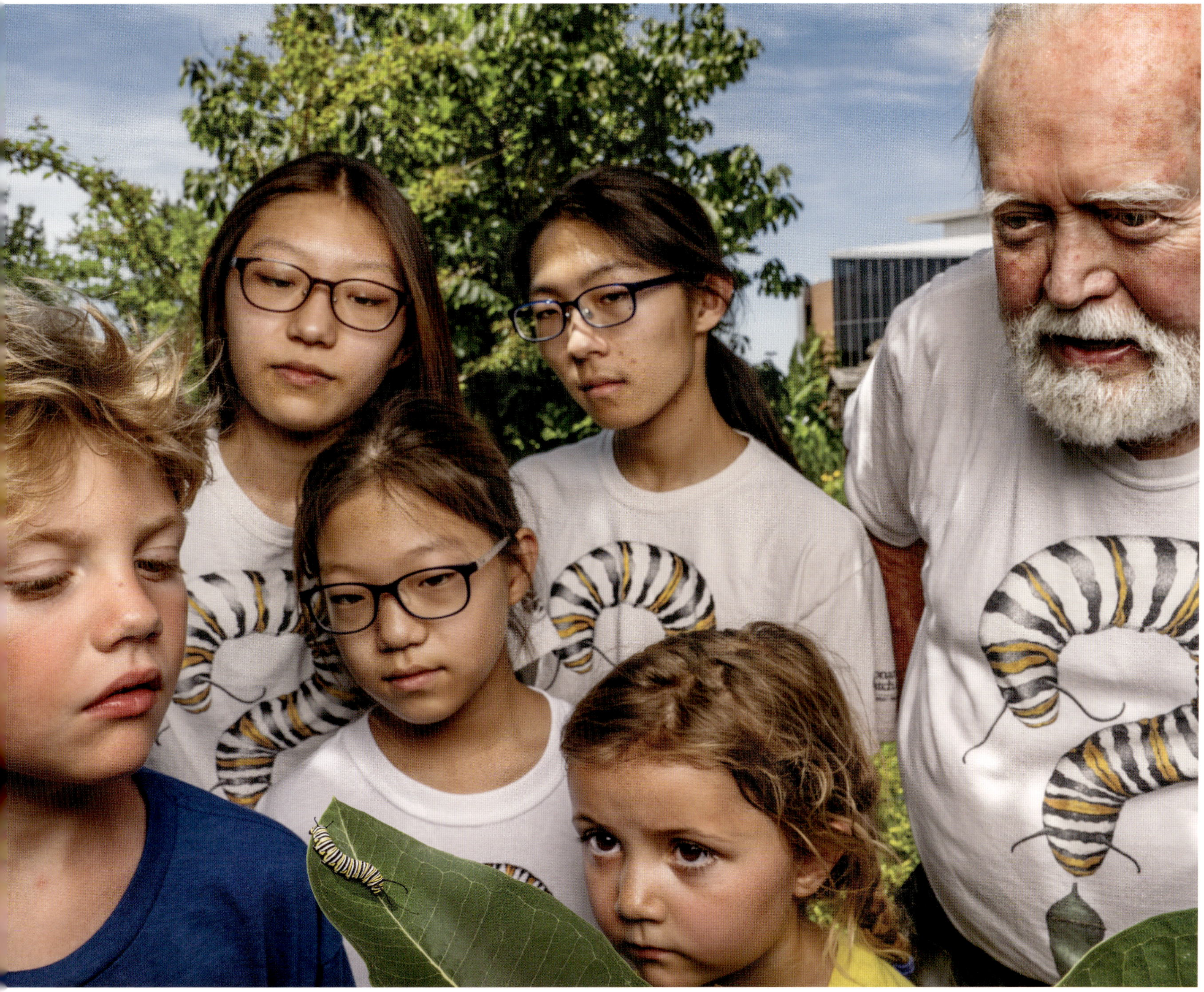

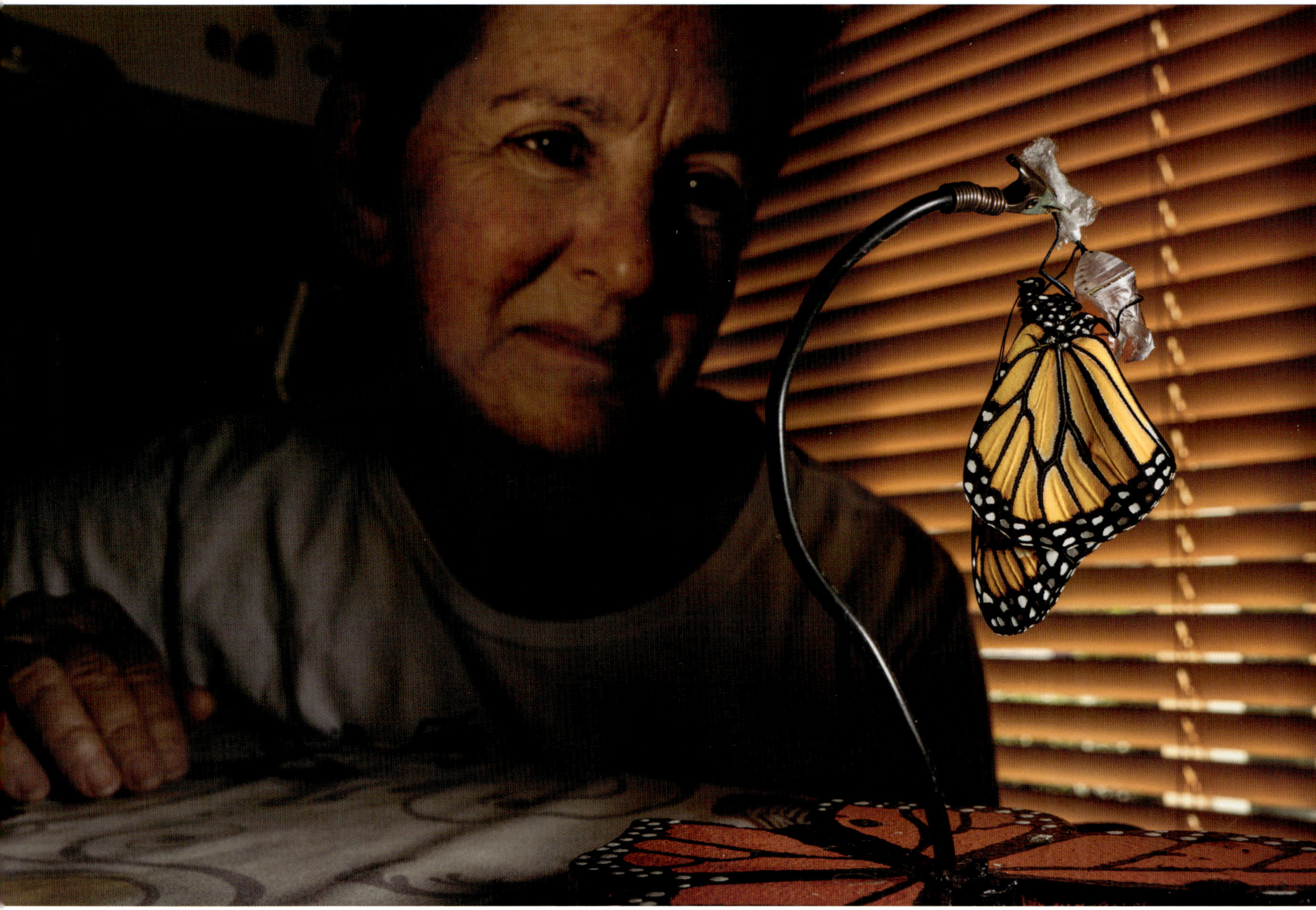

Carol Pasternak watches a monarch butterfly emerge from its chrysalis at her home in Toronto, Ontario.

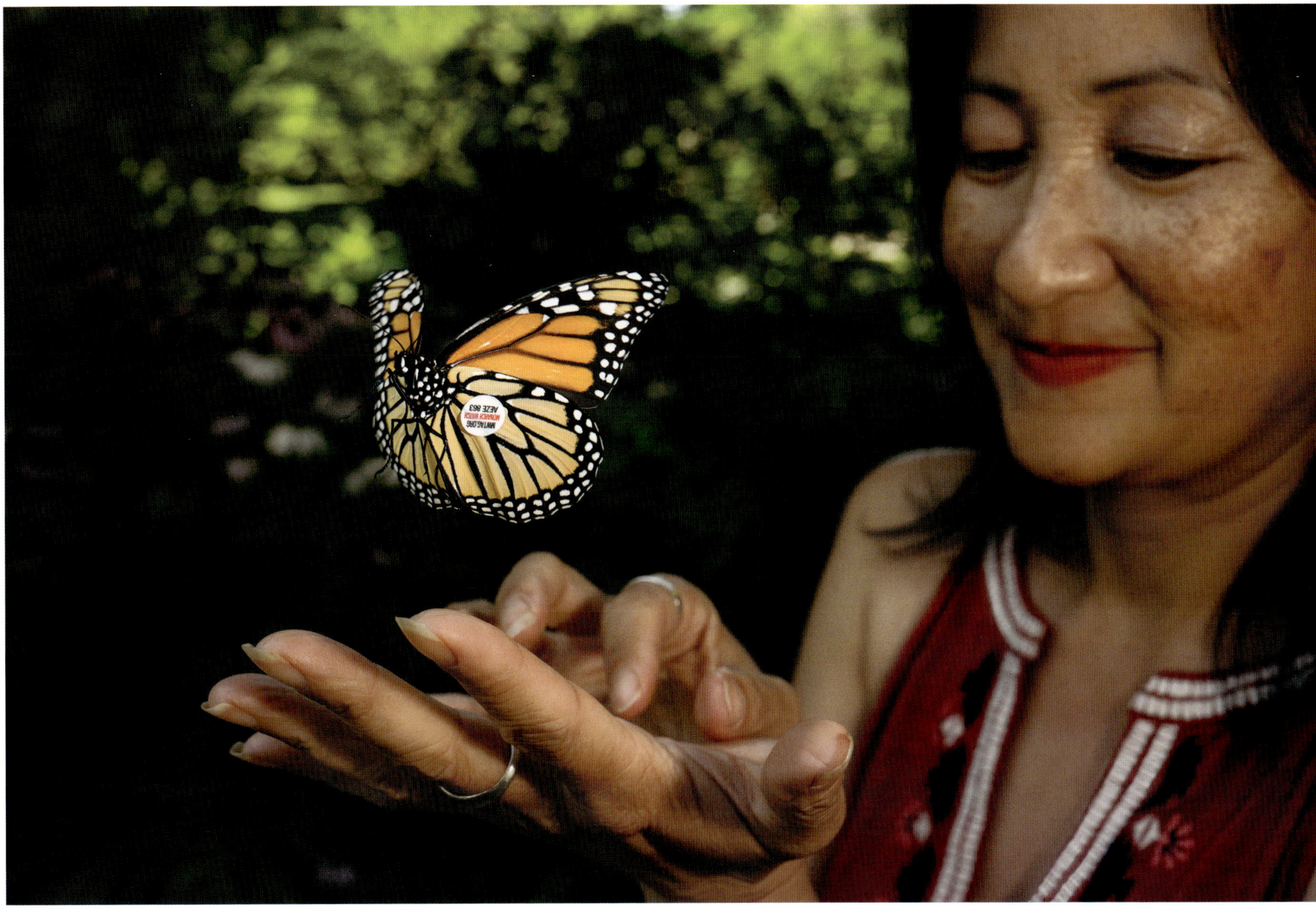

Volunteer Ann Gill releases a tagged monarch butterfly during a tagging event in Scarborough, Ontario.

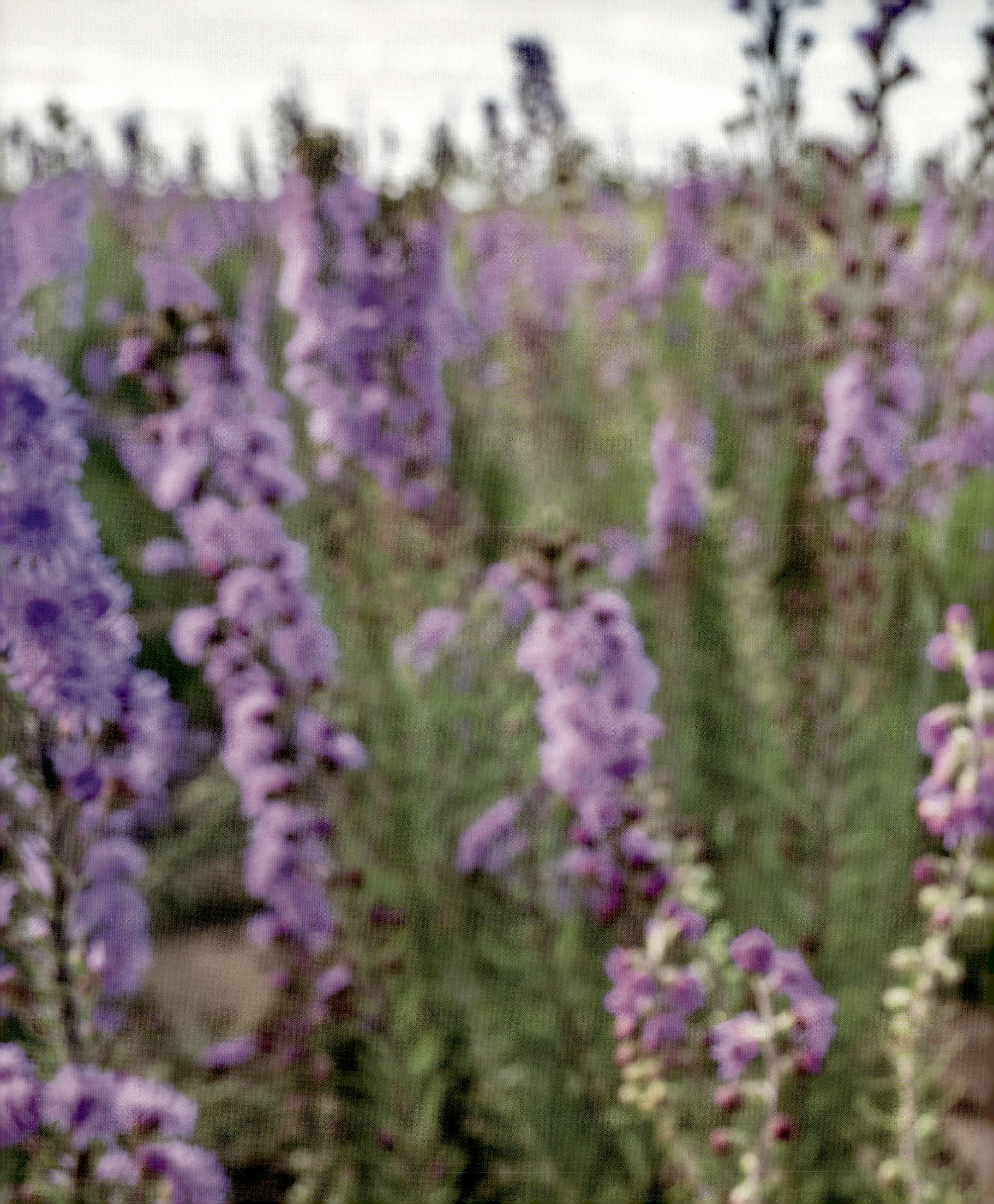

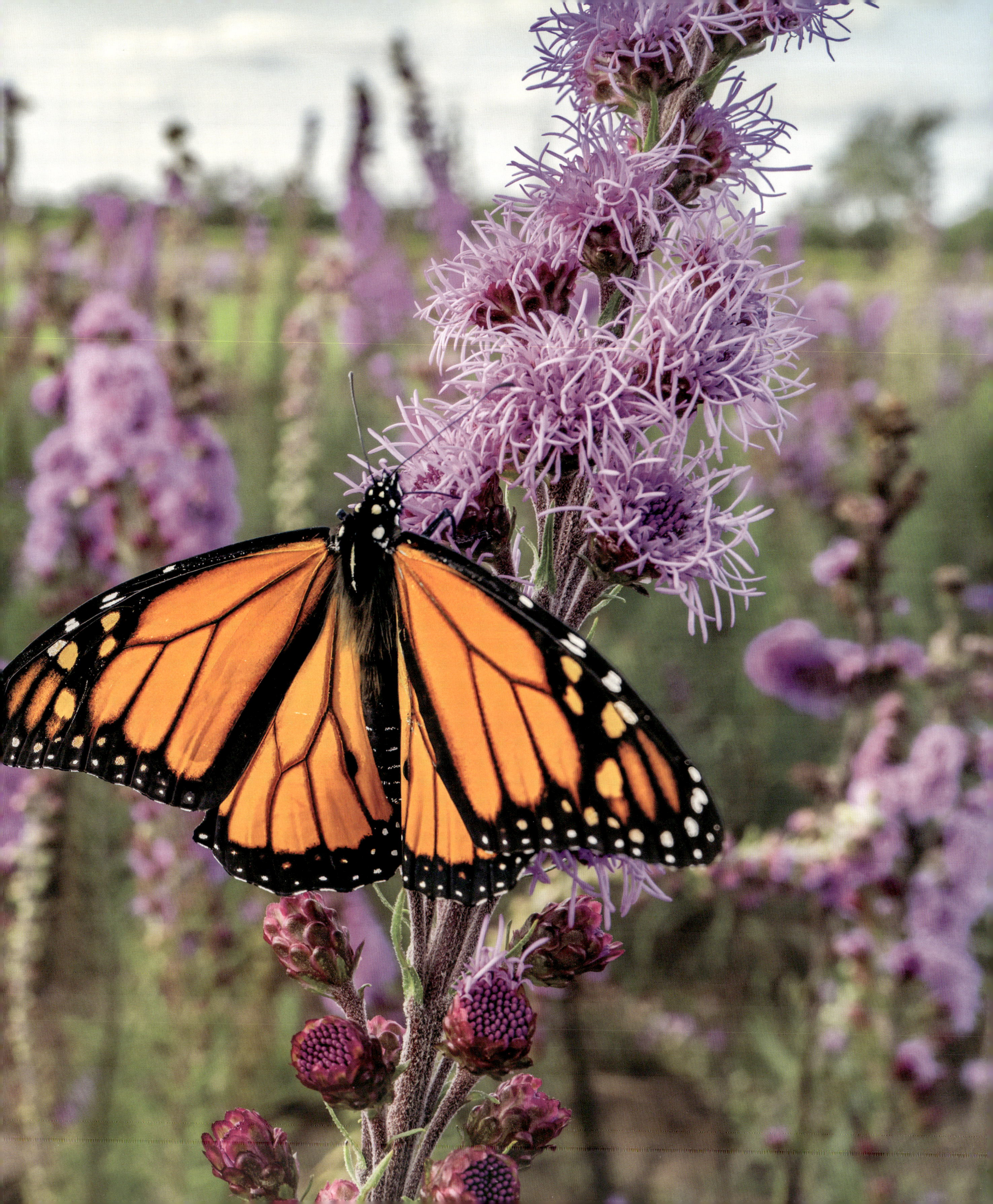

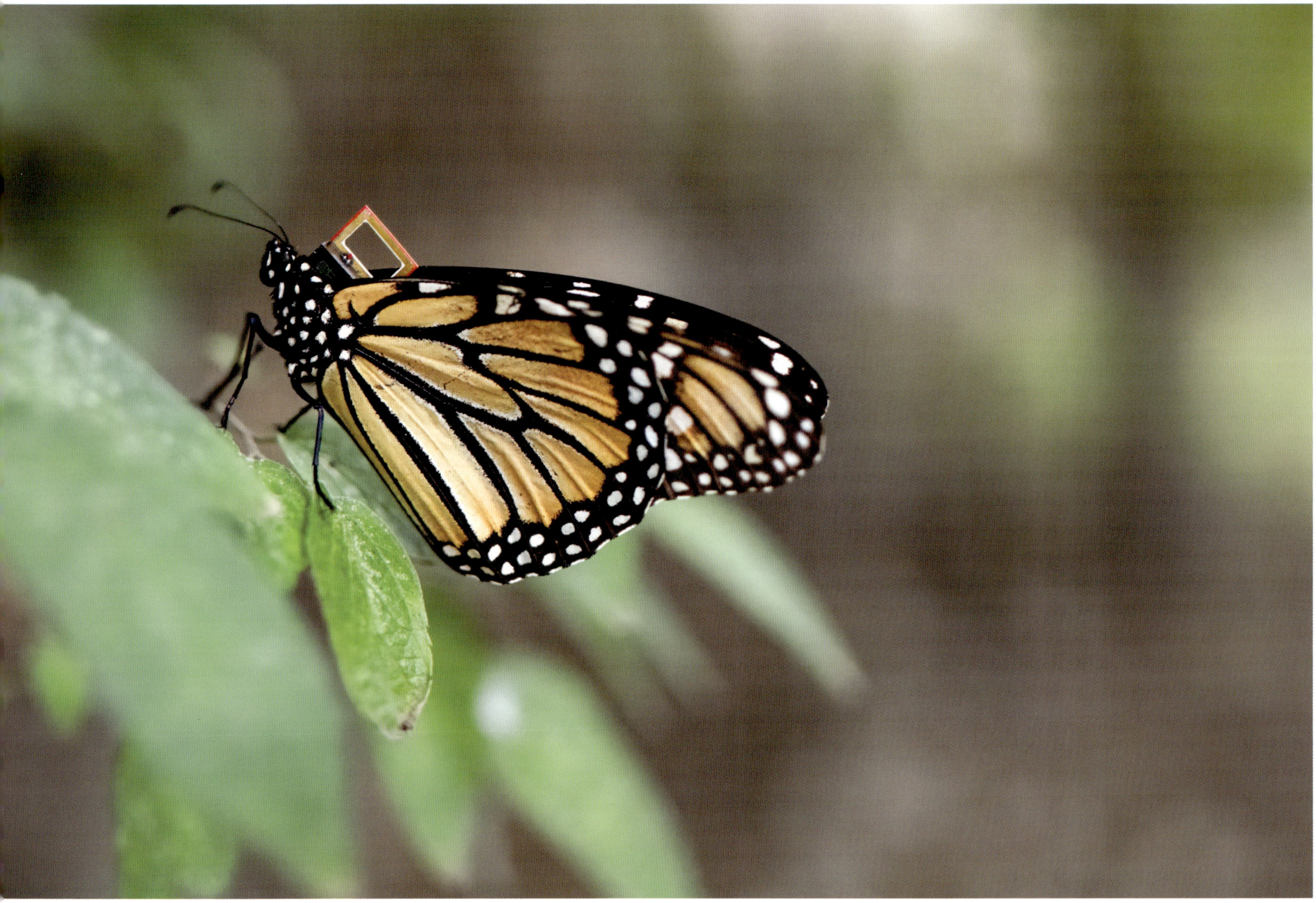

A recently tagged butterfly carries a miniature sensor designed by D. André Green II and his team from the University of Michigan. Its movement data, enhanced with AI, will provide valuable information about how monarchs navigate the landscape.

PAGES 136-137: A male monarch butterfly feeds on blazing star in Foley, Minnesota. As daylight shortens and temperatures cool, monarchs enter reproductive diapause, storing energy from late-summer blooms like these to fuel their long southward migration to the mountains of central Mexico.

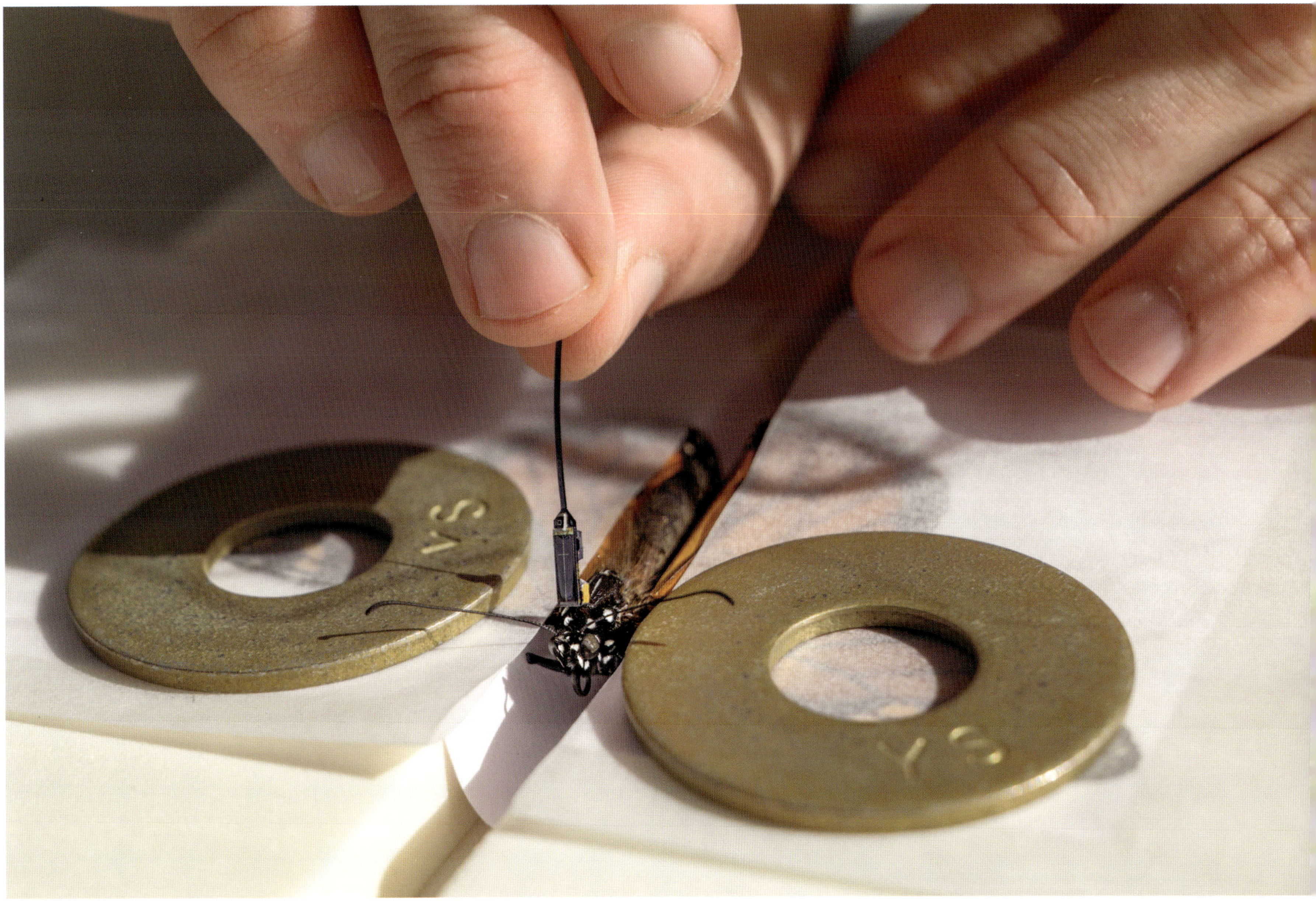

Ashley Fisher, a biologist with the Xerces Society for Invertebrate Conservation, carefully attaches a BlūMorpho sensor, a groundbreaking technology that is already enabling scientists to follow the remarkable monarch migration in real time with unprecedented detail.

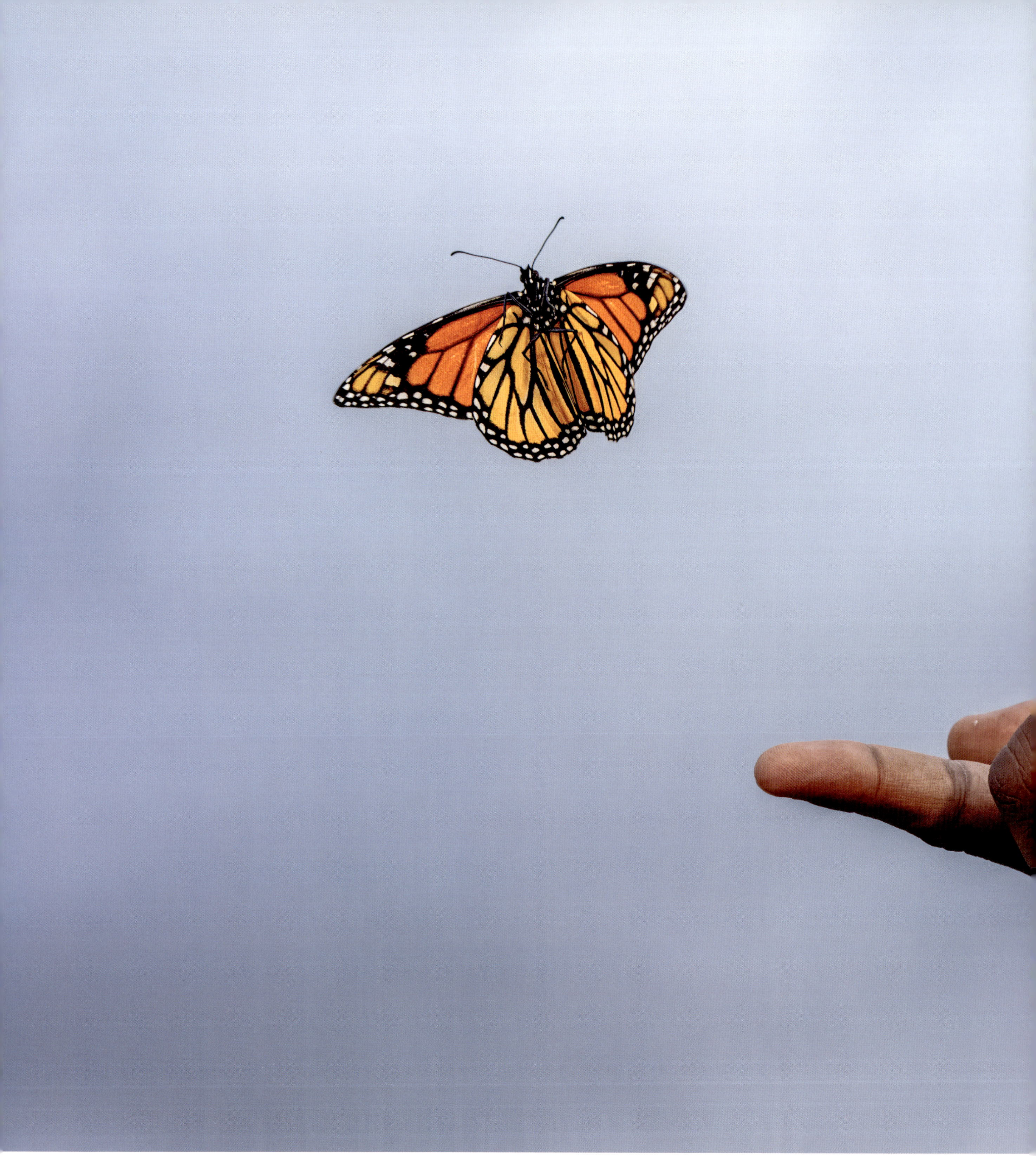

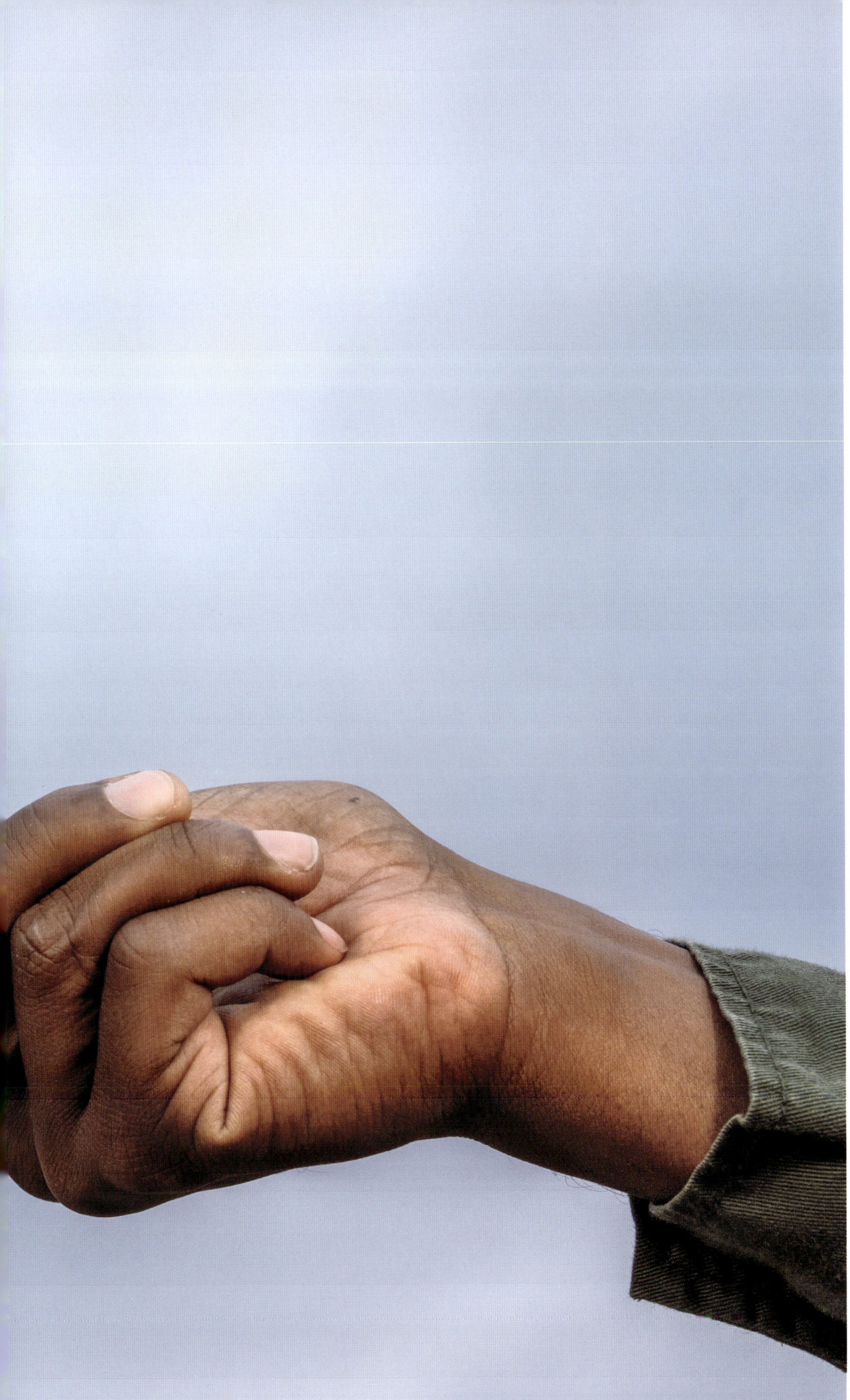

A recently tagged butterfly takes off in Junction, Texas. A new generation of scientists is pushing the boundaries in monarch-tracking science in ways that were unthinkable a few years ago.

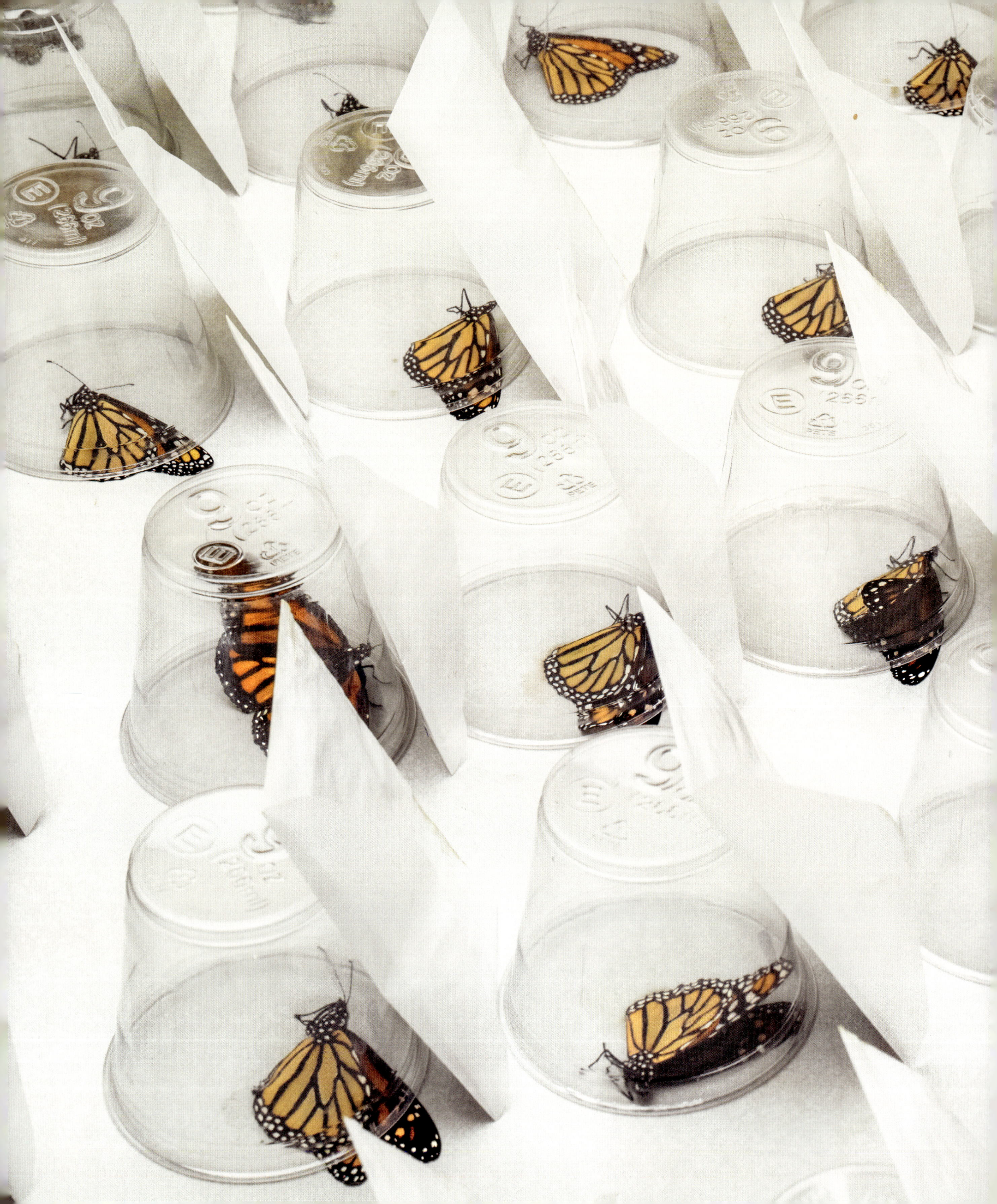

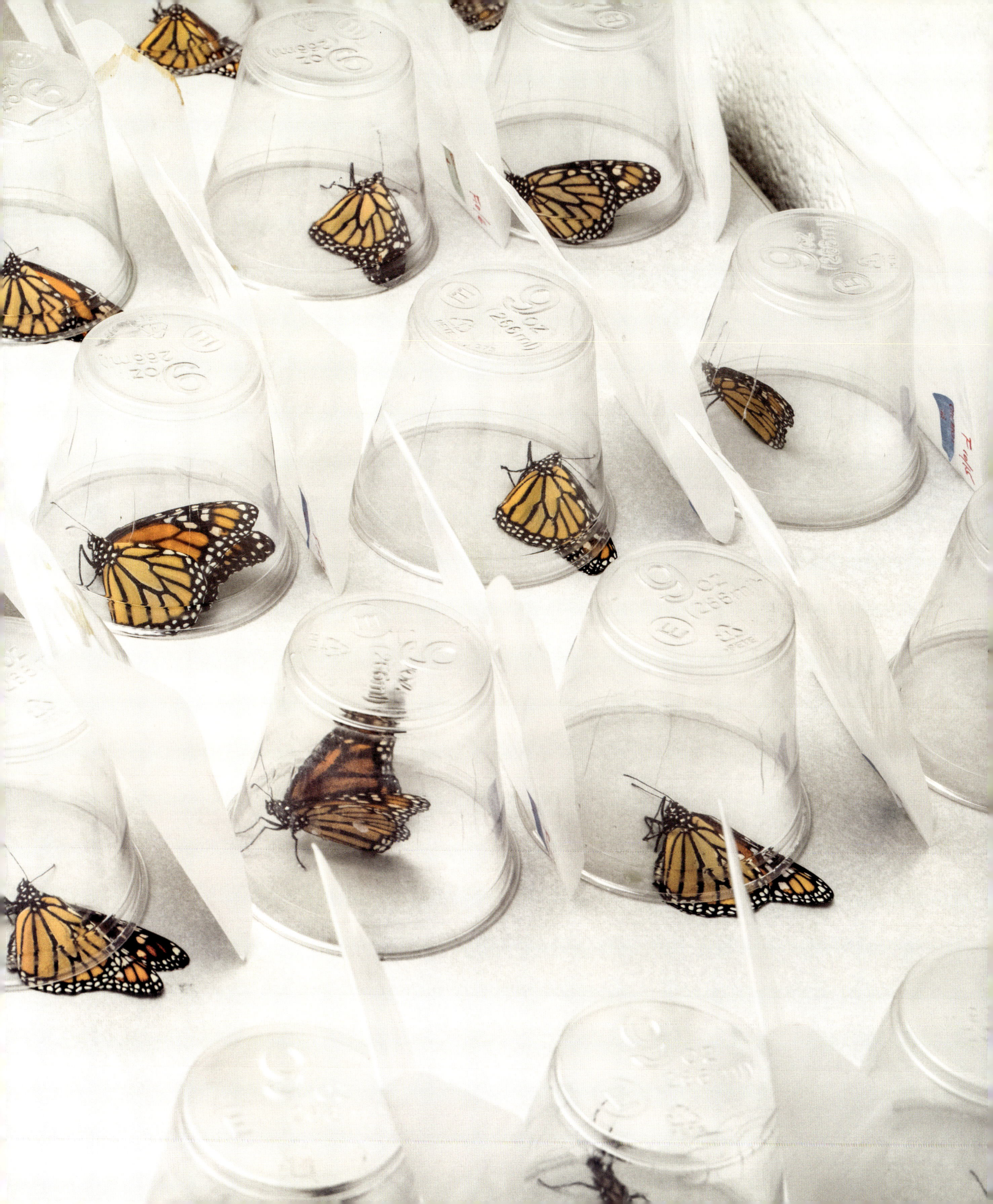

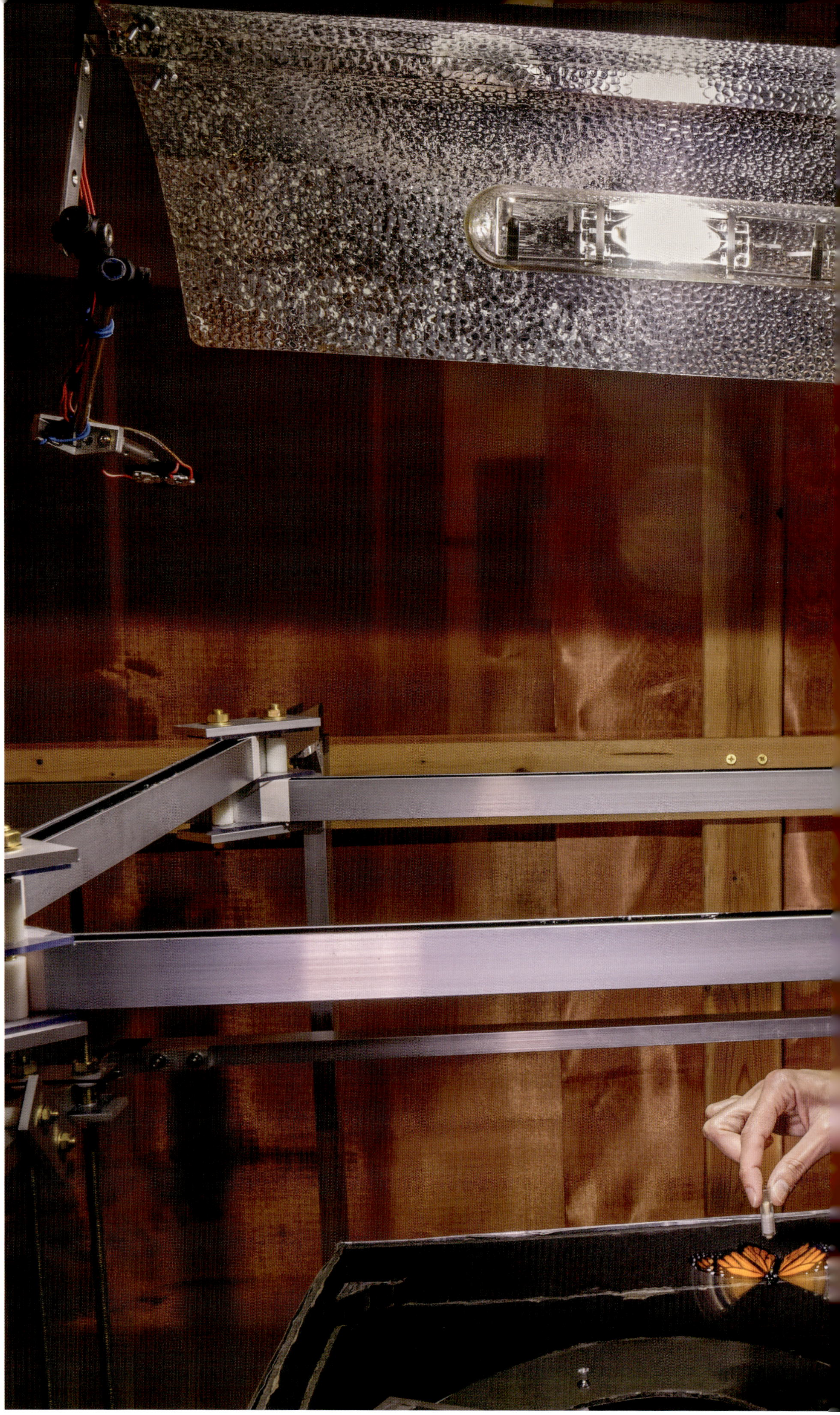

At Texas A&M University, Christine Merlin uses a flight simulator placed in a Faraday cage to test the flight direction of a monarch butterfly from her lab-raised colony—part of her research on how these insects sense Earth's magnetic fields.

PAGES 142–143: These lab-raised monarch butterflies, used in Christine Merlin's chronobiology experiments, will help reveal how circadian rhythms and internal biological clocks influence the monarchs' ability to orient and navigate during migration.

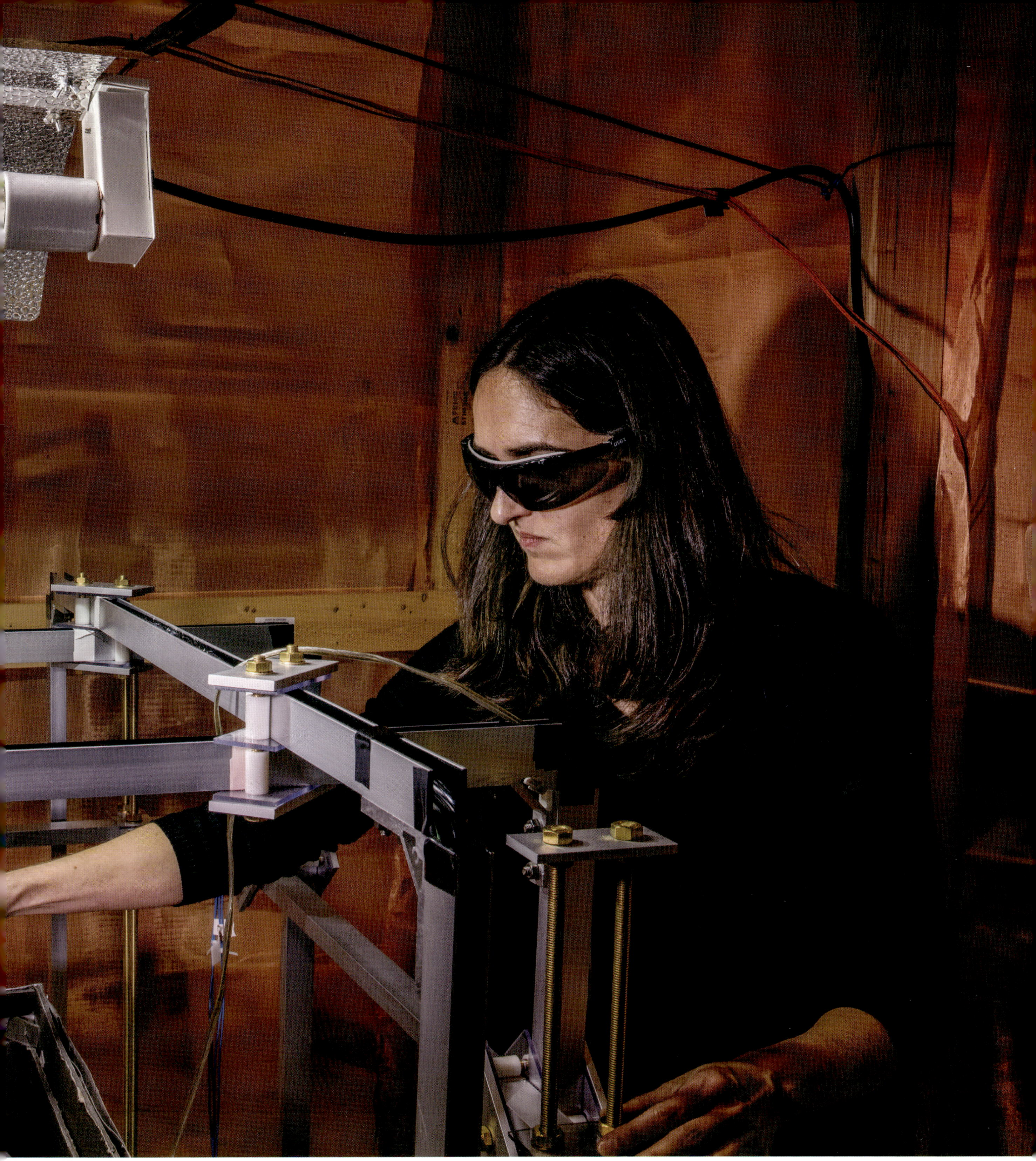

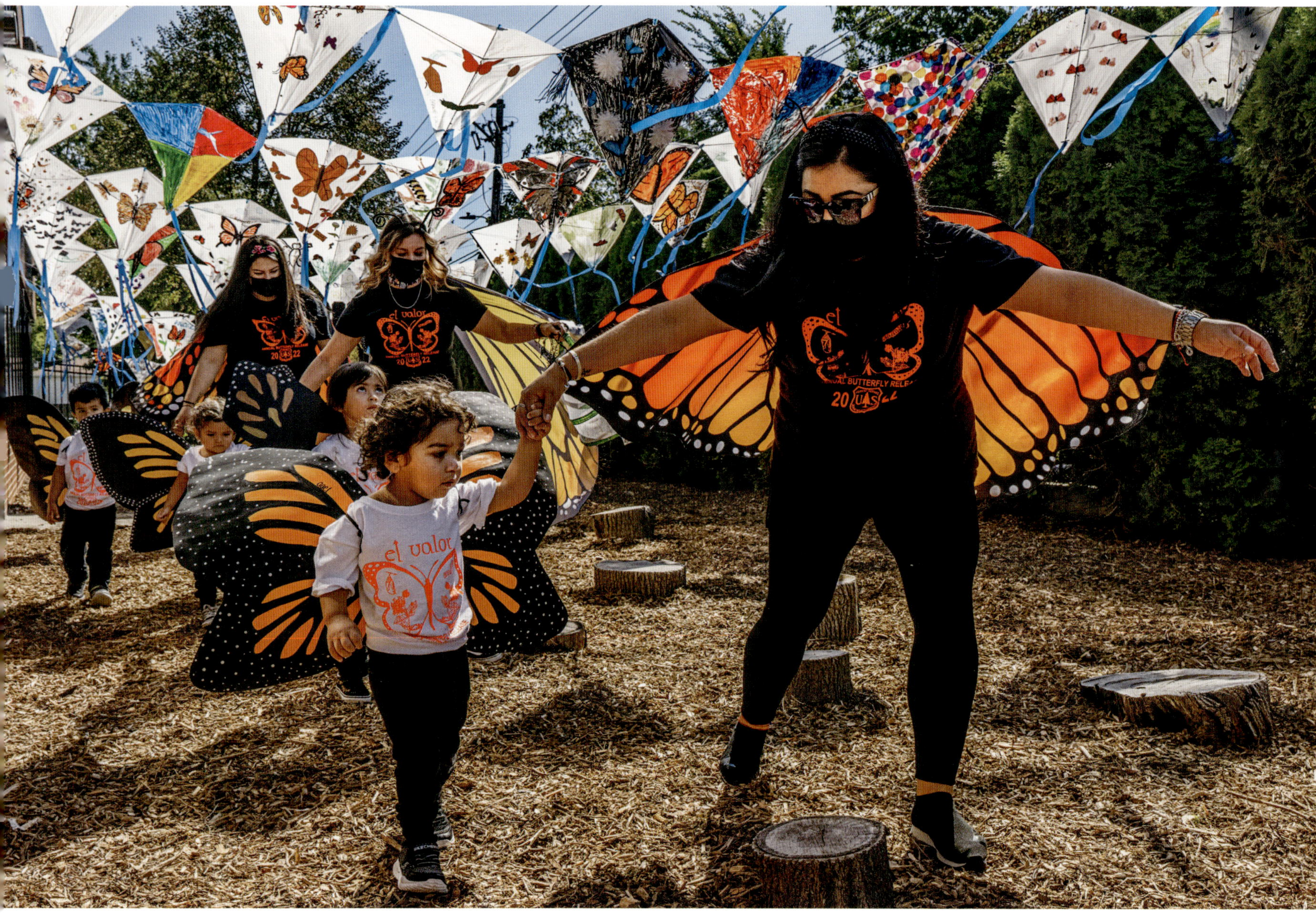

ABOVE AND OPPOSITE: Each autumn at El Valor, a nonprofit community center serving Chicago's Latino community, students release monarchs reared in classrooms during spring and summer as part of the center's environmental education program—always a celebration.

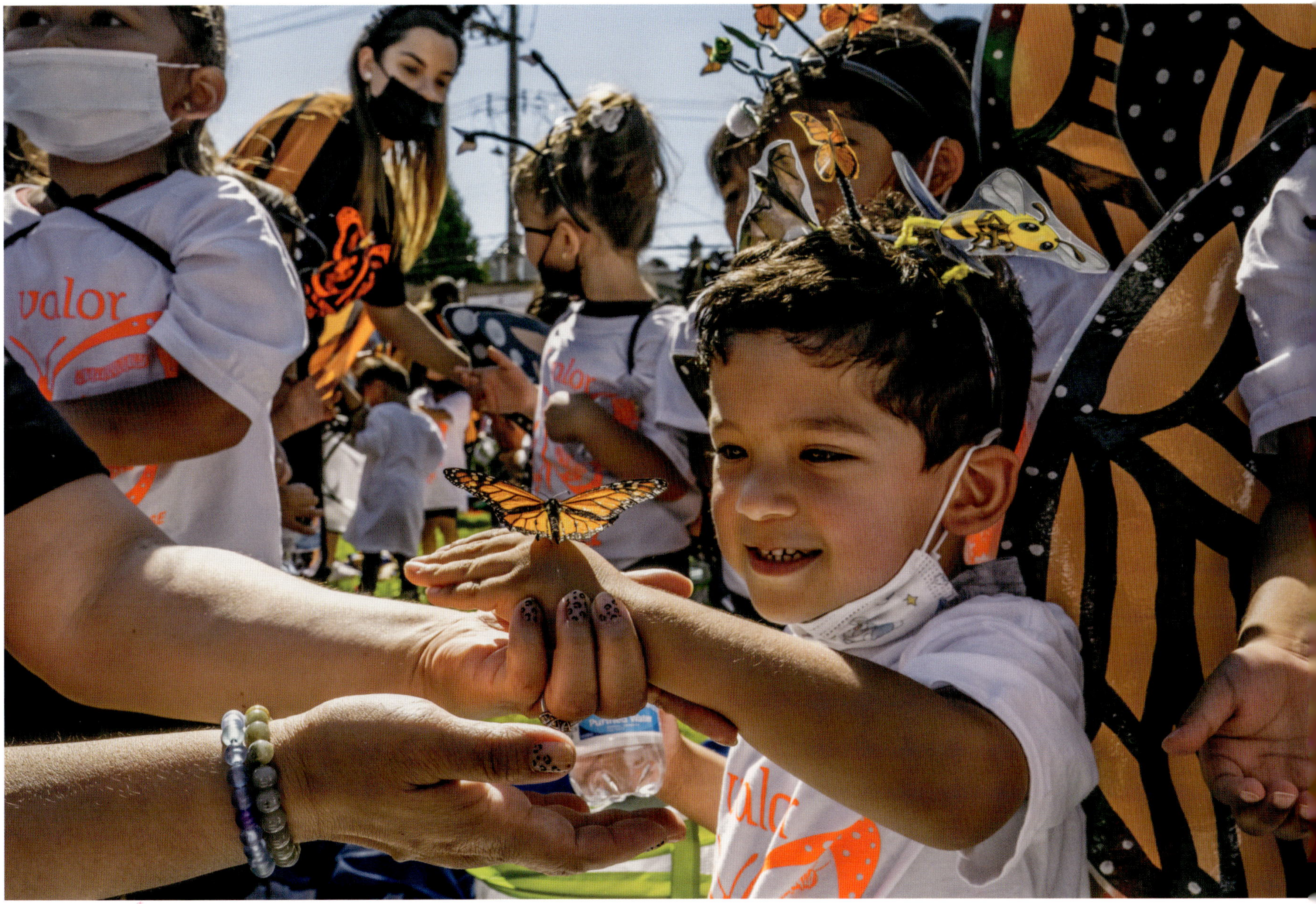

PAGES 148–149: Monarch butterflies roost in the foothills of La Huasteca near Santa Catarina, Nuevo León, Mexico, along one of the main highways of the Monterrey metropolitan area.

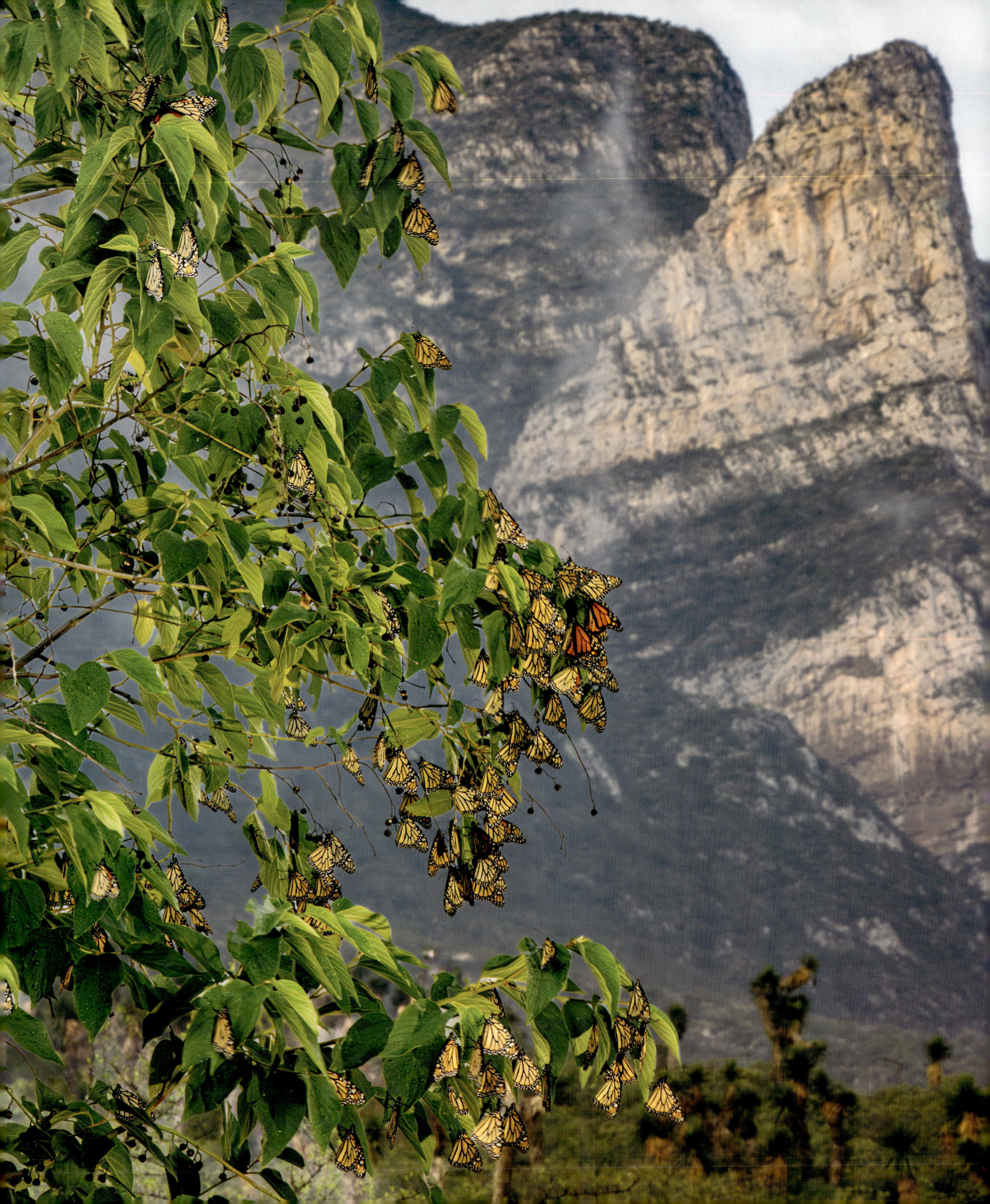

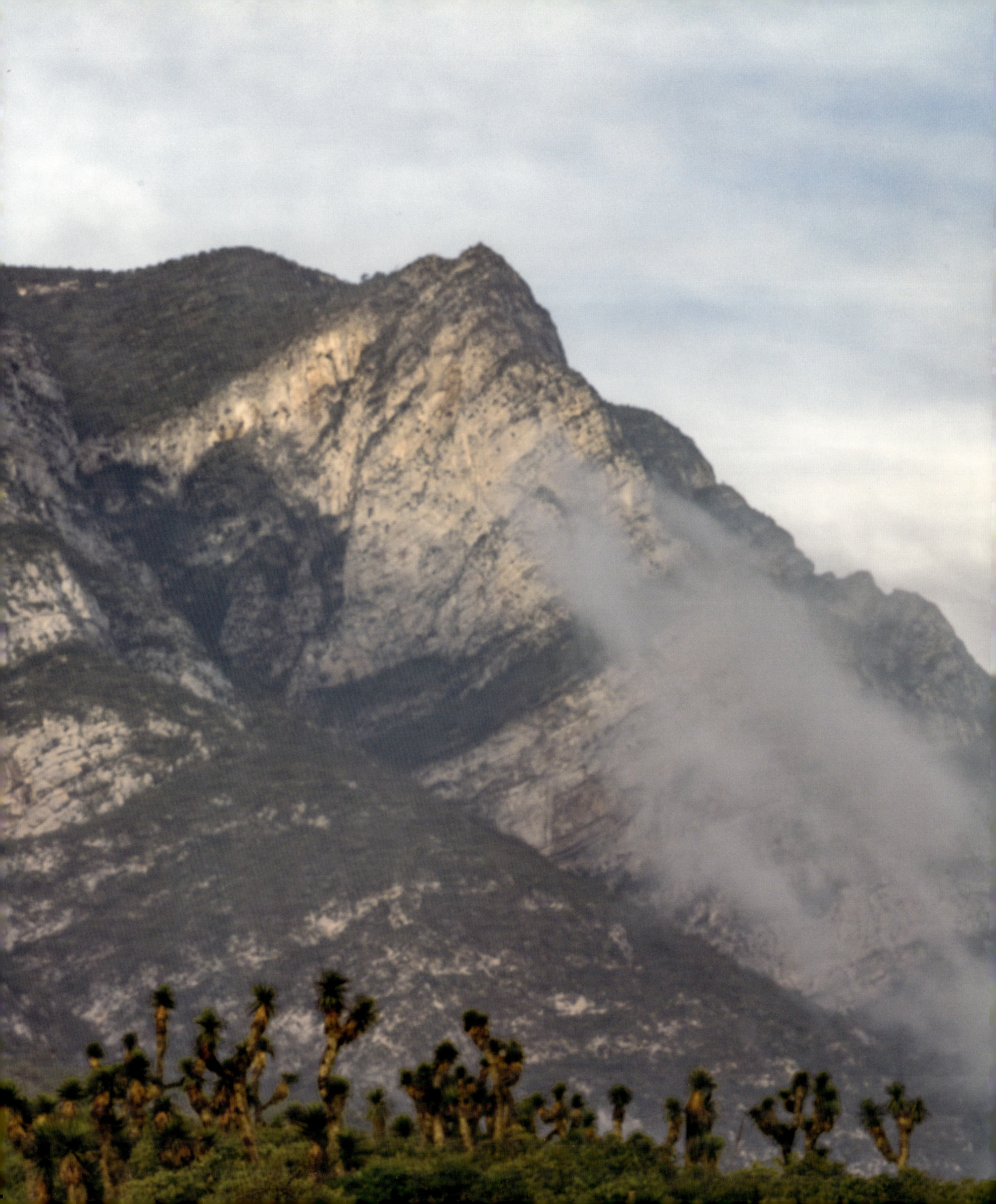

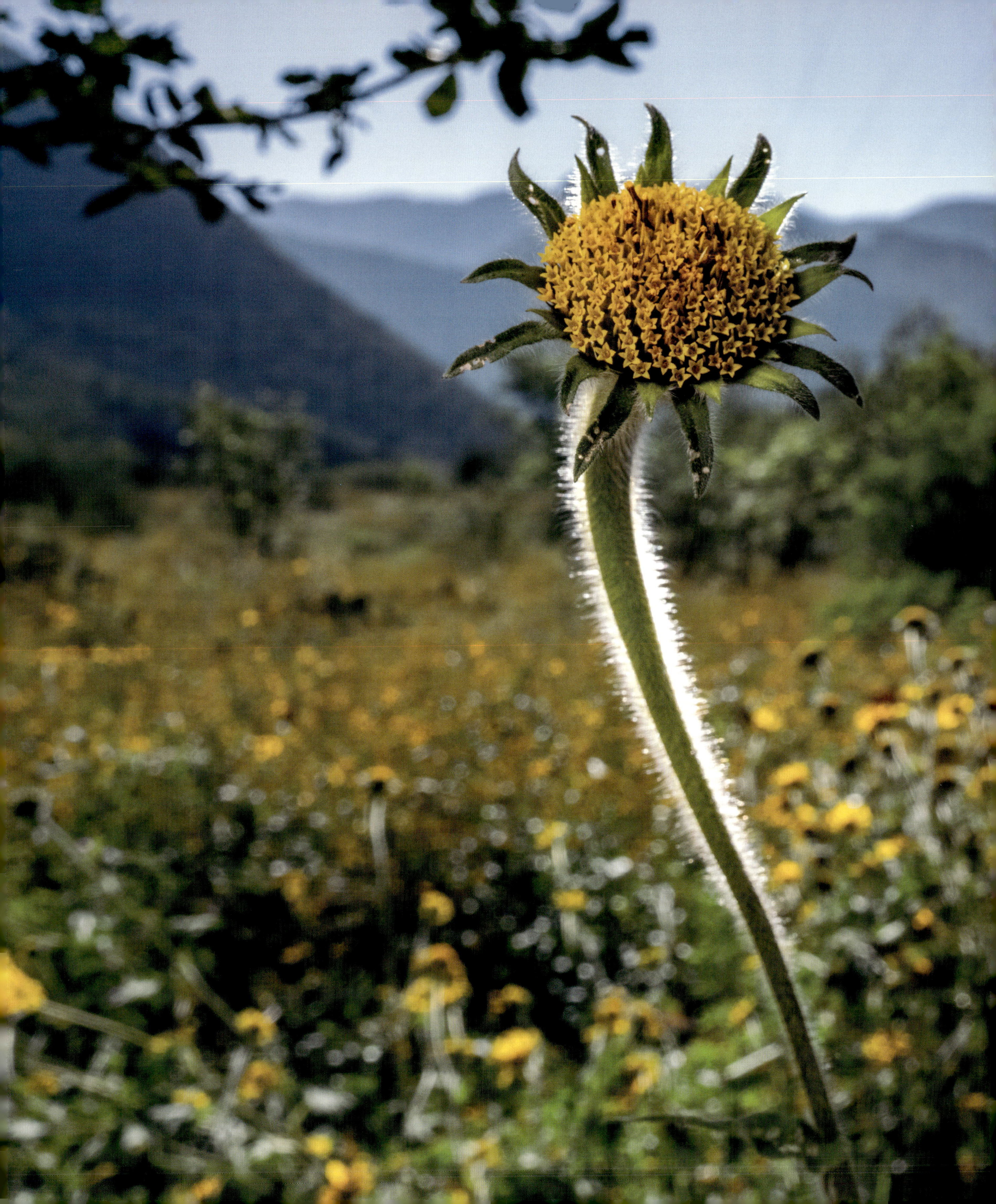

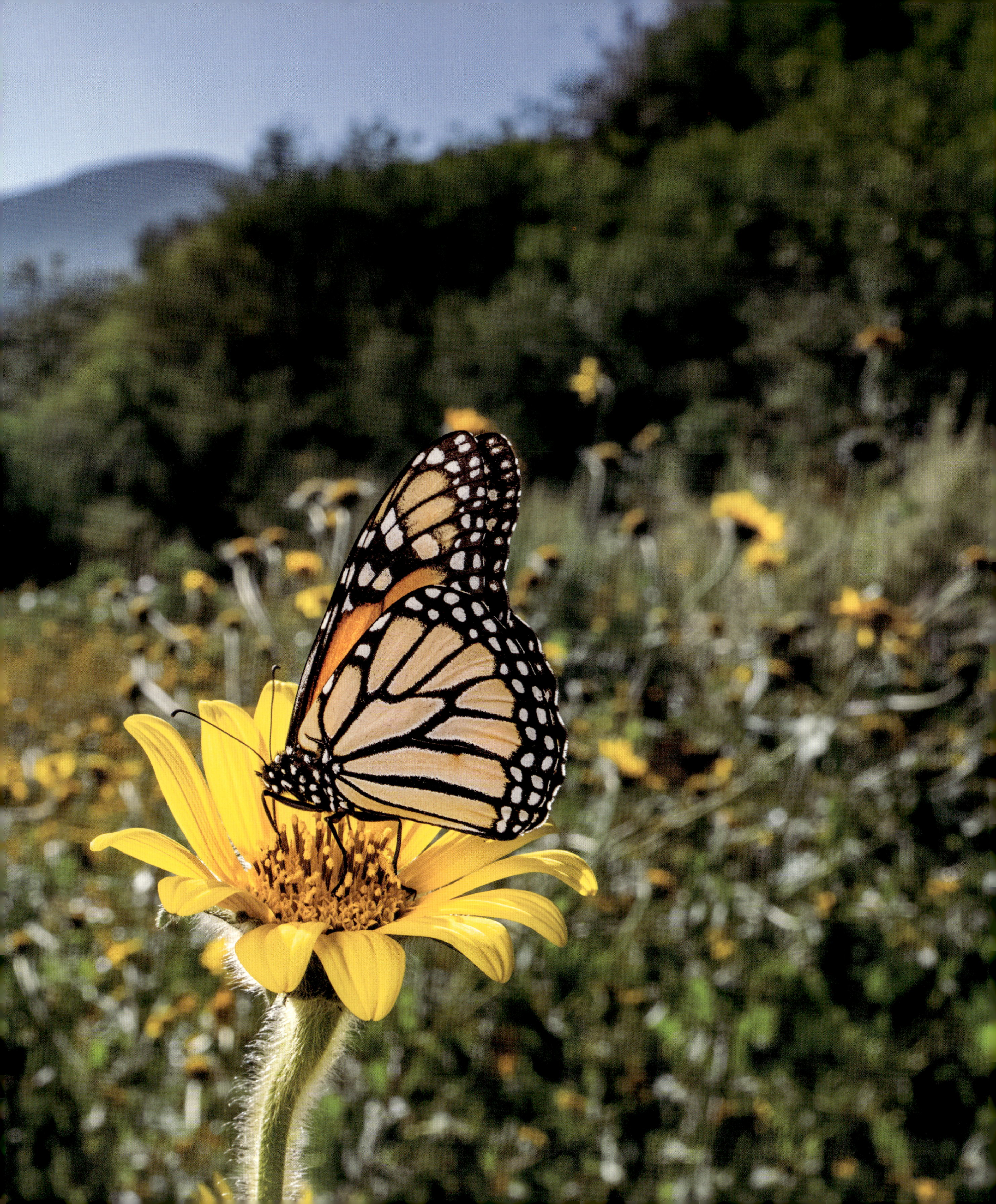

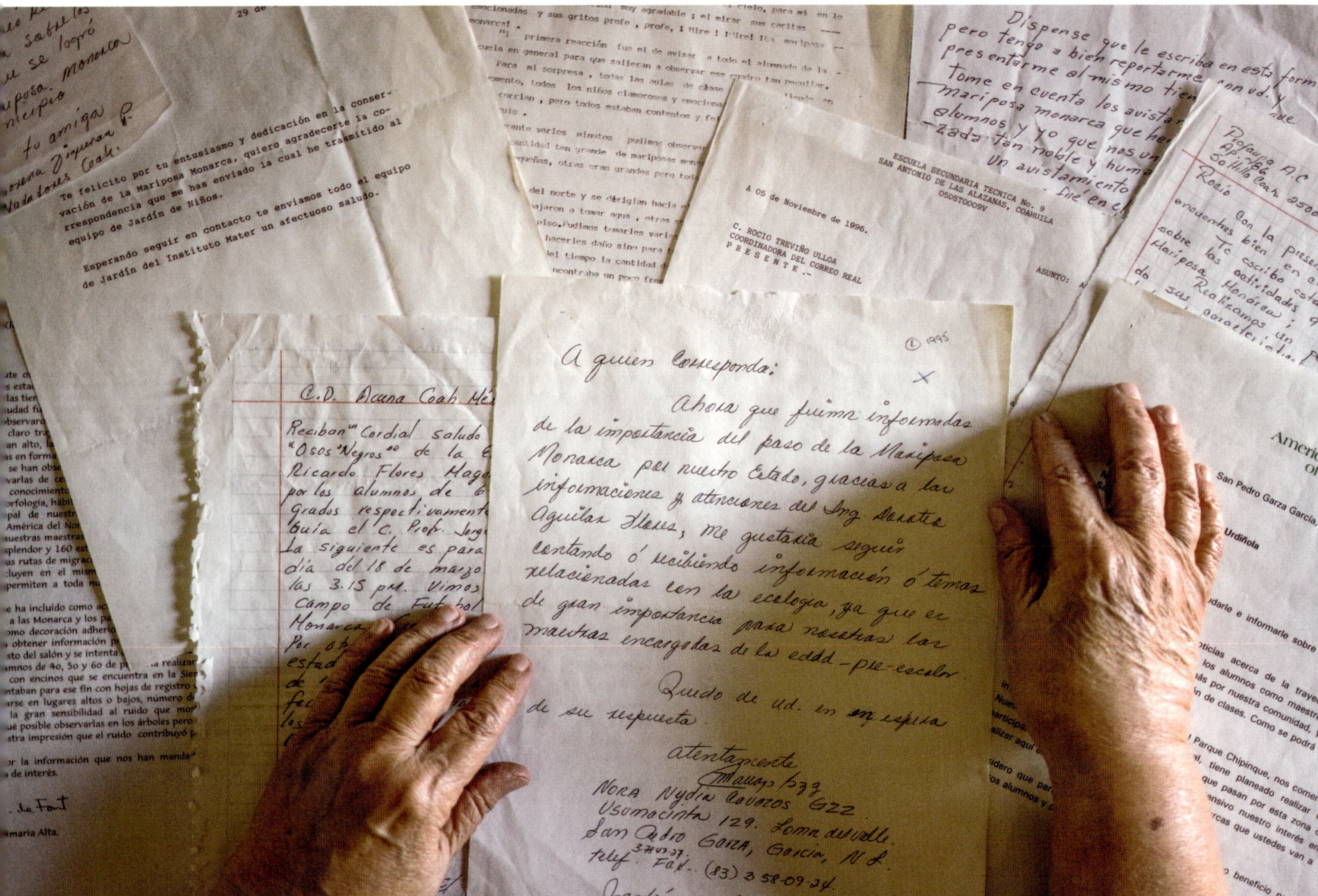

Long before email and social media, monarch lovers in Mexico reported their observations in writing. Those letters have been instrumental in mapping the monarchs' migratory route.

PAGES 150–151: Monarch butterflies feed on native flowers during their fall migration through San Lorenzo Canyon in Zapalinamé State Park, Saltillo, Coahuila, Mexico. Recent droughts have reduced flowering and thus nectar availability, leaving monarchs with less fuel for winter survival.

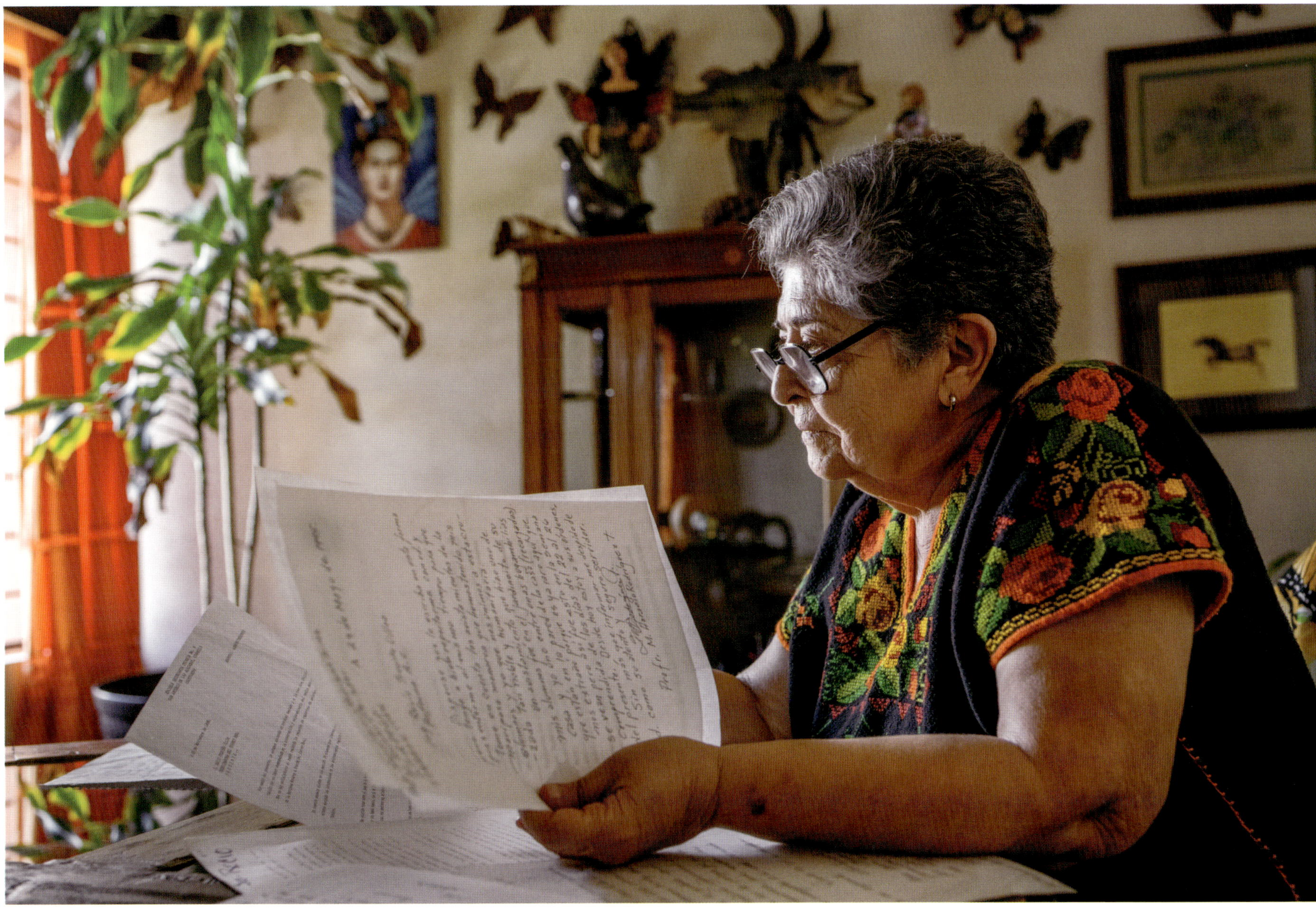

Rocío Treviño, founder of Correo Real—a 30-year-old citizen science program documenting monarch migration across northern Mexico—reviews some of the earliest letters submitted by volunteers.

PAGES 154–155: Monarch butterflies perch on trees along their migratory route near a major highway outside Monterrey, Nuevo León, Mexico. Vehicle collisions are an underrecognized threat, and light pollution can disrupt monarchs' circadian rhythms.

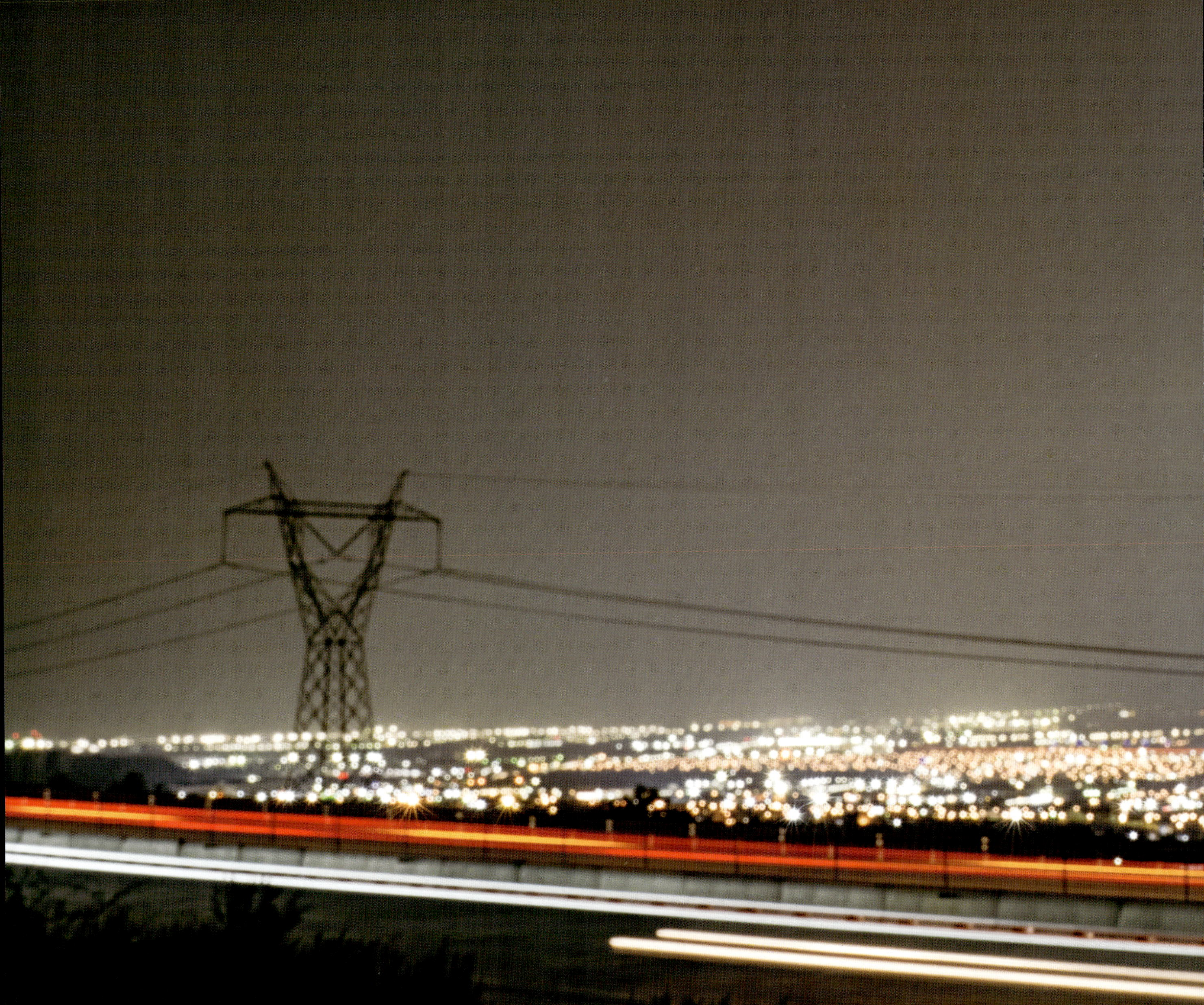

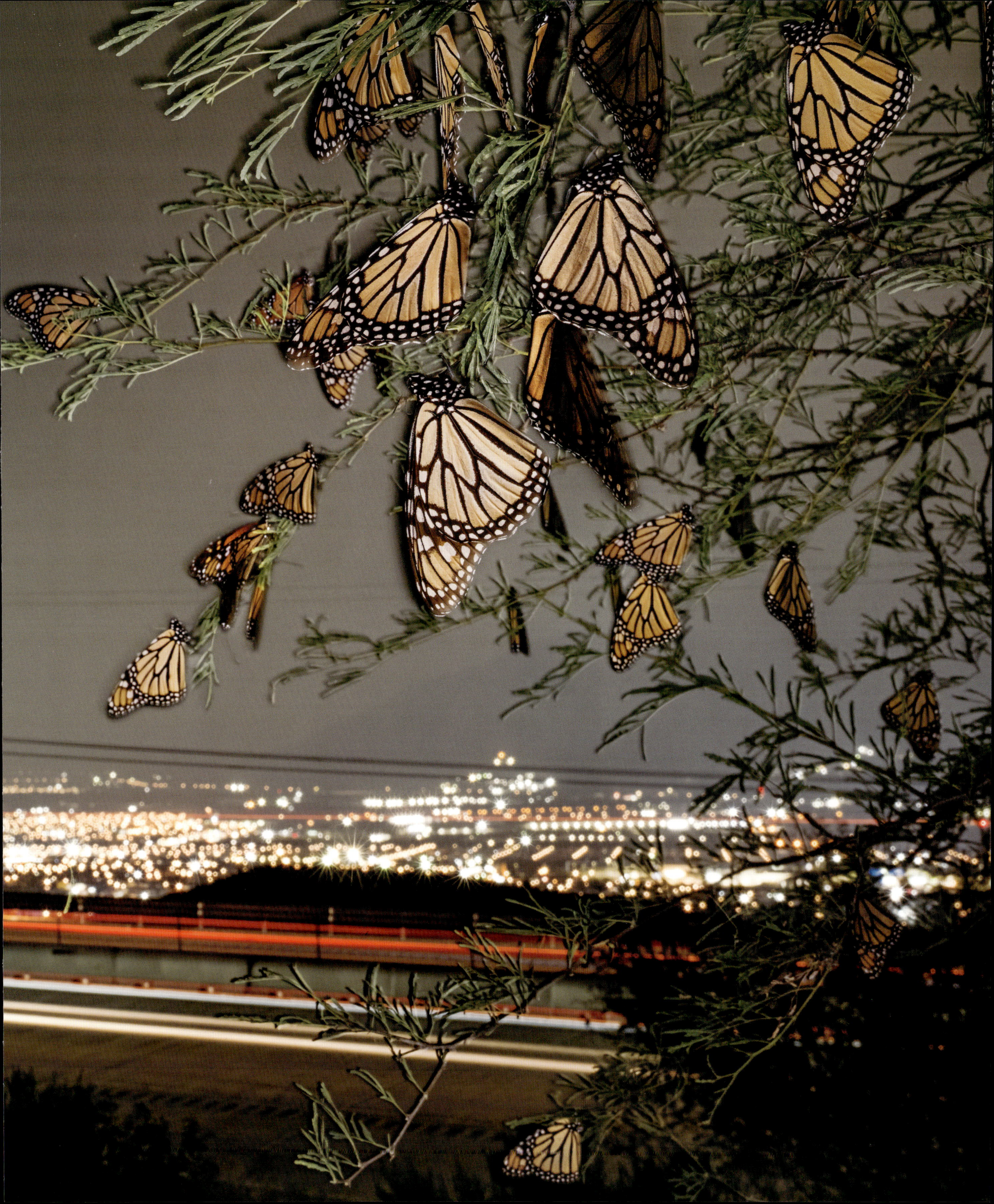

CITY LIGHTS

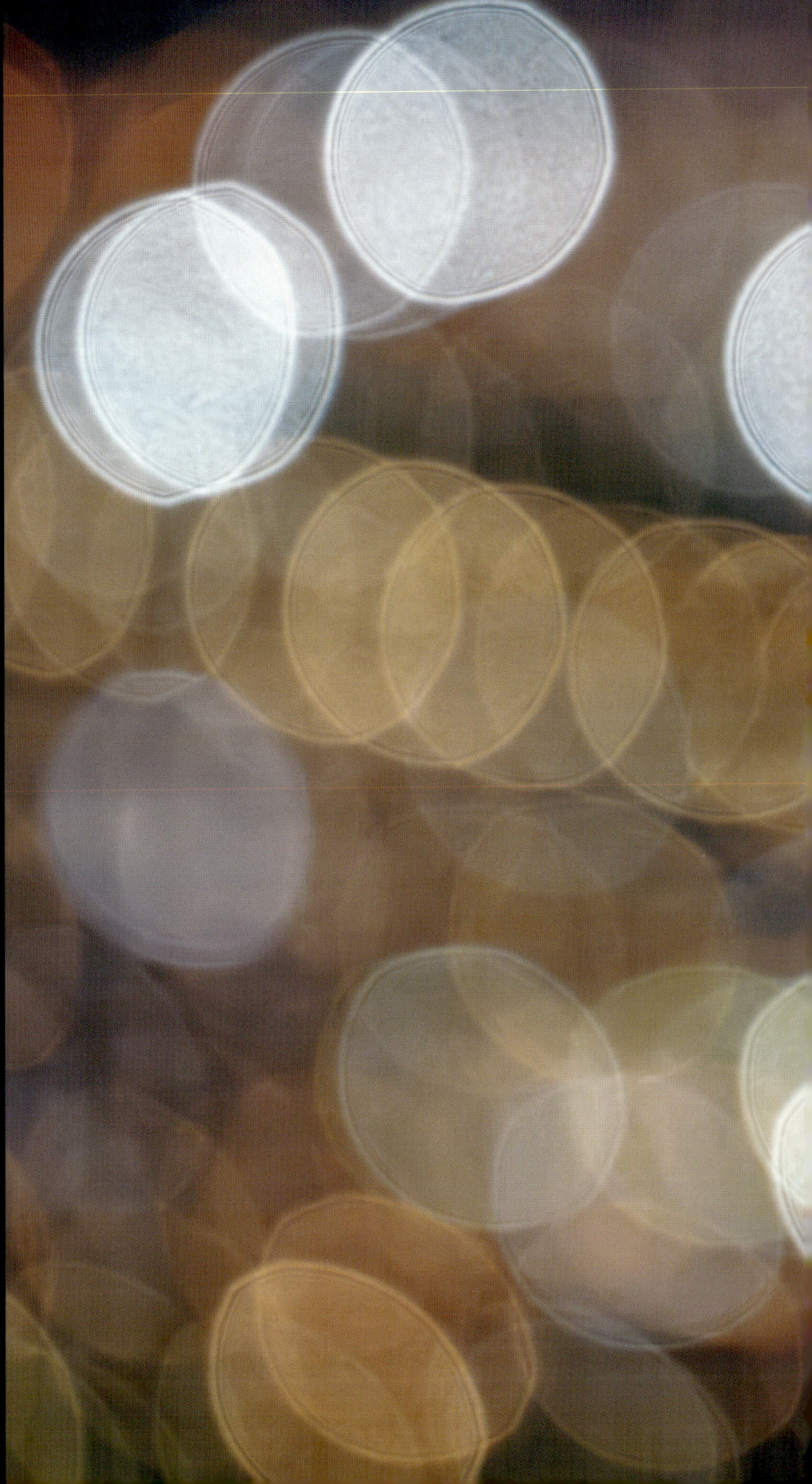

SOME PHOTOS REVEAL THEMSELVES in the most unexpected situations. I came to Monterrey in northern Mexico because in past years I had seen reports of massive accumulations of road-killed monarchs around Santa Catarina, a suburb along Highway 40D, a major road that cuts across much of Mexico and through one of the most important corridors for monarchs on their way to the wintering sanctuaries. The pixelated images I had seen on social media were staggering: almost two feet of dead monarchs piled in the ditches along a specific stretch of highway. It was an important photo for me to capture.

Security is always a concern when working in some regions, and I didn't want to go alone into an area I didn't know. My colleagues at Profauna México, a conservation nonprofit based in northern Mexico, recommended I work with Omar Franco, a Colombian photographer who lived nearby. Omar often biked through the very area I wanted to explore, and every year he had seen monarchs in the same spots. We met the next day for a quick scout, and he proved to be a charming, enthusiastic companion.

From afar, the highway cuts through two very different worlds: behind us, the dramatic limestone cliffs of La Huasteca rising above foothills covered in treelike yuccas; in front, the sprawl of Santa Catarina and the skyline of Monterrey. The first scout with Omar was unsuccessful. We didn't see a single monarch, dead or alive. The branches that should have already been filled with clusters hung bare, swinging gently in the breeze. The next day we found a few thin clusters in the trees but no signs of monarchs in the ditches. I spent three days in the area scouting, testing ideas with traffic and slow shutter speeds, trying to salvage what was looking like a failed shoot.

The next day I had to go to Saltillo, two hours north of Monterrey, so I thanked Omar and left. But we kept in touch, and two days later I got an excited text from him: The first swarms of monarchs were arriving. I decided to give it one more try.

We arrived in the afternoon, just in time to watch the last light climb the limestone walls of La Huasteca with the hum of traffic in the background. Then I saw it: a flicker of light here, flashing police sirens there, the highway lamps switching on. The city was entering night mode. We had never stayed so late, so we waited a bit more. I found clusters of monarchs with the abstract background of city lights and started testing. I switched to a telephoto for an even more blurred, almost painterly background, and then I added a teleconverter, a lens attachment that forced the blur even more. Then it happened. The picture emerged: sharp, clear, and beautiful monarchs silhouetted against a blur of city lights.

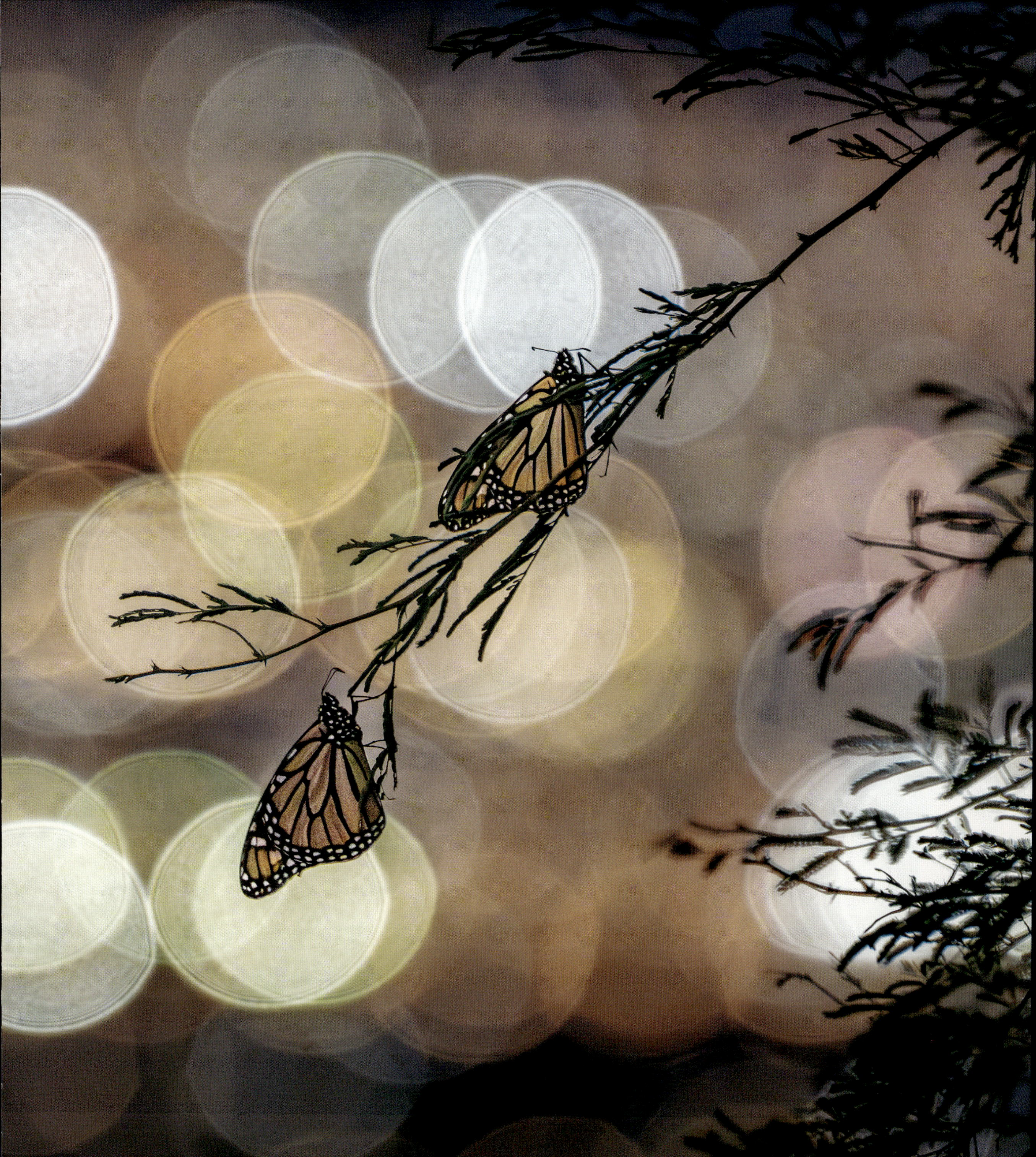

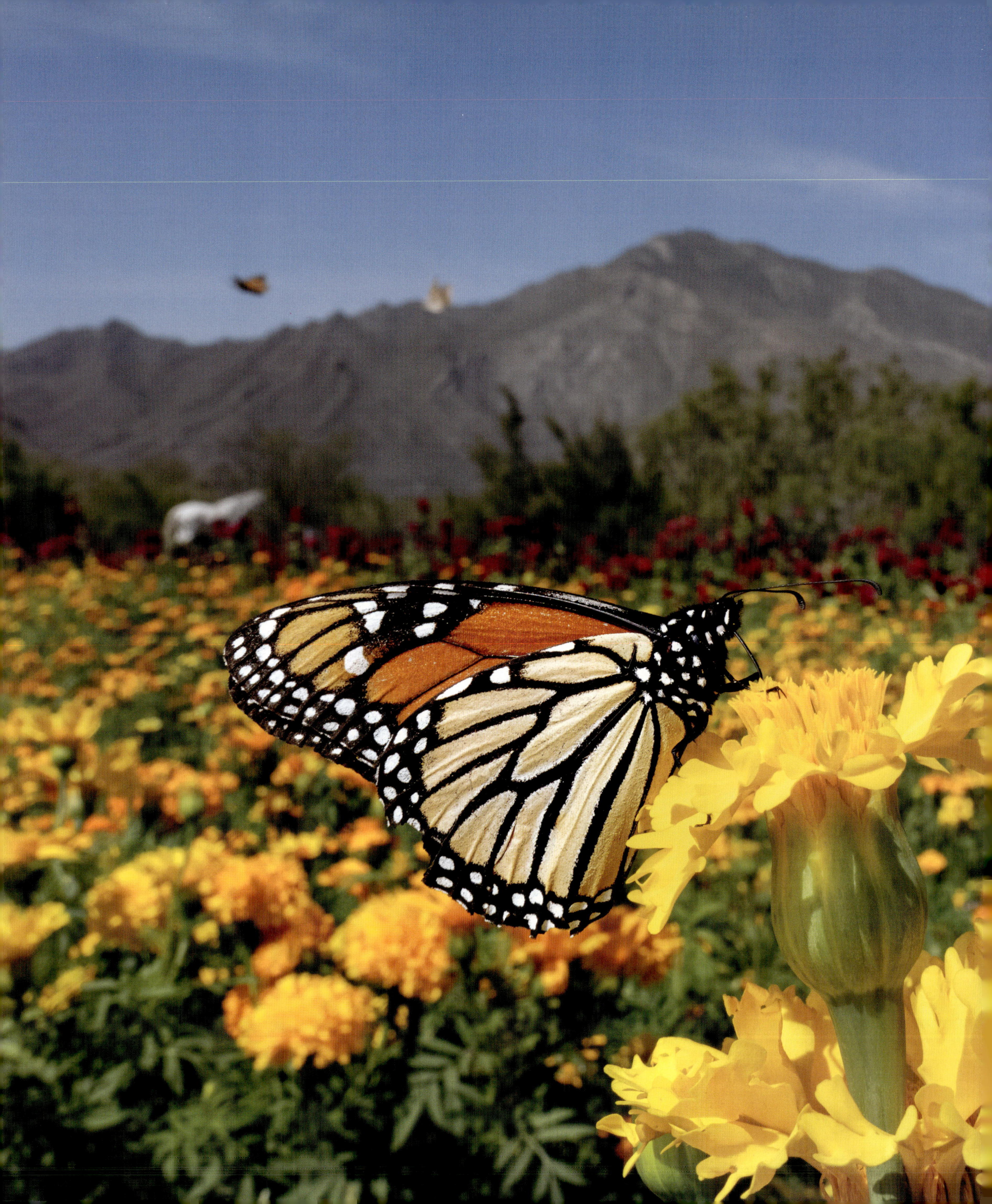

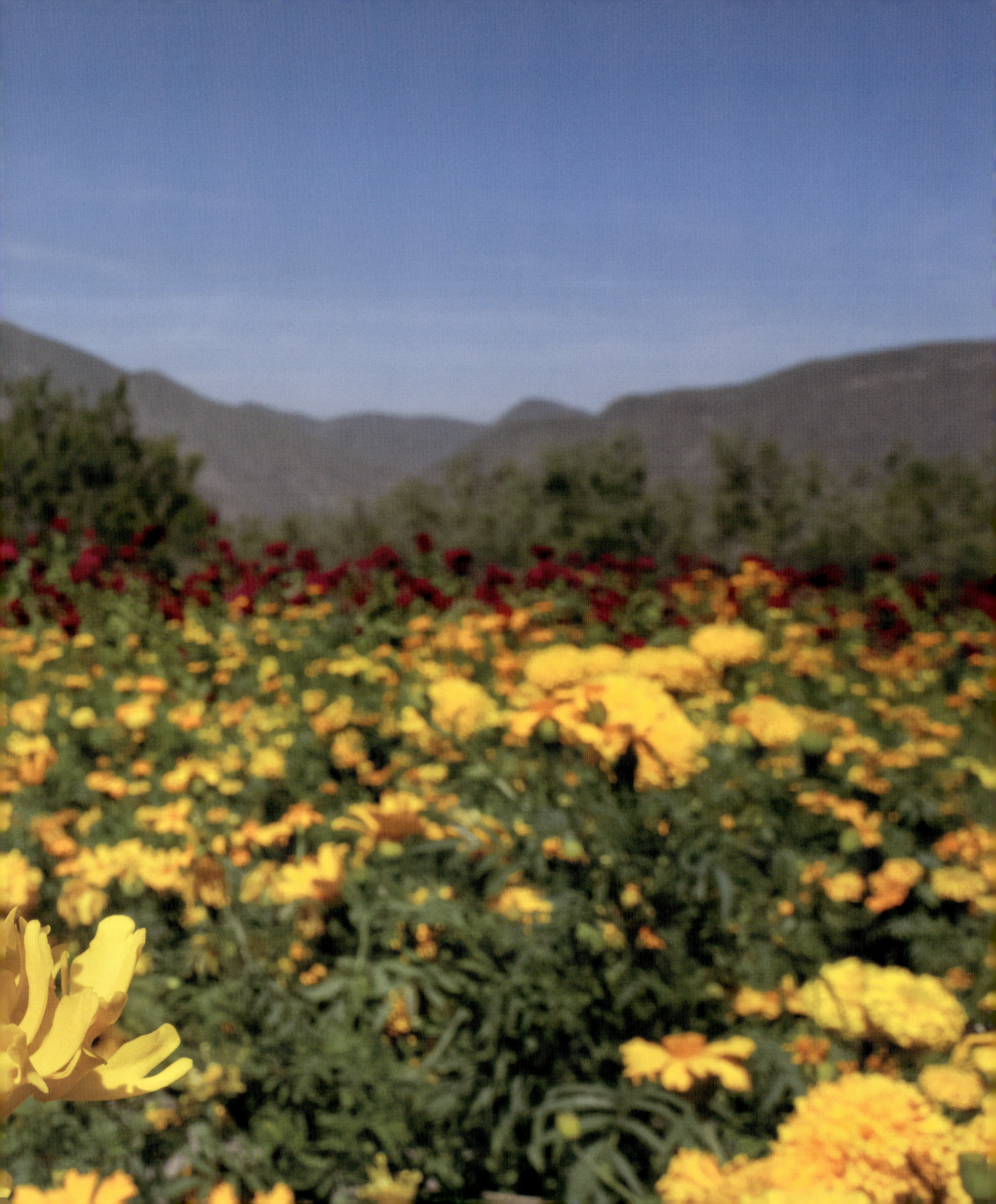

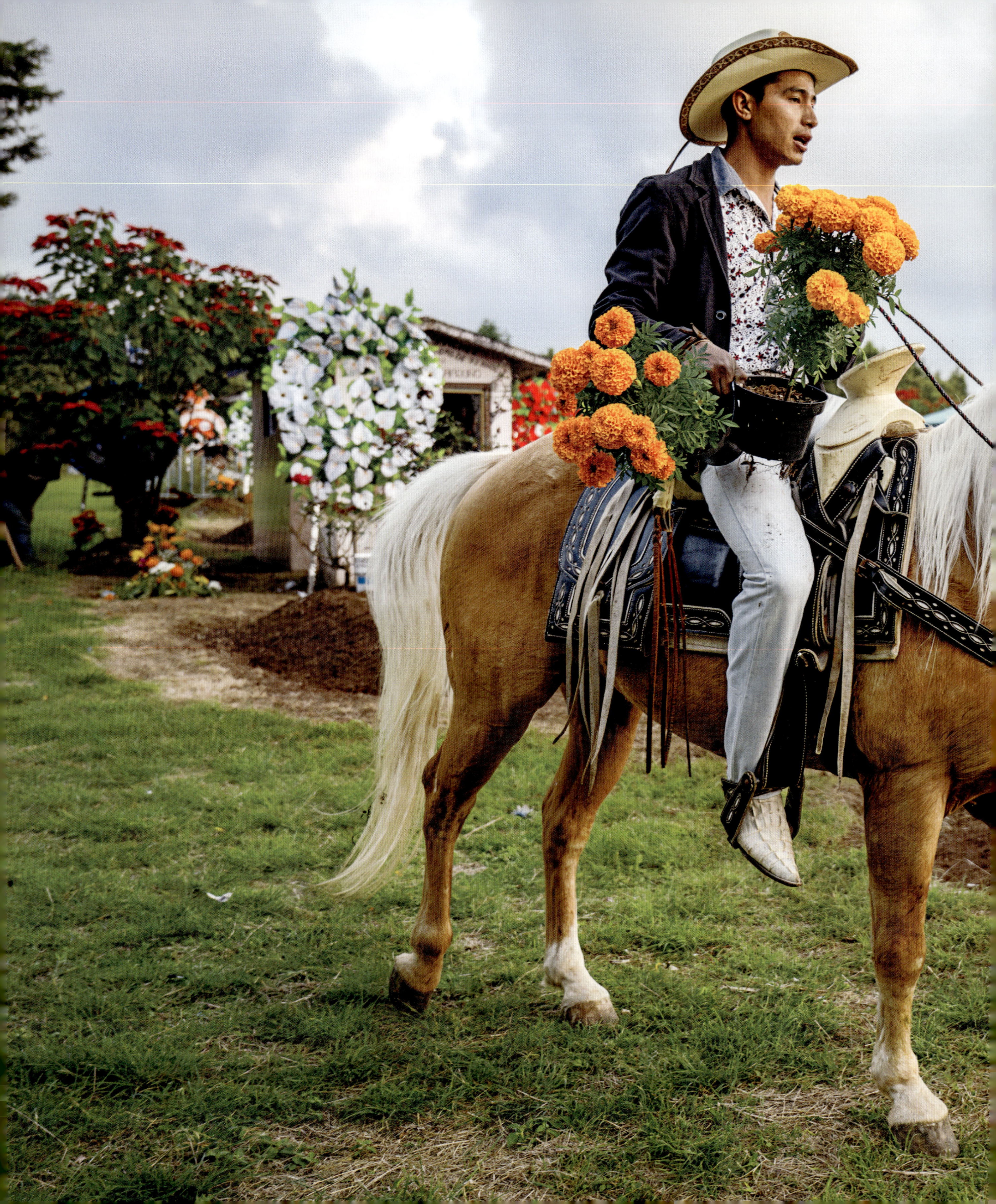

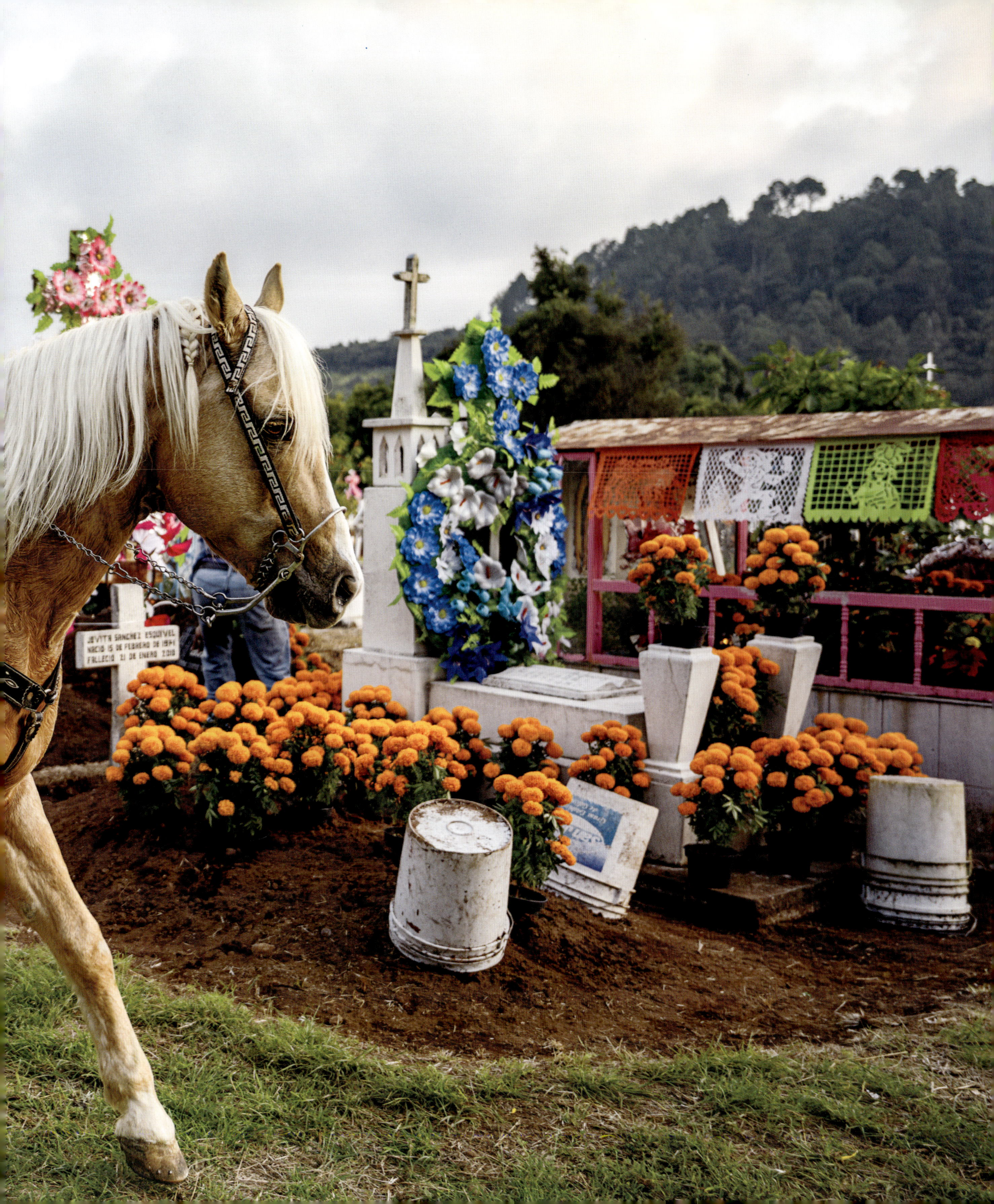

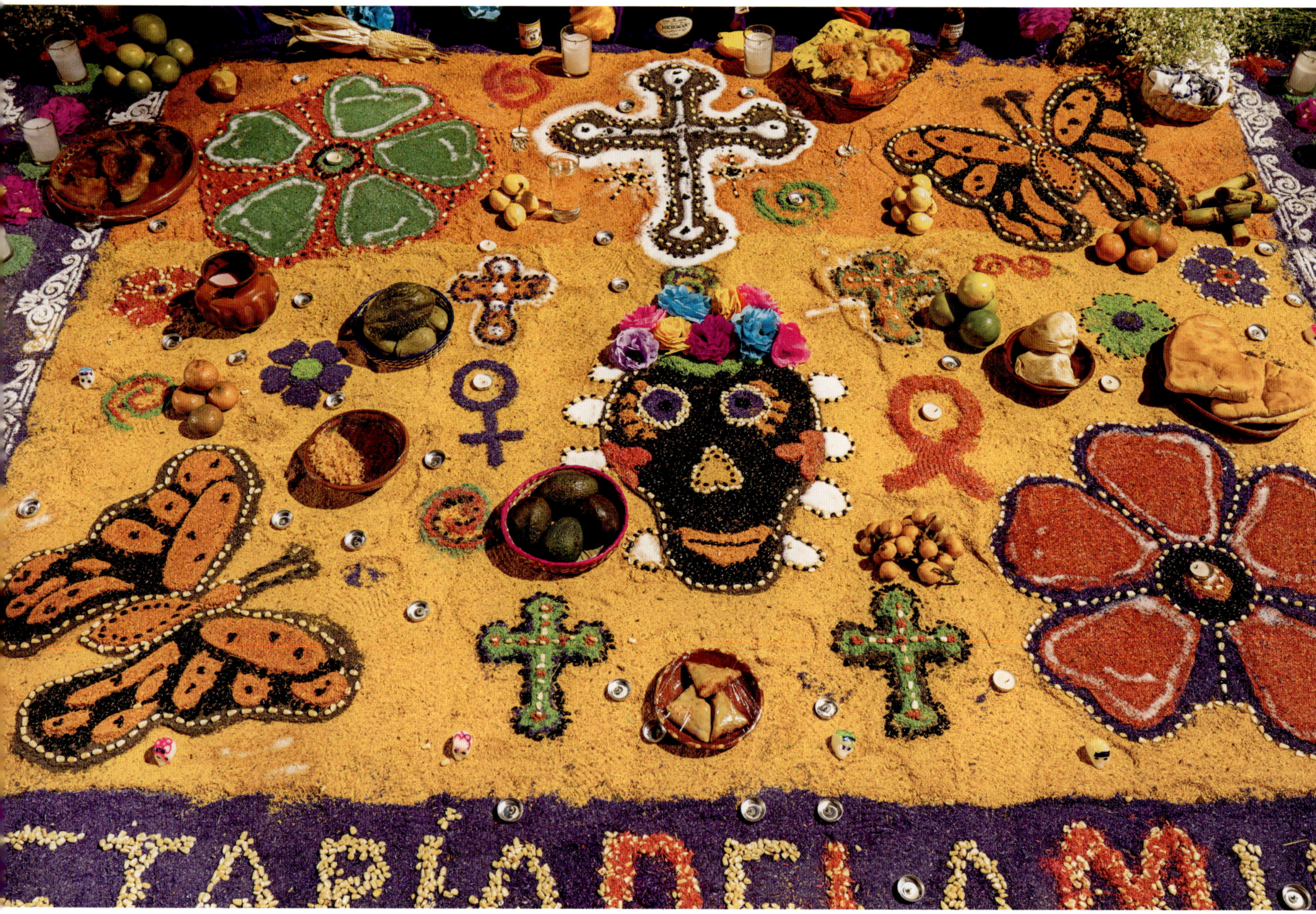

An altar decorated with monarch butterfly motifs during the Day of the Dead in Zitácuaro, Mexico, a small city in the heart of the monarchs' wintering region

PAGES 158–159: A monarch butterfly rests among *cempasúchil* flowers—the marigolds traditionally used for Day of the Dead altars—at Rancho Rivera in La Candelaria, Nuevo León, Mexico.

PAGES 160–161: Carrying cempasúchil flowers, Héctor Colín rides to the cemetery in Carpinteros, a small community just outside the Monarch Butterfly Biosphere Reserve.

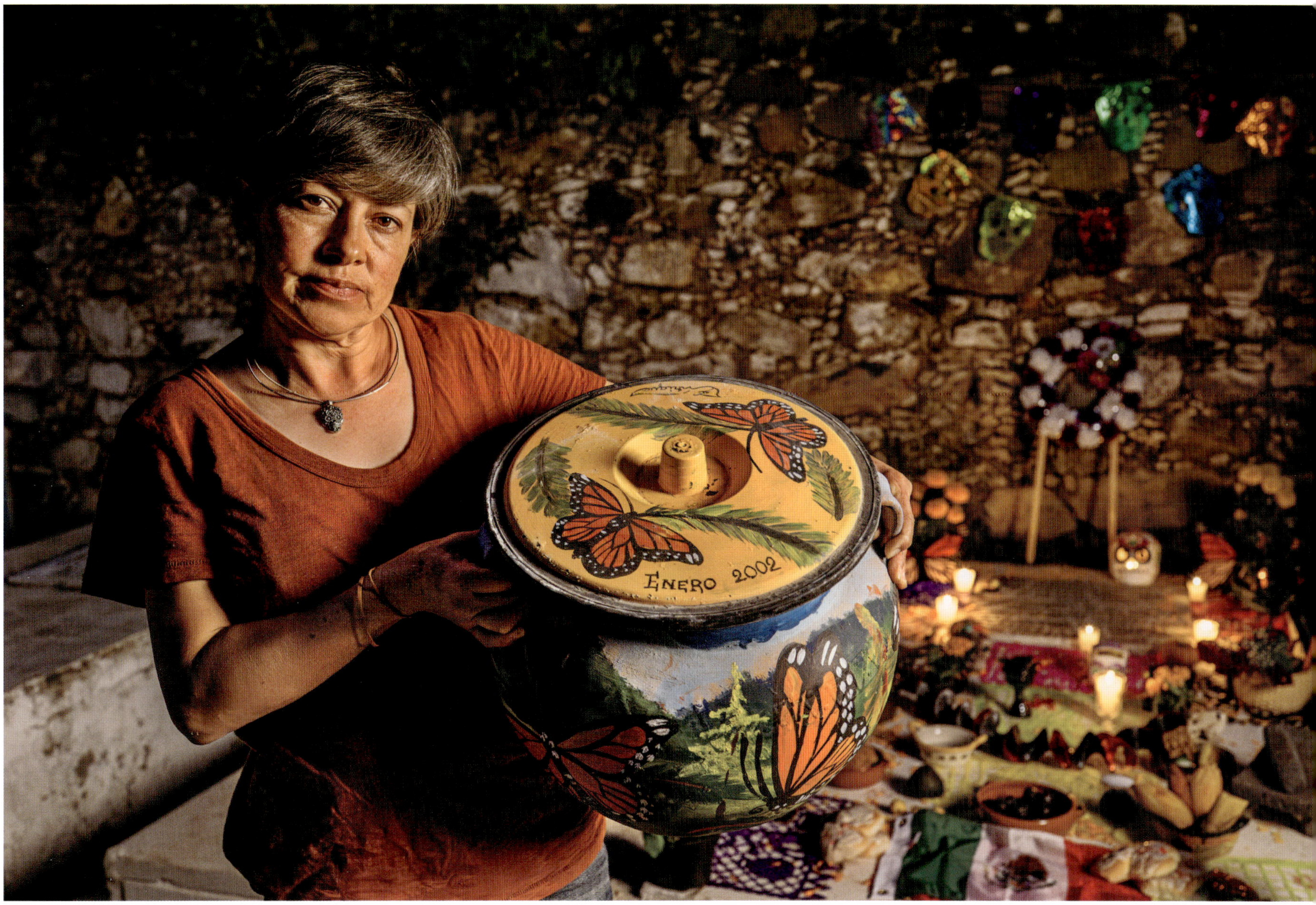

At the cemetery in Angangueo, Michoacán, Estela Romero—a Journey North collaborator who reports on the status of monarch colonies in Mexico—holds a funerary urn that commemorates the thousands of butterflies that died during a 2002 snowstorm.

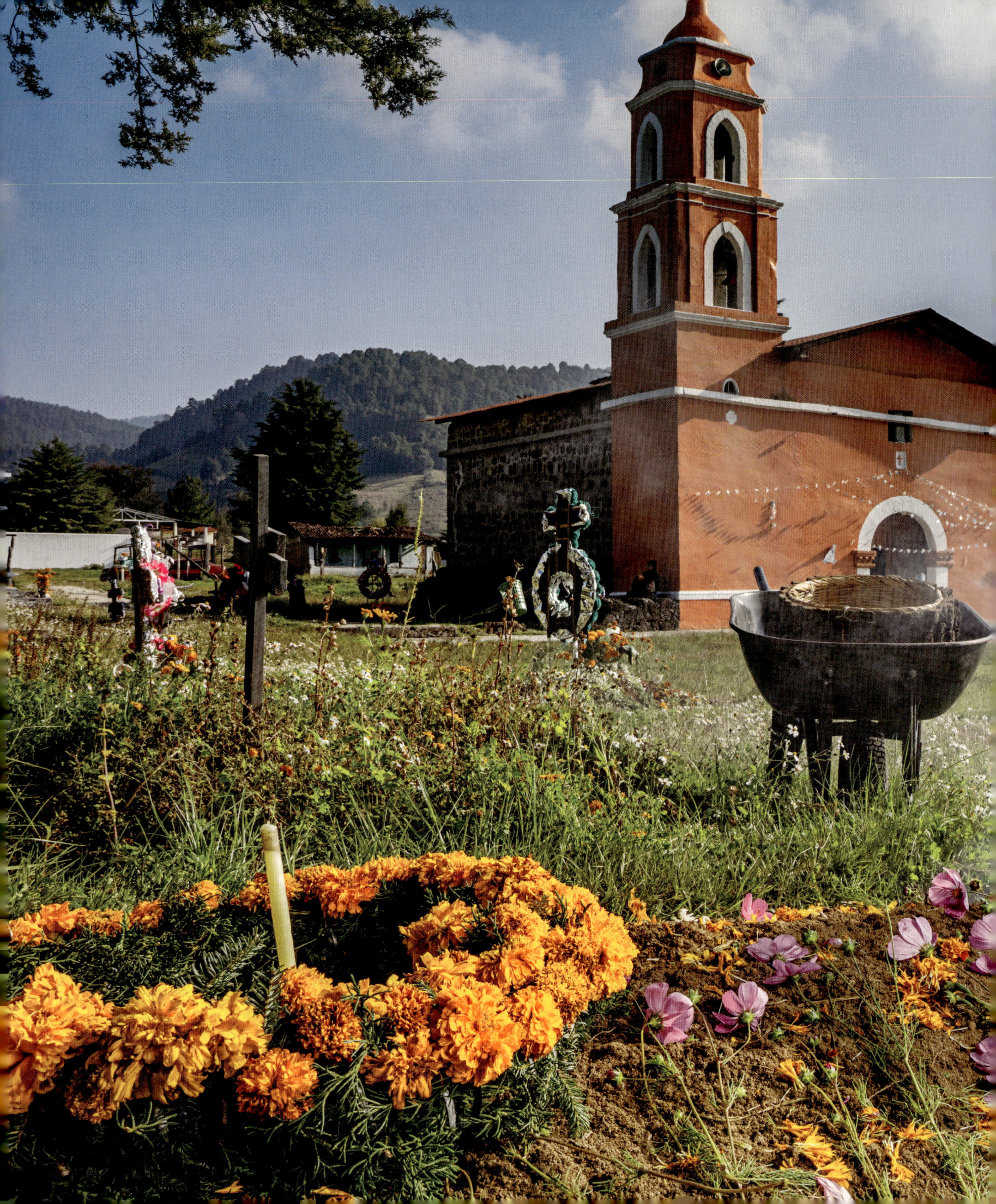

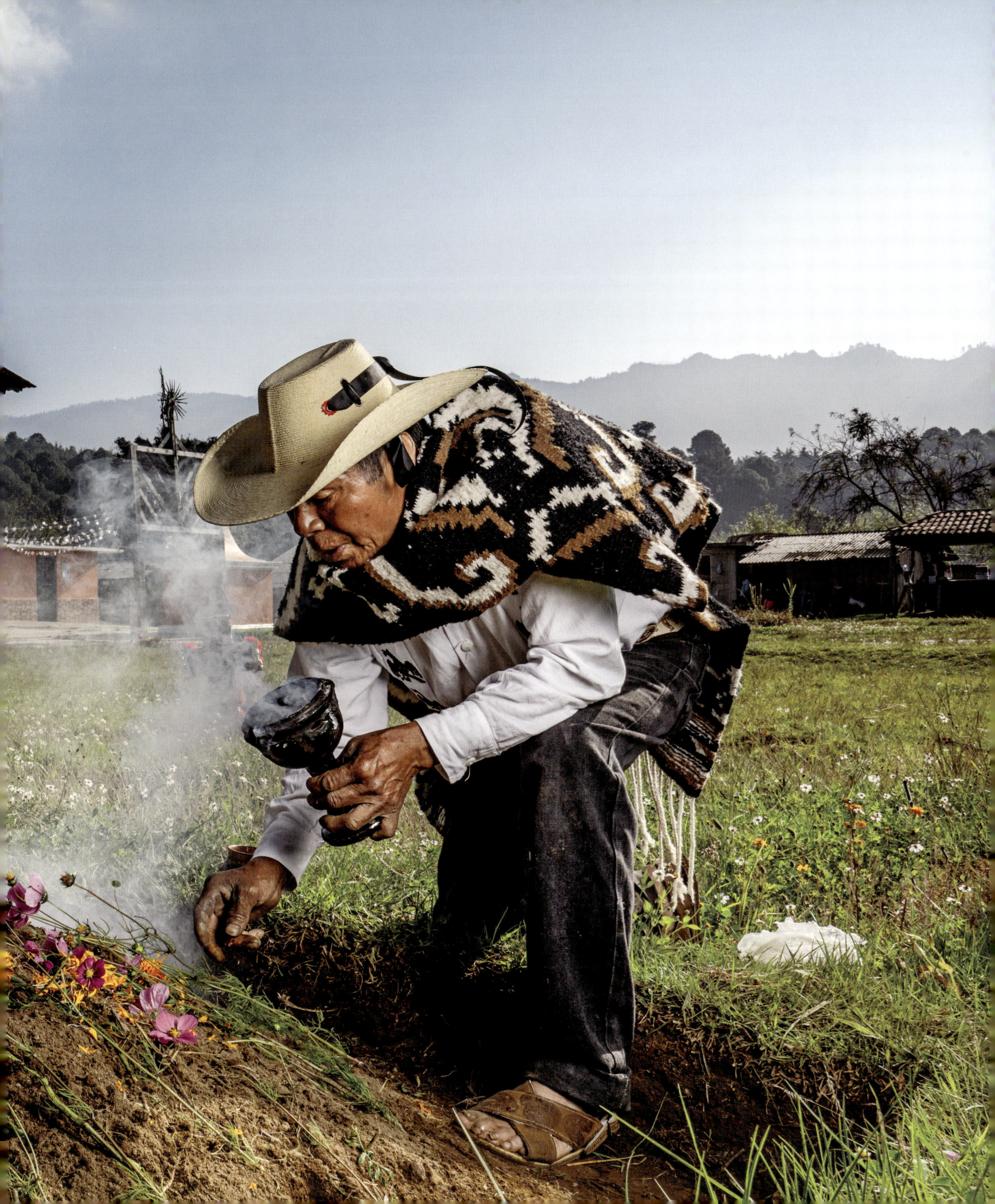

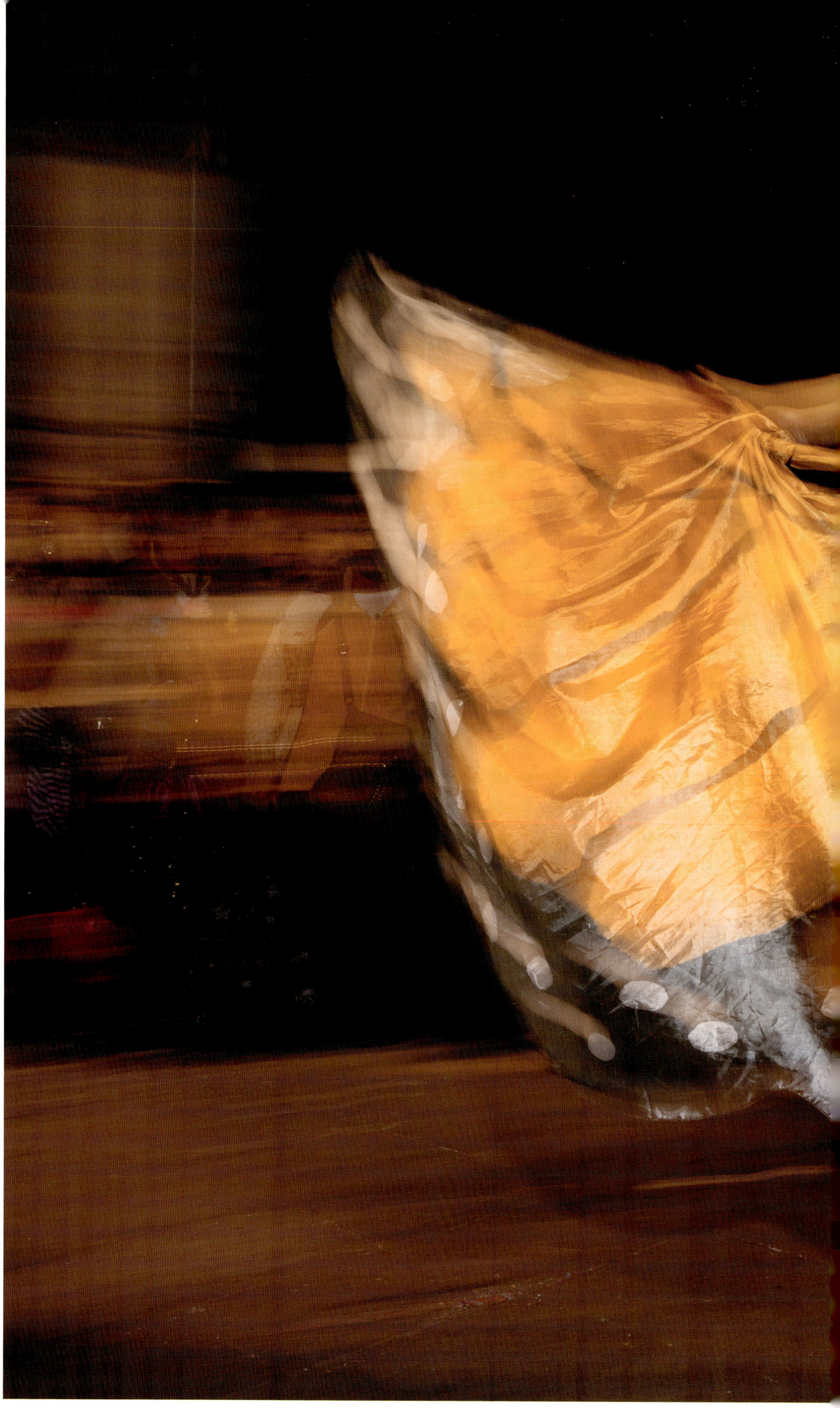

A young woman dressed as a monarch butterfly dances during the Day of the Dead student parade in Zitácuaro, Michoacán. Like many cultural traditions, the celebration continues to evolve, and the monarch has become one of Mexico's most recognizable contemporary symbols.

PAGES 164–165: To celebrate the annual Day of the Dead, Sabino Marín Reyes cleans and decorates the graves of his relatives in Ejido Francisco Serrato, a Mazahua Indigenous community in Michoacán. The Mazahua believe that monarchs are the souls of their ancestors.

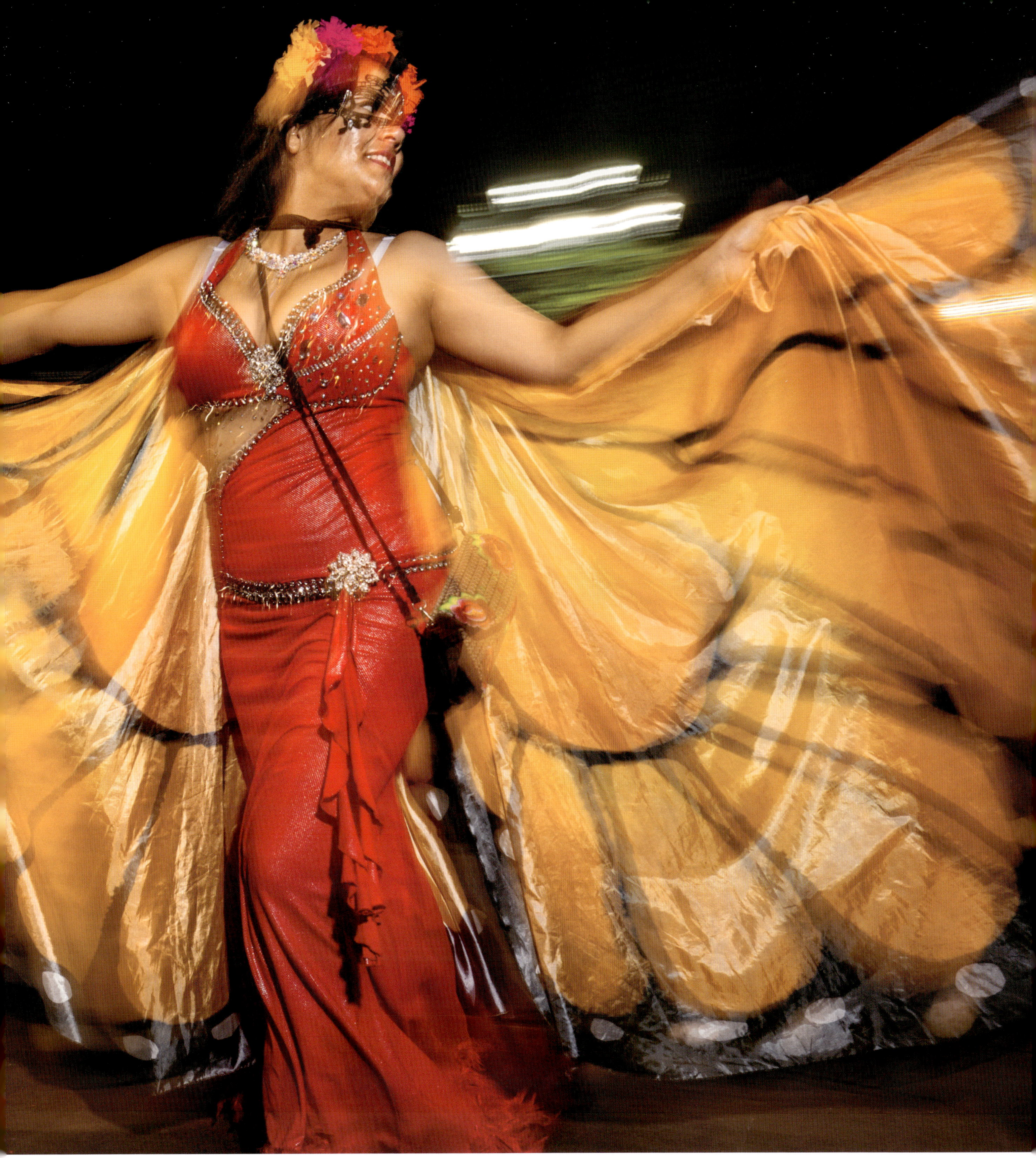

WINTER

FOREST HOME IN MEXICO

I HAVE NEVER BEEN THERE to see it with my own eyes, but I like to imagine it's like Christmas morning when I was a kid: You go to bed with an empty tree, and the next morning—surprise!—it's full of monarchs (or presents, however you want to see it). Of course, I know that's not how it happens. When I ask the local people, what they tell me is that it all unfolds gradually, almost without notice—until one day the monarchs are simply there, as if they had never left.

I also like to imagine the season's first butterfly that sets foot on an oyamel branch. Because there has to be an absolute first, right? How does it choose which tree among the thousands of oyamels? Why that one and not this one? Will it make a pleasurable relief sound in the language of the butterflies? Apologies in advance for this anthropomorphism; I should know better than that at this point in my career. But it is so tempting and easy to get lost in these kinds of musings, and surely I am not alone. The mystery of the monarchs is just too alluring.

Otherwise, how do you explain the collective obsession of a whole continent with these butterflies? Every year, the monarchs "are back," as we say, and we all celebrate it. I use the quotation marks with a lot of intention. Because if you think about it carefully, they can't be back—because these individual butterflies have never been in these forests before. Yet it is impossible not to think that this Mexican forest is their home as much as the prairies of North America. What other animal is welcomed home in two different places in the same year? As an expat from Spain, having lived many years in Mexico, I can definitely see myself in that mirror.

A WINTER REFUGE LIKE NO OTHER

Oyamel fir forests create the perfect microclimate for overwintering monarchs (pages 184–185). At elevations above 10,000 feet, the dense canopy shields them from rain and wind and keeps temperatures cool enough to slow their metabolism but not so cold that the monarchs freeze. Sunlight filtering through gaps gives just enough warmth for occasional flight. Think of the canopy as working in three ways: a blanket, trapping warmth; an umbrella, shielding from rain and dew; and a hot-water bottle, with the trunks holding heat through the night. These simple effects can mean the difference between survival and death.

Monarchs are tough. If they stay dry, they can withstand temperatures well below freezing. But when they are wet and cold, their

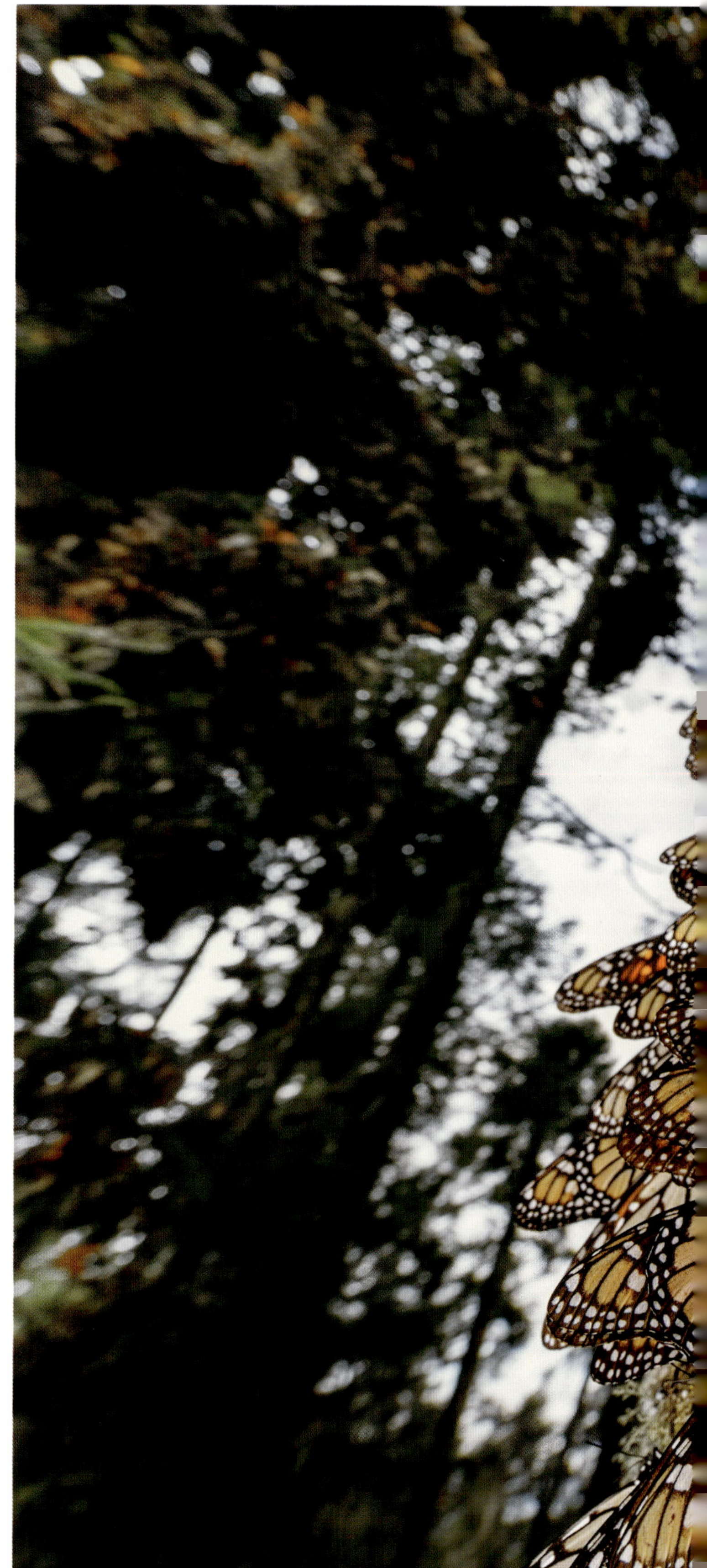

Monarch butterflies cluster on the trunk of an oyamel fir in Sierra Chincua Sanctuary, part of Michoacán's Monarch Butterfly Biosphere Reserve.

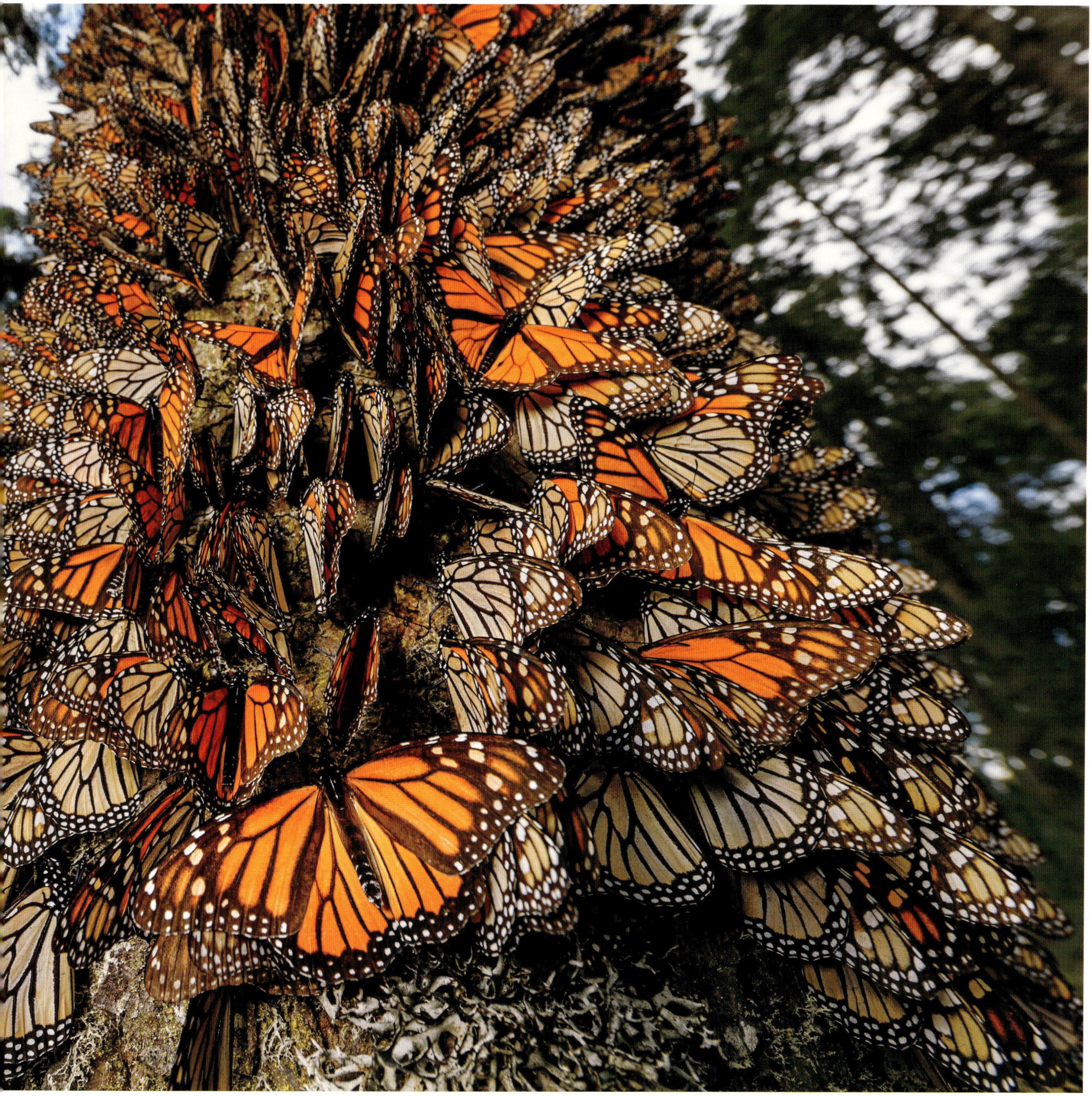

chances of surviving a freeze plummet, especially if they've been knocked down to the forest floor by wind or rain. And sometimes, when the oyamel canopy is thinned by logging or drought or pests, it can lose its protective superpower altogether.

In March 2016, a historic snowstorm hit the monarchs' wintering grounds right as the colonies were preparing to travel back north, their fat reserves depleted after the long winter stopover. Scientists estimated that between 30 and 35 percent of the butterflies were lost in a single night. I remember hearing the news. I was in a small roadside restaurant in the desert of Oaxaca, where I was filming military macaws. I dropped everything, packed, and headed to Michoacán. I had to wait three days before I was granted access to the sanctuary. Chaos reigned: The authorities were struggling to manage the swarm of TV crews covering the story. Each team was given a 10-minute slot to enter the colony, gather their visuals, and exit.

I had been waiting in line since early morning, but the light was harsh—patches of blinding sunlight mixed with pitch-black shadows on the snowy ground—so I let everyone else go first. By mid-afternoon the light had softened, and I finally walked the narrow trail that rangers had cleared, carefully moving fallen monarchs by hand so that visitors wouldn't step on them.

What I saw was heartbreaking. Thousands of butterflies lay on the ground, piled around the oyamel trunks like autumn leaves in a messy garden. The sun was going down, and with it the light was changing fast. For a few seconds, golden rays filtered through the canopy, and I caught a beautiful photo—but it lacked context. It was lovely light, but it didn't explain what I was witnessing. I needed something flatter, something clearer, something that showed the full scale of the loss.

It was getting late, and the ranger was asking me to leave. I begged for a few more seconds. I knew that as soon as the sun dropped below the horizon, the forest would turn into a flat, dull scene—normally something I would avoid, but exactly what I needed that day. I wasn't allowed a tripod, so I climbed onto a rock, balancing, trying to keep my camera steady in the low light. Most frames were blurry. At last, I held my breath and shot a rapid sequence of photos, hoping. One of those frames became the image I had in mind (pages 246–247)—one that later changed my life, as I describe in the final chapter.

In Michoacán, winter storms like this are rare, though with climate change they may become less so. But the opposite, and greater, dangers are quieter, and they unfold in slow motion: Droughts and higher temperatures can weaken or kill the oyamels themselves, and land-use changes alter the forest forever. These trees did not evolve for such pressure, and these forests are dwindling. Without them, the monarchs will lose the only refuge that makes their extraordinary journey possible. To capture the full scale of the problem in my photographs, I needed a monarch's-eye view of the forest.

MONARCH'S-EYE VIEW

I knew that the best way to capture the complexities of conservation in the Monarch Butterfly Biosphere Reserve landscape was from the air. I had flown over the sanctuaries once before, in 2006—a helicopter ride courtesy of CFE, Mexico's federal electricity company. The experience was eye-opening, but the photos from that day have always been a frustration. It was noon, the light was harsh, and it was my first helicopter flight ever, so I made plenty of rookie mistakes.

So, almost 20 years later, with many more hours of aerial photography behind me, I decided to try a helicopter flight again. My friend Moisés Acosta, whom I had met more than a decade before, was instrumental in arranging the flight and the permits. He joined me, along with Ganesh Marín, who once again was assisting me on this trip.

Like a giant, noisy hummingbird, the chopper hung suspended over Cerro Pelón, a sanctuary in the core zone of the Monarch Butterfly Biosphere Reserve. We had flown over most of the route I had carefully planned, and I was a bit disappointed. There was an annoying haze, and all my photos looked washed-out, not doing justice to the breathtaking sight in front of us. But there was some wind and a thin window of clear sky on the horizon: a small chance that, as the sun went down, the light would turn magic. I asked the pilot to circle Cerro Pelón once more while I restlessly scanned the landscape for a change of light. By the time we came back to where we began, the difference was remarkable. It was now or never. We retraced our path, and I took more photos of those iconic views: the best stretch of unaltered forests, the avocado plantations around the reserve (pages 190–191), the sunset light over Cerro Pelón. The images were finally working. In that extra half hour, I achieved more than I had in the previous two hours.

The mountains of the Trans-Mexican Volcanic Belt stretch east to west, linking the two Sierra Madres. They hold some of the highest peaks in the country, well over 15,000 feet, yet the mountains the monarchs choose for their wintering grounds are not particularly remarkable. At just 9,000 feet, they may seem big on a local scale, but they are easily lost in the broader geography of the range. Most of those mountaintops are now protected by the Monarch Butterfly Biosphere Reserve, but that didn't happen overnight.

In 1980, soon after the overwintering colonies were revealed to the world, the Mexican government created a special reserve to protect small patches of forest around the best-known roosting sites. Those early protections were fragile agreements between landowners, scientists, and officials—often more symbolic than effective. In 2000, under pressure from conservation groups and international partners, the protected area was greatly expanded to cover more than 138,000 acres of oyamel forest, and it was renamed the Monarch Butterfly Biosphere Reserve. The expansion came with a new conservation fund to support local communities and pay for forest protection. On paper, it was a triumph of science and diplomacy, turning a handful of secret groves into what would later become a UNESCO World Heritage site. On the ground, it was more complicated: Boundaries overlapped ejidos and other communities; people who had lived with the monarchs for generations now had to adapt to a model where conservation, politics, and money were all closely tied together.

I first arrived in Mexico in 2004, and the official story was still that the monarchs' biggest problem was deforestation there. And to be fair, that had been true for a long time. Illegal logging in and around the monarch wintering grounds had been devastating. But in the late 2000s and early 2010s, satellite data showed that the rate of deforestation was finally slowing. Community-led efforts, government action, and international support had started to pay off. As that success story took hold, the conversation began to shift. Some critical voices had already been warning that prairie loss and milkweed decline along the breeding grounds and migration pathways had serious impacts on the monarch population, but that narrative didn't really take off until the mid-2010s, when two things happened: Karen Oberhauser and John Pleasants published a turning-point paper showing the link, and the 2013–2014 wintering population hit the lowest level in recorded history—precisely at a moment when deforestation was in sharp decline. Since then, the influence of prairie loss and milkweed decline on monarchs has become a dominant conservation message, alongside climate change and deforestation in Mexico, which still pose threats. Small-scale clandestine logging, usually for firewood or modest income, invisible from the air, and salvage logging—the commercial removal of downed trees after storms (pages 200–201)—are still going on, and both weaken the canopy and delay forest recovery.

But even as logging slowed, new threats emerged. In the late 1990s, the United States lifted its ban on Mexican avocados. The only state authorized to export them was Michoacán—the very heart of the monarchs' wintering grounds. Since then, demand has exploded. Today, most of the avocados eaten in the United States come from this state. Flying over the reserve, I could see the patchwork clearly: dark green blocks of avocado orchards pressing against the ragged edges of the oyamel forest. Avocados are thirsty trees and need much more water than the oyamels and other native species. Entire hillsides have been reshaped for plantations. From above, it was easy to see how two global cravings—one for guacamole, the other for monarchs—collide in the same landscape. And in a warming world, with water already scarce, that tension will only grow sharper.

On the way back to the airfield in Zitácuaro, I watched the mountains fade into the distance, like small picturesque hills in the background of a painting. It seems almost impossible that monarchs can find them year after year. But the attraction of these particular oyamel forests must be strong in the butterflies' genes. Every November they return, oblivious to the enormous social complexities that have unfolded over the past three decades to ensure that these mountains remain protected.

BIRD PREDATION

Back on land, back on sure foot, I went once again after a photo that had haunted me for years. My nemesis. I have already mentioned how monarch caterpillars feed on milkweed and absorb the plant's toxins,

(Continued on page 174)

SOON AFTER THE OVER-WINTERING COLONIES WERE REVEALED, THE MEXICAN GOVERNMENT CREATED A RESERVE TO PROTECT ROOSTING SITES.

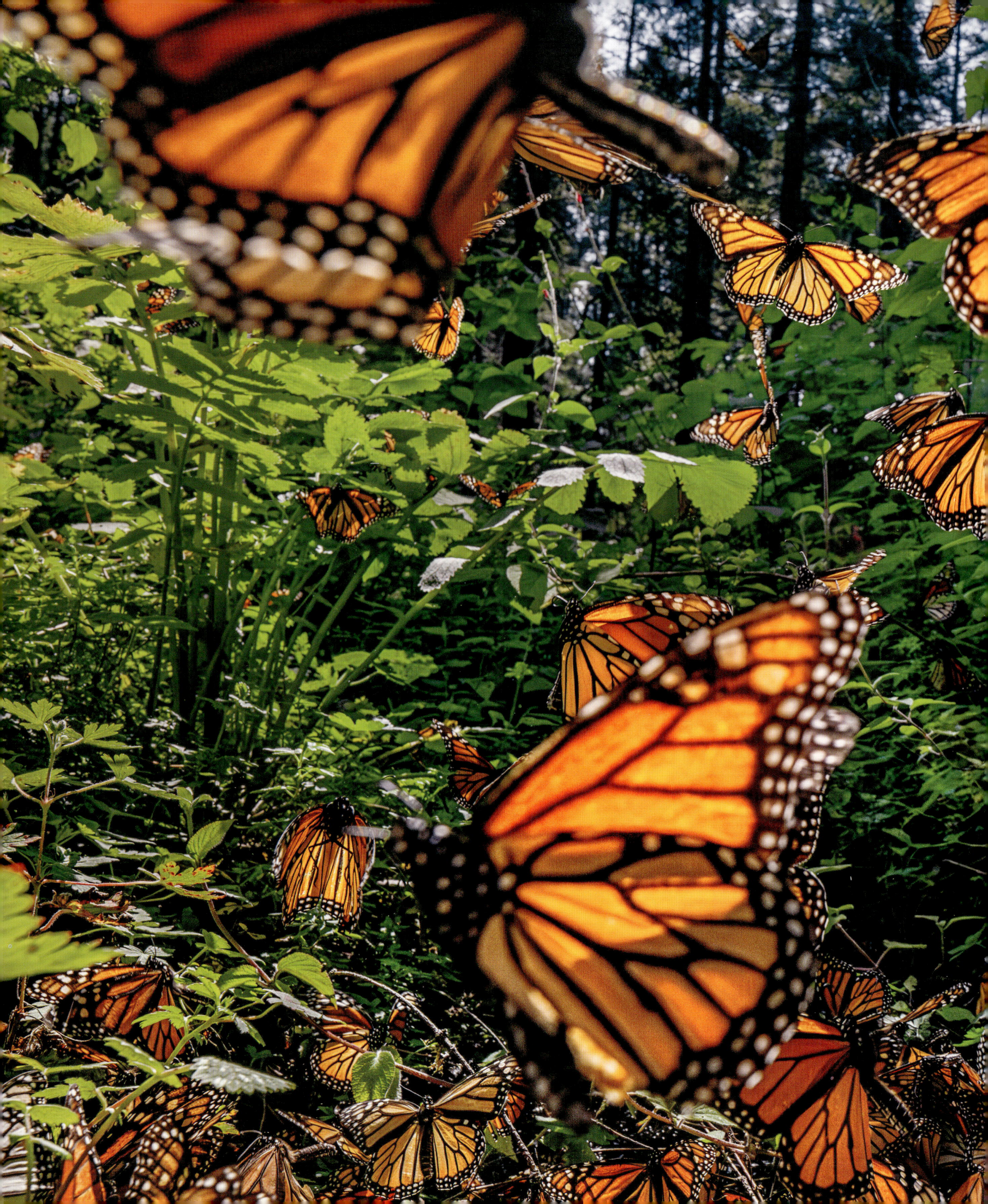

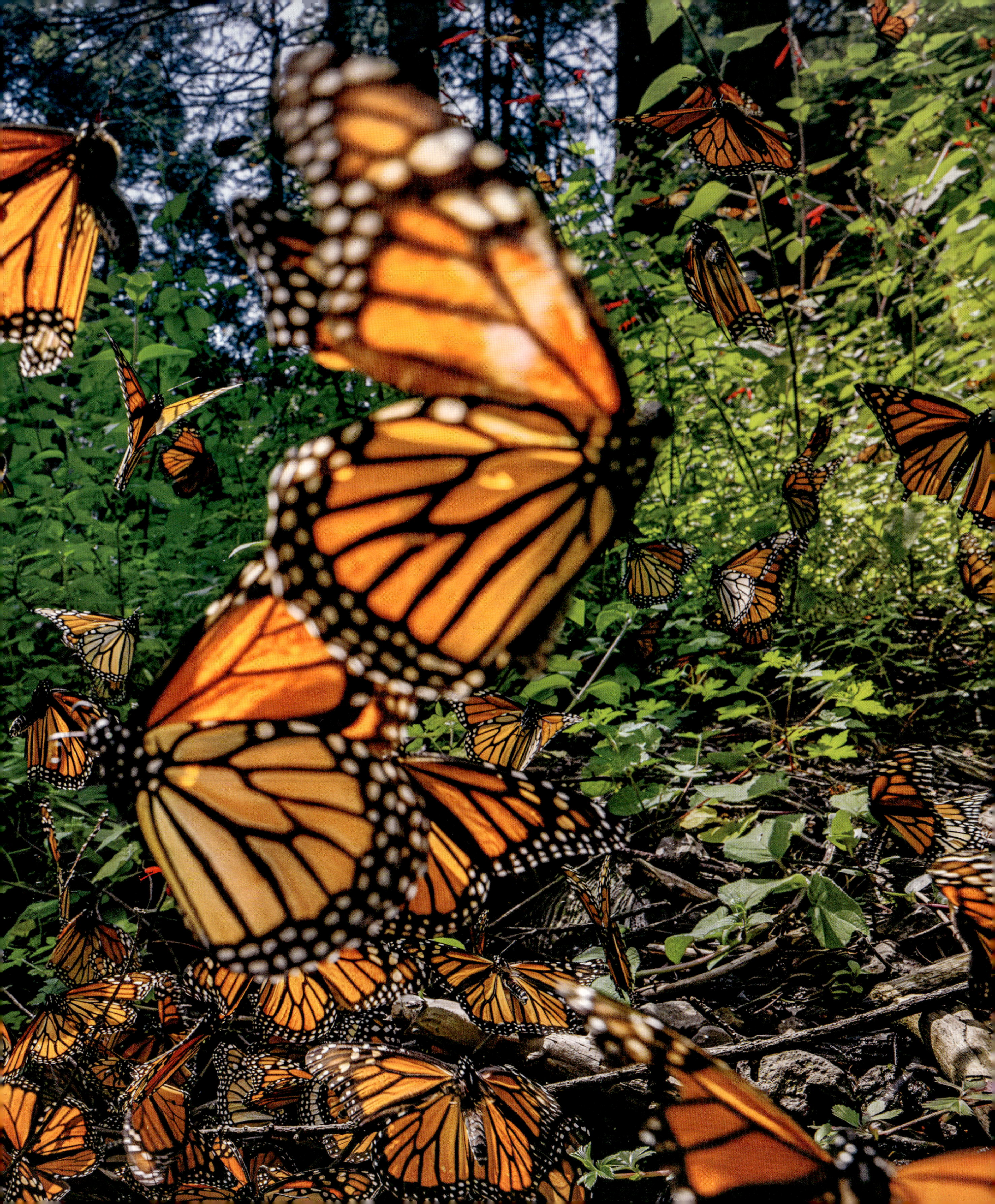

I LIKE TO SAY THAT MY BIGGEST SKILL AS A PHOTOGRAPHER IS PERSISTENCE. I CAN BE STUBBORN TO THE POINT OF OBSESSION.

PAGES 172–173: Monarch butterflies in El Rosario Sanctuary in Michoacán's Monarch Butterfly Biosphere Reserve can be quite active in the heat of the day. To quench their thirst, hundreds swarm to the ground near a small stream, where they sip liquid and important minerals from the wet soil—a behavior called puddling.

(Continued from page 171)
keeping them in their bodies as a defense. As adults, their orange-and-black pattern is a warning signal—aposematism is the technical word—telling predators they're poisonous. And it works. Few birds dare to eat them.

But at the sanctuaries in Mexico, certain kinds of birds—mostly grosbeaks, tanagers, and orioles—have developed a tolerance to the toxin and can eat monarchs. It is a wonderful example of coevolution. What's really interesting is that different bird species use different strategies. The shape of their beaks—strong Darwinian reminiscences—gives a hint as to each species' technique. Grosbeaks use brute force, their massive beaks tearing monarchs apart, eating the abdomens whole, consuming not only the fat but also some of the scales on the wings, which store the toxins (pages 196–197 and 199). Orioles and tanagers, with their sharp beaks, are the surgeons of the forest (page 198). They target the butterfly's fat-rich abdomen, accessing it by making a quick incision and eating only what they need.

The consequences of these strategies are obvious. Grosbeaks start strong, gobbling one monarch after another, but after a while the small amount of toxins they inevitably ingest kicks in. You can see them step back and sit aside. I like to imagine them burping, in a slightly diagonal position, suffering indigestion like the weird great-uncle who drank and ate too much at Christmas. Orioles and tanagers, however, can keep their monarch buffet going for much longer. These few bird species account for nearly all natural predation in the monarch colonies. Mice, and occasionally even squirrels, make up the rest.

Predation happens when the butterflies are most vulnerable. At dawn, in the cold, their metabolism is so slow that they hang motionless in the canopy and can't react. I don't envy those monarchs. What a horrific sight to see a grosbeak approaching when you can't even move. But the scene from the ground is different. Usually you don't even see the birds. What you see is the consequence: a sudden rain of monarchs dropping from the branches. By the time you reach that spot, the culprits have flown away. The ground is littered with wings, or abdomen-less butterflies, bright orange against the dark forest floor, like confetti after a parade.

For years I had tried to capture this moment and failed. The problem was light. It all happens early in the morning, in near darkness, and normally very high up in the canopy, where it is difficult to see, and so it is hard to frame a photo. My early cameras simply couldn't handle it, and all I got was a sad collection of blurry frames. But then digital sensors improved, ISO sensitivity got better, and suddenly the impossible felt within reach. Now I was ready to give it another try. I like to say that my biggest skill as a photographer is persistence. I can be stubborn to the point of obsession. This time I was totally committed: 20 days in a row, waking up at 4 a.m., hiking in the dark, setting up my tripod, scanning the canopy with binoculars, waiting. Most days I went back to the hotel empty-handed.

Then, two days before I had to leave, it happened. It was cloudy that morning, which meant the birds started a bit later, when there was already enough light in the forest. I was watching a branch that I'd noticed they favored, one with a natural opening that gave me a clear angle, with a beautiful shape and a dense texture of monarchs. And right on cue, as if planned in a script, two black-headed grosbeaks landed there. Seconds later they began tearing into the monarchs, the forest quiet broken by the soft sound of butterflies falling apart. I knew I had just a few seconds. My heart was pounding. I locked the lens, snapped a quick run of shots, and for once the frames came out sharp. After years of failure, I had the photo. And the irony? It was never published! So you can imagine how happy it makes me to see it now in print (pages 196–197). But to be fair, the behind-the-scenes video we made about my numerous attempts did really well on National Geographic's social media, and it became one of the most watched pieces of the whole project. The paradox of outreach in the 21st century.

HOPE

I used to be a serious birder—or at least that's how I liked to see myself. But now that I've packed enough years in my bag and met some true birders who travel the world chasing what they call lifers (new sightings that they add to their life list), I see it was pretentious to call myself that. Real birders come to these forests only to watch birds; for them, monarchs are just a side benefit. The Trans-Mexican Volcanic Belt is full of endemics, and when I first came to the sanctuaries in 2004, I had some species on my wish list.

But on my first visits I was shocked to discover that almost any songbird was broadly referred to as *jilguero*—a word commonly used to name the brown-backed solitaire—and that was it. Ecotourism was still in diapers. Today it's grown up, with all the contradictions of a teenager. Some guides can now name every bird species found in these forests in four languages. Visitor numbers, and the infrastructure to serve them, have grown fast. What used to be a few isolated creaky wooden cabins is now a fully assembled food court with a parking lot. Where an old lady once served only quesadillas, vendors compete fiercely for business with a rich menu of assorted options. Yes, there are downsides to this growing tourist industry—crowds, erosion, noise—but ask anyone locally, and the response is overwhelmingly supportive of this newfound bonanza.

Hundreds of thousands of people visit the sanctuaries in a single overwintering season, and in towns like Angangueo, tourism has boosted the local economy. It's not a cure-all, but tourism has become a lifeline for many families once trapped in poverty. And this has a direct effect on monarch conservation. When families see that a standing forest can bring in more income through tourism than a forest cut down for timber or firewood, the incentive to protect trees is real. Former loggers are now guides, horse handlers, or food vendors. Local cooperatives reinvest part of the profits in trail maintenance and community forest patrols, making the monarchs' presence an economic asset worth safeguarding. Of course, many challenges remain around poorly managed mass tourism, but overall, this trend has created a powerful—and at times fragile—balance between livelihoods and conservation.

All things considered, I have always been critical of the "economic alternatives" approach in conservation. I understand the pragmatic need to use economic value as a tool, but I have always found it morally conflicting. Call me a romantic, or a privileged observer, but something as remarkable as the monarchs has an enormous intrinsic value, and I constantly wonder whether there are other ways of protecting them without falling into the money trap. Engaging local communities in conservation beyond tourism is difficult, and even harder to quantify in economic terms, but I am convinced it is part of the solution. The benefits are enormous and expand over time like ripples in water. I work on long-term projects where I return to the same places over and over and witness change across years. Sometimes those changes are discouraging, especially deterioration in the condition of the land, but one common denominator is clear: When local communities are engaged, when there is a sense of ownership and commitment to this elusive, abstract idea of conservation, the outcomes are stronger. And in the Monarch Butterfly Biosphere Reserve, it is no different. In just two decades, the change in attitude and the growing participation and commitment of the nearby communities have been extraordinary. Yes, tourism and money play a role, but so do local leadership and bottom-up approaches.

Every year, local communities in and around the Monarch Butterfly Biosphere Reserve join forces with the National Commission of Protected Natural Areas (CONANP) and World Wildlife Fund in Mexico (WWF-Mexico) to monitor the size of the wintering colonies that year. Monitoring takes place several times each winter season, and then the results are averaged. It is a repetitive and almost mechanical process, but in that simplicity lies its beauty. In an age of LiDAR and AI, the monitoring is still done the old-fashioned way—with a measuring tape, a compass, and GPS. Precisely because it has been done this way since the 1990s, it provides valuable points of comparison across decades. I joined Eduardo Rendón Salinas, the lead monarch scientist at WWF-Mexico, in one of these monitoring sessions in San Pablo Malacatepec, Michoacán (pages 210–211), and I can testify that it isn't easy work. You have to climb, bushwhack, and side-slope like a madman; you need to know your math and be rigorous. But the result is worth it. Every March, when WWF-Mexico and CONANP announce the numbers, everyone in the monarch community holds their breath. The data they announce will set the pace for the next news cycles about monarchs.

Ecotourism and community-led conservation are essential, but there's still a missing link. In the middle ground lies education—the foundation for decision-making, both now and in the future. A few years ago, I visited the Papalotzin Environmental Education Center in Zitácuaro. It's an initiative established by Moisés Acosta, who had helped me with the helicopter flyover. That week a local school was visiting. It was May, long after the monarchs had left Mexico, so I was

(Continued on page 178)

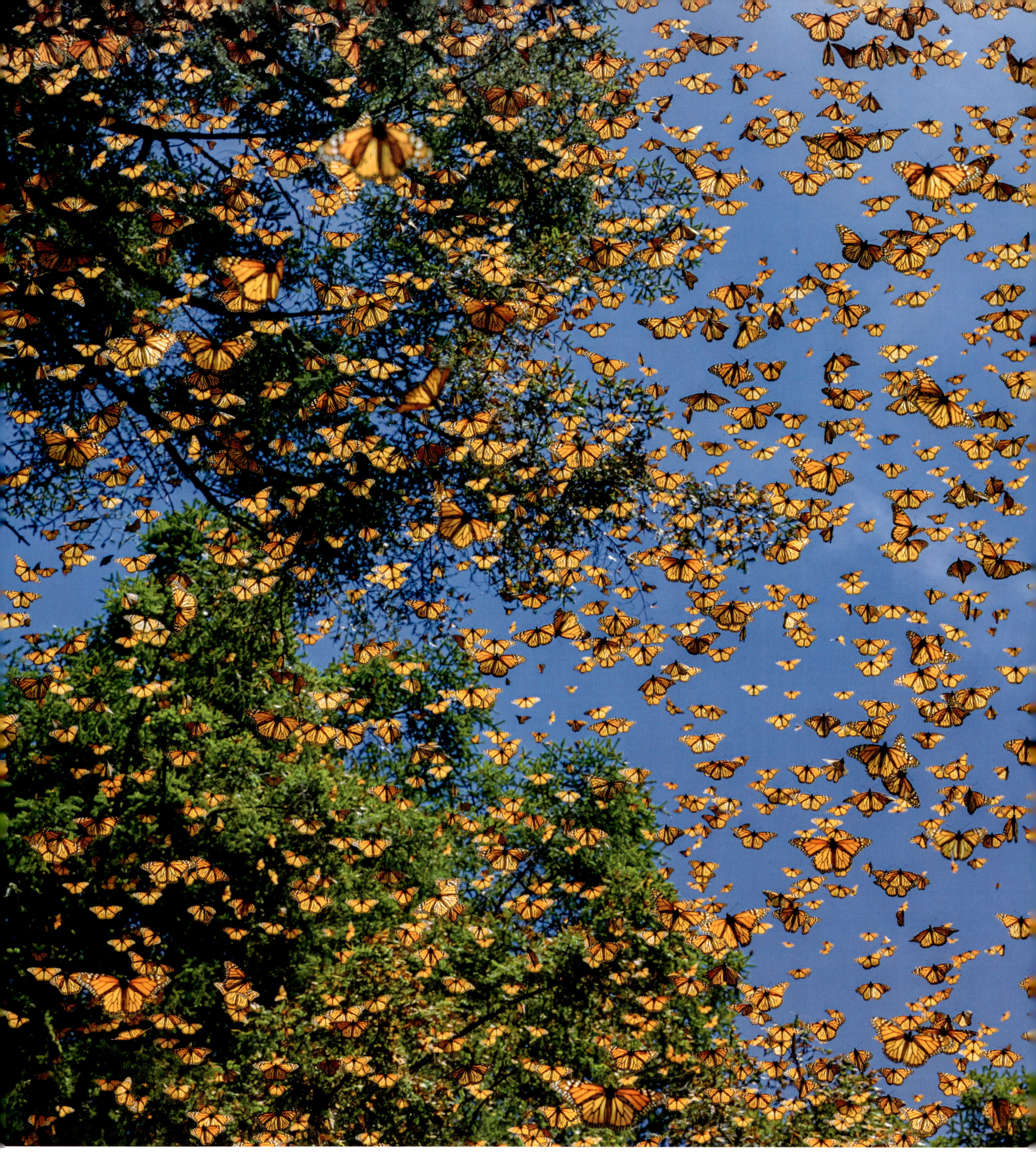

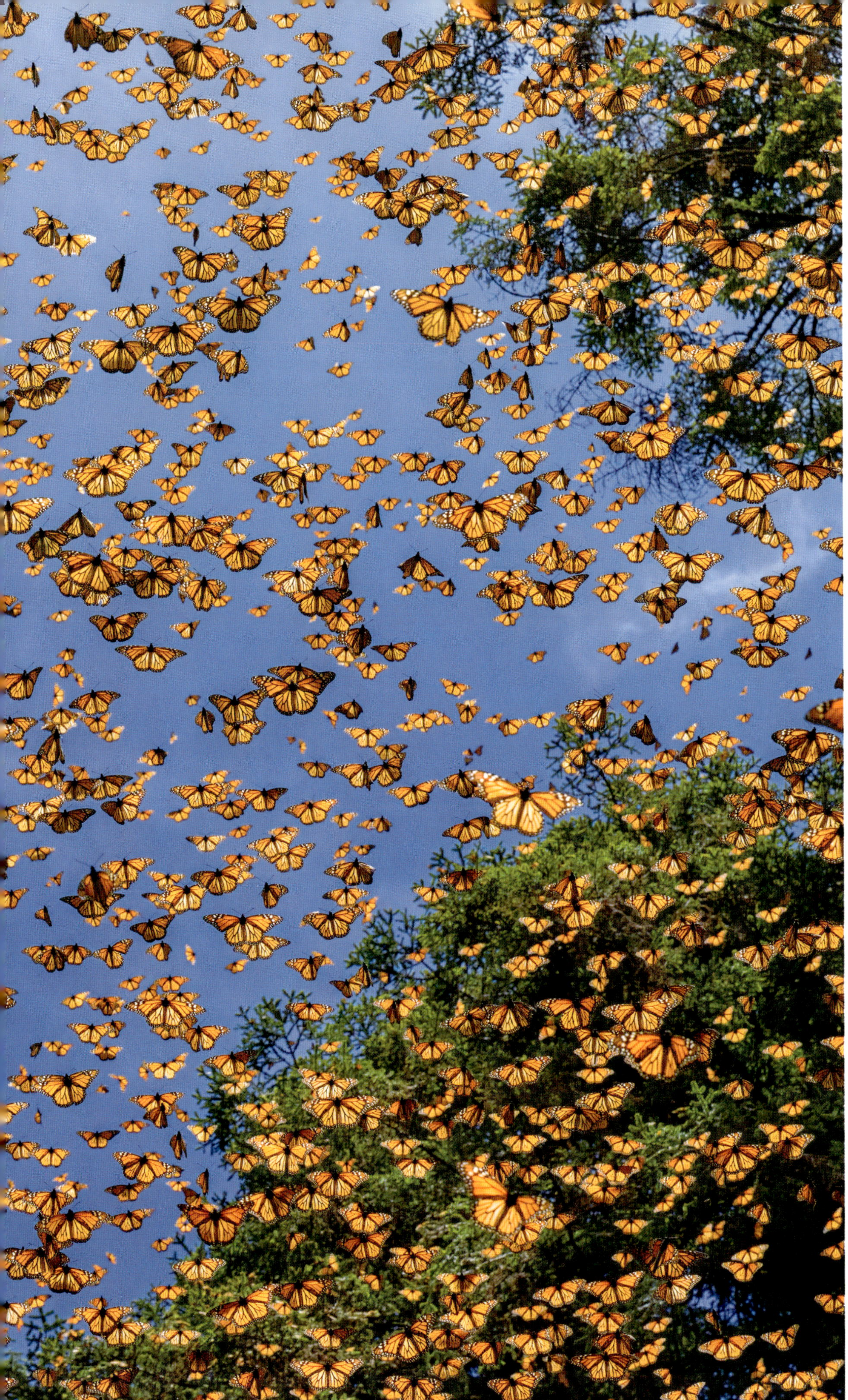

Monarch butterflies stream through the oyamel fir forest at El Rosario Sanctuary. Shaded mountain slopes there create the cool, humid microclimate that allows millions of monarchs to survive the winter.

(Continued from page 175)
curious to find out what he had prepared. He promised, mischief in his tone, it would be something the kids would never forget.

I arrived and was greeted by Moisés and his wife, Verónica Velazquez. Soon the bus arrived, full of noisy primary schoolers, and chaos took over. After some chasing around, and patient guidance by the teachers, the children crammed into a small classroom, where Moisés quickly grabbed their attention (pages 204–205). On the table he had resin models of the monarch life cycle, live caterpillars feeding on milkweed, and real chrysalises. At his feet sat a cardboard box, which he carefully kept covered. As he moved on, the kids grew more engaged, and Moisés—like a skilled magician—prepared his final act.

He asked for a volunteer, and dozens of hands shot up. He picked a small boy with big, curious eyes and neatly combed black hair plastered with gel. Moisés asked the rest of the students to turn around. He pulled a set of costumes from the box, and before our eyes he staged a live metamorphosis. The boy, dressed as a caterpillar, crawled into a green cocoon, wiggled, and moments later—with help from Moisés and plenty of suspense—emerged, beautifully dressed as a monarch, standing on top of the teacher's desk. Everyone in the room cheered and clapped.

In 2023, Moisés told me he had to close the Papalotzin Environmental Education Center. After much struggling, he had run out of funding. That made me sad, and it made me reflect again on the economic-value approach. What is the price tag of those wide-open eyes full of wonder?

I don't have empirical data, but as a father who has seen the same wide-open wonder in my daughters the first time they looked up at a blue sky filled with streaming monarchs—sending me straight back to my own moment in 2004—I am a strong believer in the power of education, especially when combined with hands-on experiences in nature. Those kinds of experiences can change you for a lifetime. Their ripple effects are unpredictable but always powerful.

The experience can be so transformative that many a visitor who comes to the sanctuary becomes, in a way, an ambassador for the monarchs. I've met dozens of seasoned monarch enthusiasts who still keep long-standing friendships that began on a pilgrimage to the sanctuaries. And I use the word "pilgrimage" deliberately, because the whole experience carries an aura of spirituality and mysticism. Many people cry when they see the monarchs in Mexico for the first time. Name another insect capable of stirring such intensity—and join me in crowning the monarchs the true rock stars of insect biodiversity.

INCREASED ACTIVITY

Like every rock star, monarchs know how to put on a show in front of a crowd, even if they don't mean to. Ask any of the visitors who gather in the sanctuaries in February what they felt when they heard the faint whisper of millions of wings, the soft currents stirred in the air, the paintbrush strokes of orange against the deep blue sky (pages 186–187).

By late winter, the days are longer and warmer, and the butterflies, extraordinarily adapted to sense their environment, become more active than ever. This is the best time to witness what's colloquially known as streaming: At noon, vast flocks of monarchs pour downhill in a clear directional flow, apparently searching for water. Once their body temperature rises above 55°F, they're ready to fly. And they do it almost without warning. What starts gradually—a certain restlessness in the branches, monarchs beating their wings, more active as they get warmer—soon becomes mayhem, and the branches seem to explode like popcorn on the stove, butterflies taking off by the thousands.

It's a classic scene, and tourists from all over the world come here hoping to see it. That's why I always recommend at least two or three days in the sanctuaries. Some days are cloudy and cold, and the butterflies barely move. Some days are modest, with sudden bursts of small streams here and there, over even before the show starts. But if you're patient, or if you are like me, visiting the same places over and over, you may witness something unforgettable.

I remember one such day. After a slow, cool morning, clouds started breaking just enough for the sun to heat the oyamel forest. I could see some activity in the branches that were more exposed, and suddenly, it happened. A torrent of monarchs cascaded downhill over me, so thick that they hit my face, my camera, even tangled in my hair. I planted myself in the middle of a trail for a clear angle and kept shooting. At first it was chaos, but gradually I found patterns and rhythms I could anticipate. After thousands—I'm not kidding, thousands—of failed attempts, I finally captured the image I had imagined: a blue sky framed by the oyamel forest with thousands of butterflies flying in the same direction, all in sharp focus, neatly distributed in the frame (pages 176–177). It may look easy, but it took me 20,000 frames. I like to call that photo "a photographer's dream, an editor's nightmare"—a dream to capture, but a nightmare to sift through all those images.

In El Rosario there is a small stream beside the stairs visitors climb to reach the colonies. The colonies shift up and down the slope as temperatures change and the season advances, but the butterflies return to this little trickle of water again and again. It is one of my favorite places in the sanctuary, and I have spent hours watching monarchs drink, pressing their delicate proboscises into the wet soil of the streambed, sipping water and minerals in a behavior known as puddling (pages 4–5 and 236–237). The forest floor turns into a living carpet of wings opening and closing in the sun. Sometimes I scan them with binoculars, looking for tags. Of the four tags I've found in 20 years of visiting monarch sanctuaries, two of them were in the puddling monarchs, and on the same day!

As the butterflies' daily activity intensifies, so does mating—the final act of epic endurance for the monarchs. Monarchs engage in aerial mating: The male seizes his chance midair, crashing into a female and dragging her down through branches and leaves until they both tumble to the ground. What follows is less a graceful courtship than a contest of persistence. The female struggles, fluttering and trying to shake the male off, but he holds on with determination. When he sees the chance, he will try to take off again, dragging the female with him up to the tree canopy, where they may remain coupled for up to 16 hours—a long, almost stubborn union that drains their last reserves of energy. It is a behavior as fascinating as it is disturbing.

I had long fantasized about capturing a photograph of it: the male flying through the forest, the female resigned and pulled behind him, their abdomens locked together. But when I set out to try it, I discovered it was far from easy. I found a secluded area of the forest on a morning when the butterflies seemed especially active at mating. First I used the telephoto lens, aiming to photograph them against the blue sky, but that led to some funky pictures (pages 228–229). I wanted a forest background to add context, but with the telephoto it was impossible, and I quickly gave up, dizzy from chasing the elusive couples through the lens.

Things got interesting when I switched to a shorter lens for a wider range of focal distance. But I was too slow to react and focus. By the time I found the butterflies in the viewfinder, it was always too late. Finally I hit the sweet spot, using my 35mm lens—a middle ground in focal distance with a very fast autofocus capability—and (here comes the secret ingredient) shooting blindly as I followed butterfly couples through the forest immediately after they took off from the ground.

I often complain that I stopped being a naturalist when I became a professional photographer, because more and more I found myself less dedicated to leisurely, quiet moments of observation and more likely busy, trying to achieve a certain kind of frame. But the truth is that to get some of those frames, you have to observe and learn to anticipate behaviors. Those habits of the patient naturalist have always been useful as a photographer. I learned to predict which couples were going to take off, and that gave me a few seconds of lead time—an advantage to anticipate, frame, and shoot a quick series of photos. After an intense morning full of failures, I started to get into a flow and finally got some very interesting frames (pages 226 and 227).

Around mid-March, the end of the season, the monarchs begin to leave. For the males, mating is so demanding that it will be the only legacy they leave; most will die in the Mexican forests. For the females, it marks the beginning of the next journey north, carrying the future inside them. There are still some clusters of butterflies flying and returning to the trees, but others will take off for the last time, never to return—a farewell from their wintering havens in Mexico that will end in the prairies of Texas or Oklahoma, where the next generation will begin and continue north. I have never stayed long enough in March to see how the monarchs leave the forest, but that sight must bring quite a mix of emotions: sadness to see a once lively forest gradually fading, and happiness to imagine that new generations of monarchs will continue the cycle, returning to this place where so many people will be waiting for them.

A TORRENT OF MONARCHS CASCADED DOWNHILL OVER ME, SO THICK THAT THEY HIT MY FACE, MY CAMERA, EVEN TANGLED IN MY HAIR.

PAGES 180–181: Monarch butterflies cluster on eucalyptus branches at Lighthouse Field State Beach in Santa Cruz, California. Although eucalyptus is not native to the West Coast, these groves offer the cool and sheltered conditions that butterflies need to conserve energy through the season.

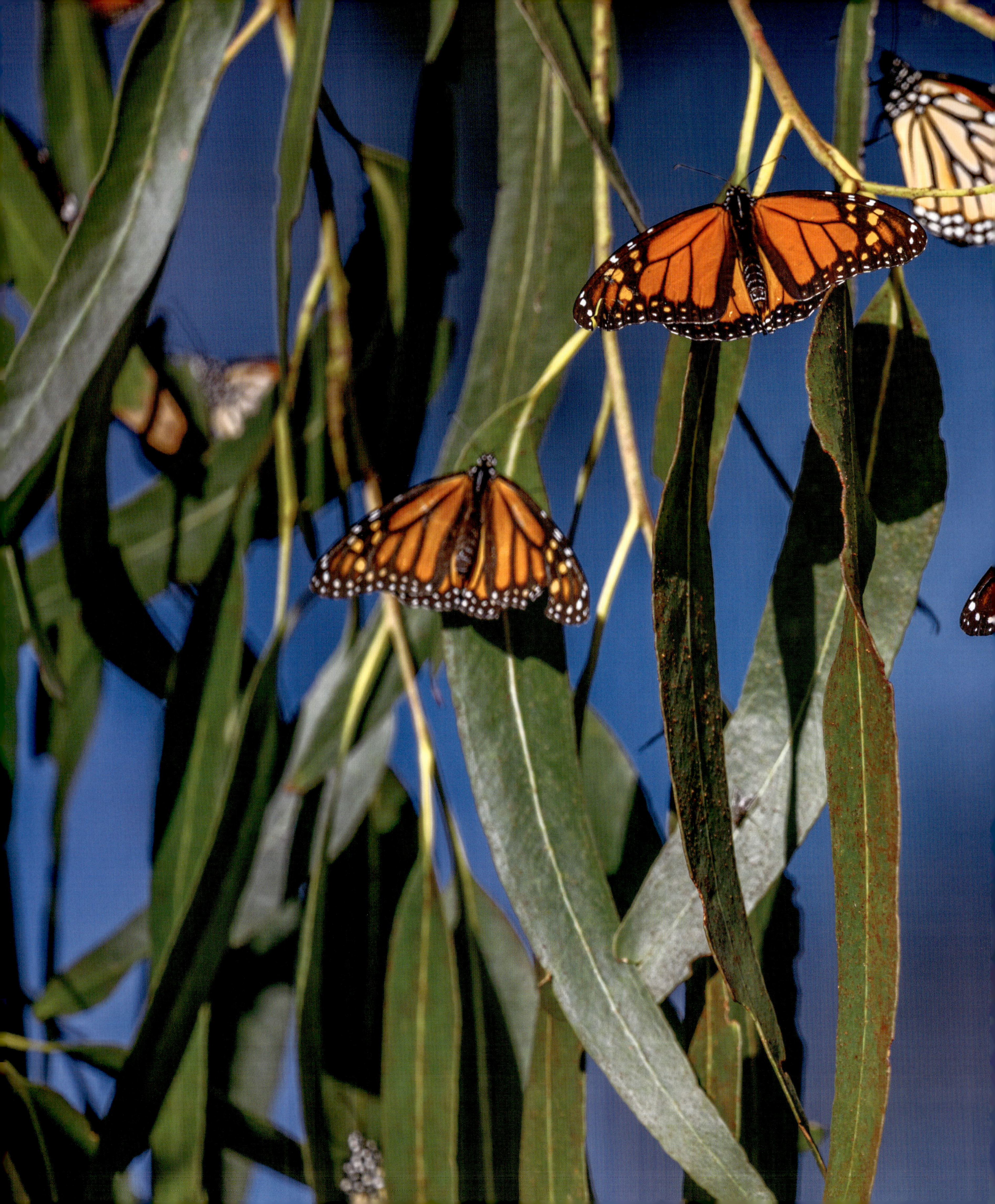

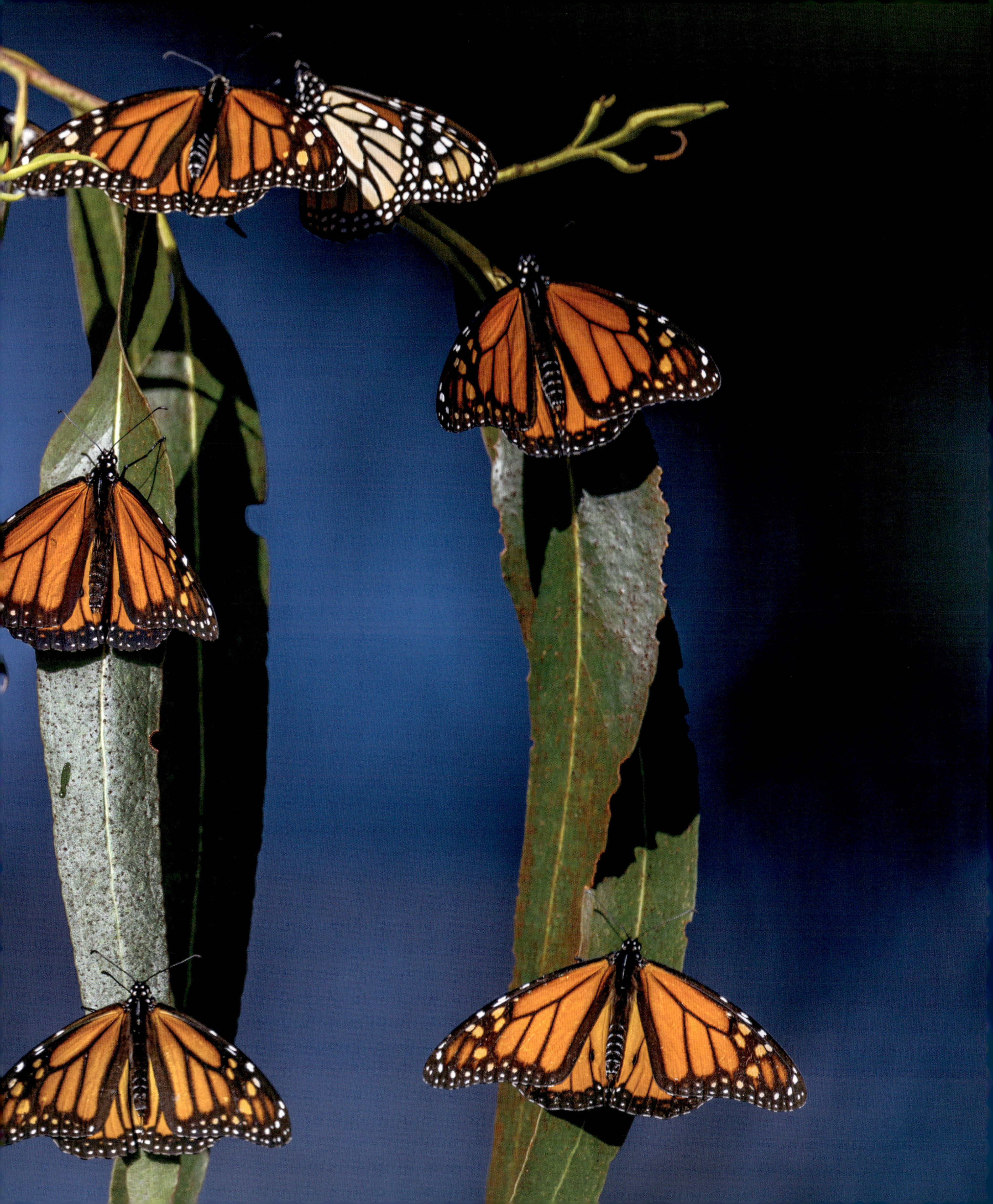

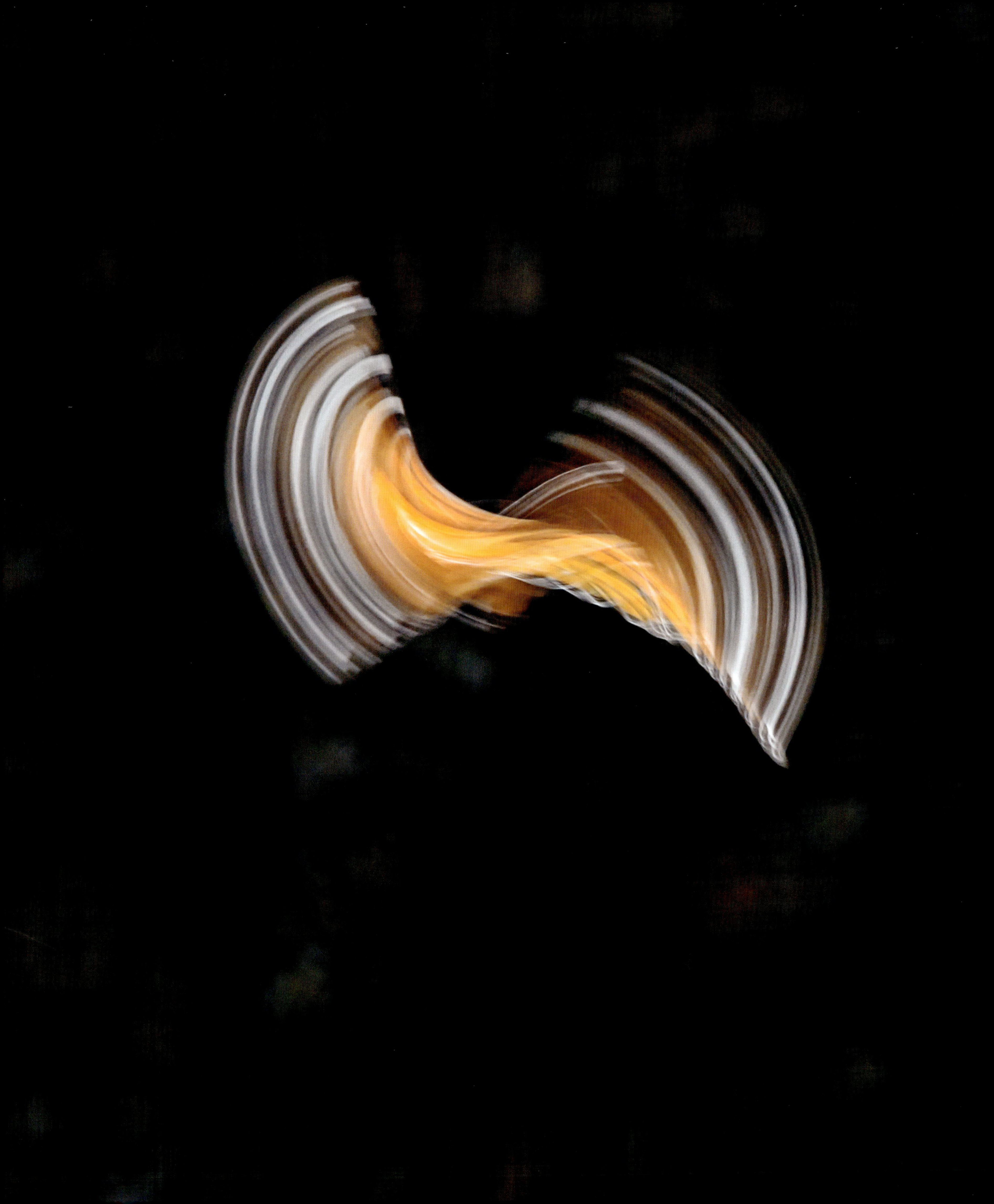

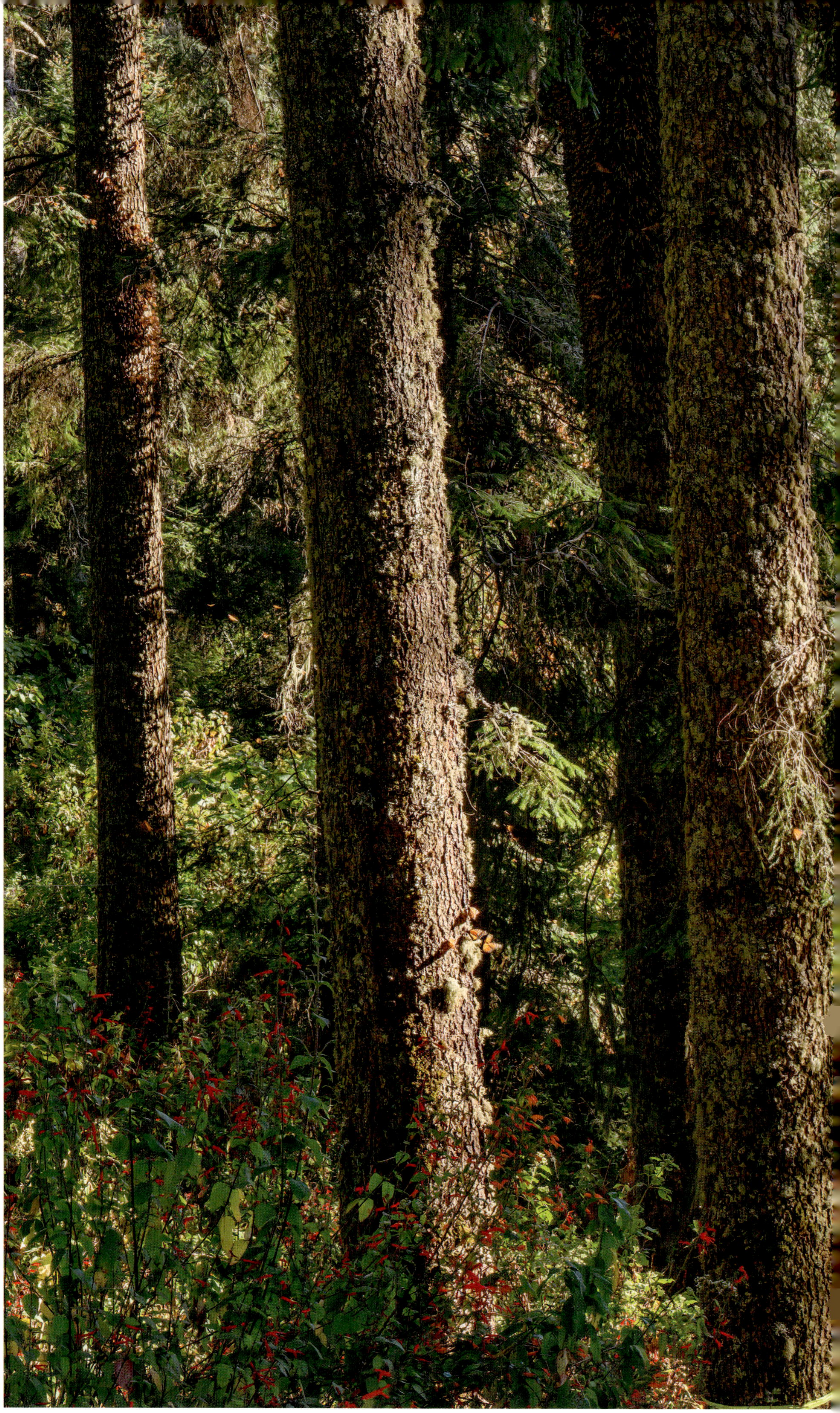

Monarch butterflies cover the branches of oyamel fir trees at El Rosario Sanctuary. The dense clusters form a living layer over the forest canopy, leaving the outlines of the trees still visible beneath the butterflies.

PAGES 182–183: A slow shutter speed captures the movement of one monarch's wings as a late ray of sun illuminates it in flight at Piedra Herrada Sanctuary, state of México, Mexico.

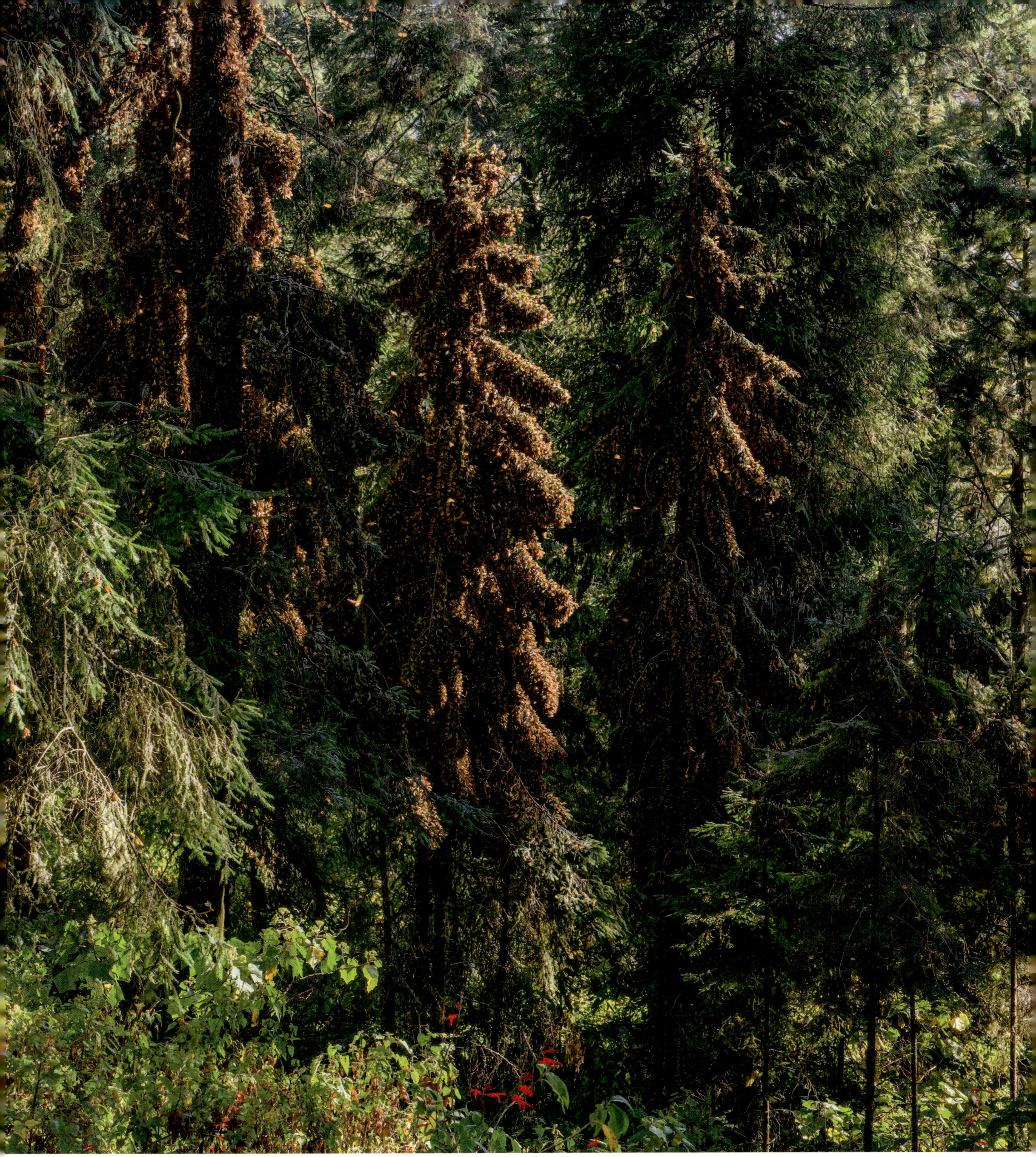

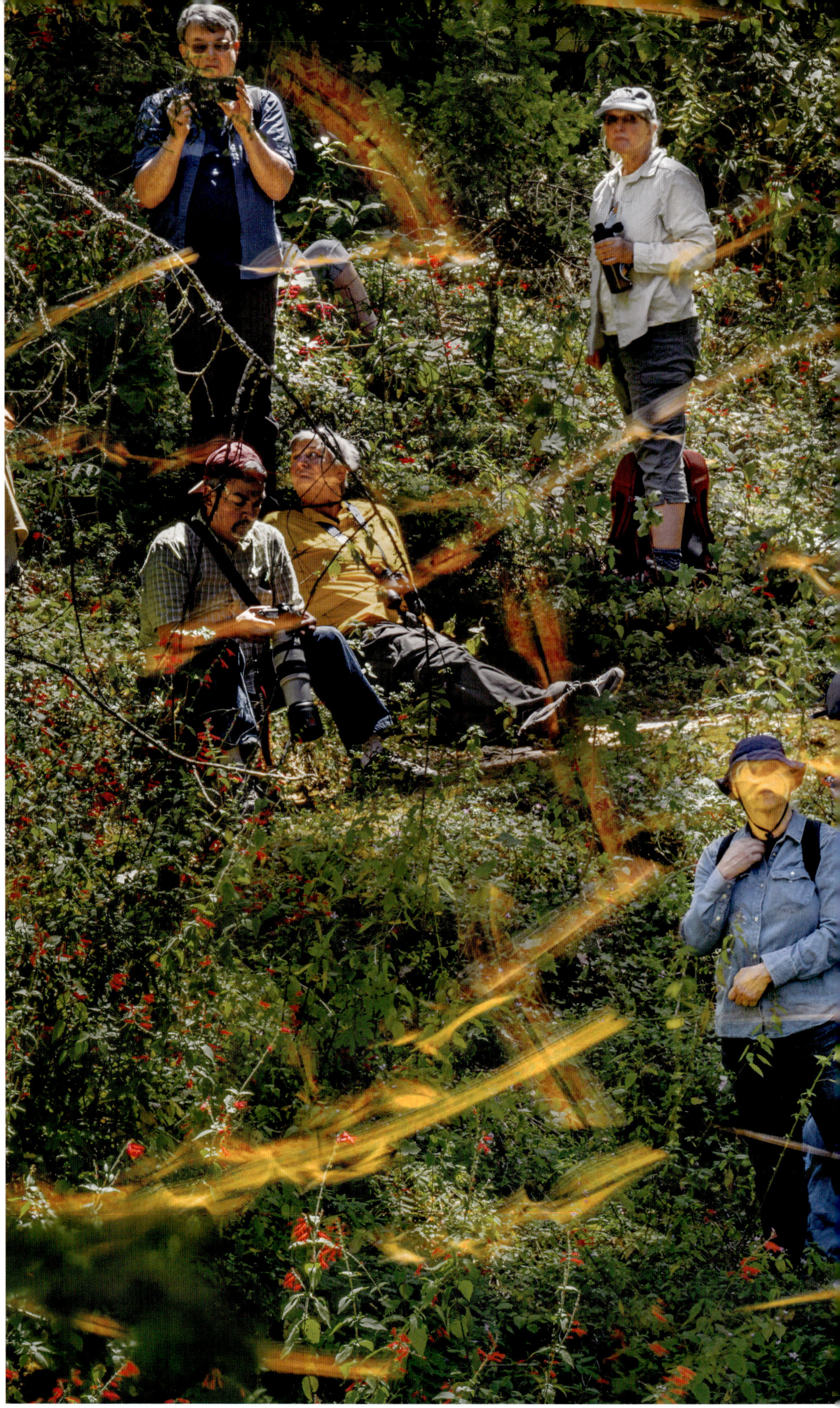

Tourists observe monarch butterflies in El Rosario Sanctuary. A slow shutter speed camera setting captures the motion of the butterflies' wings, creating soft orange trails.

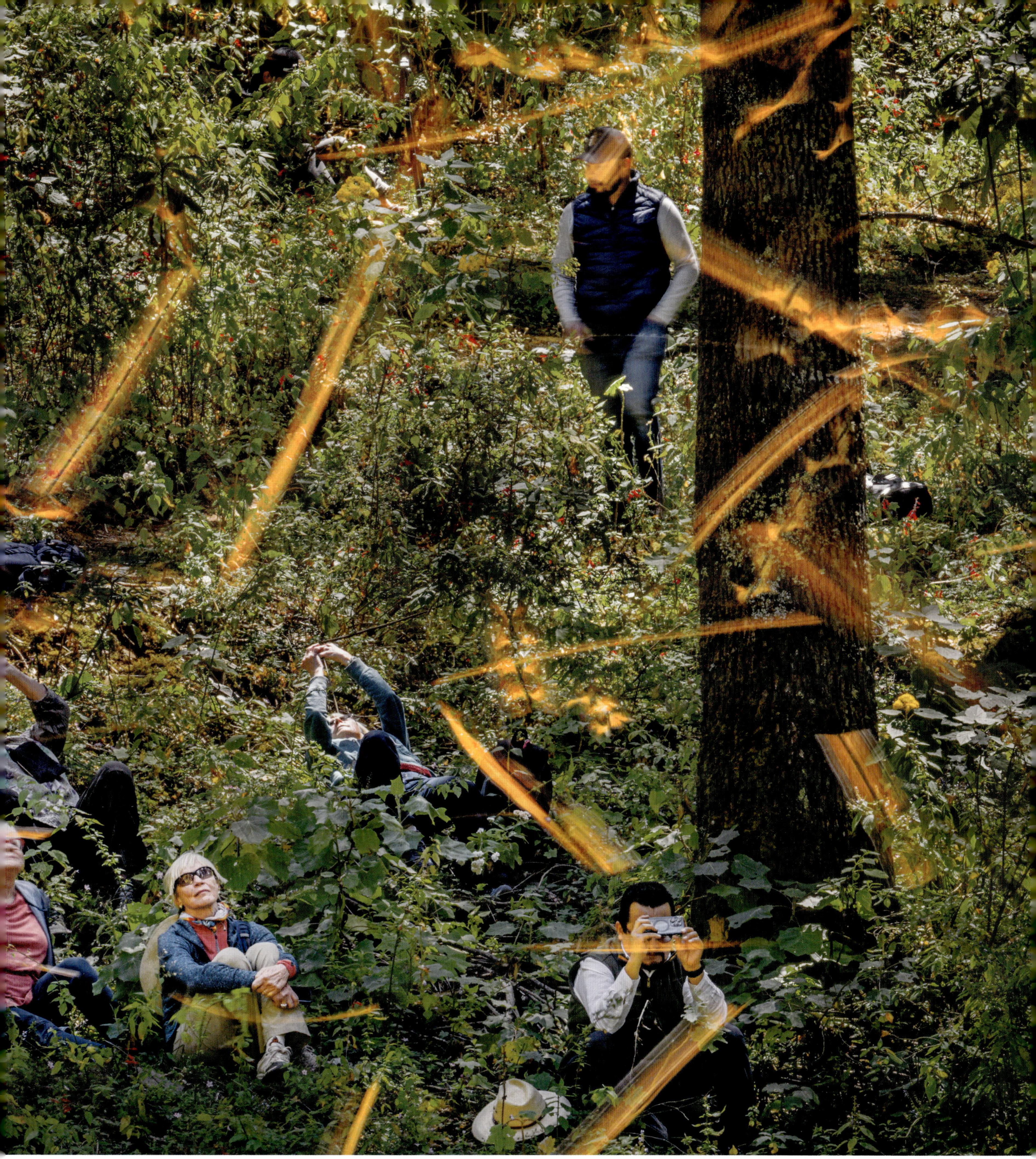

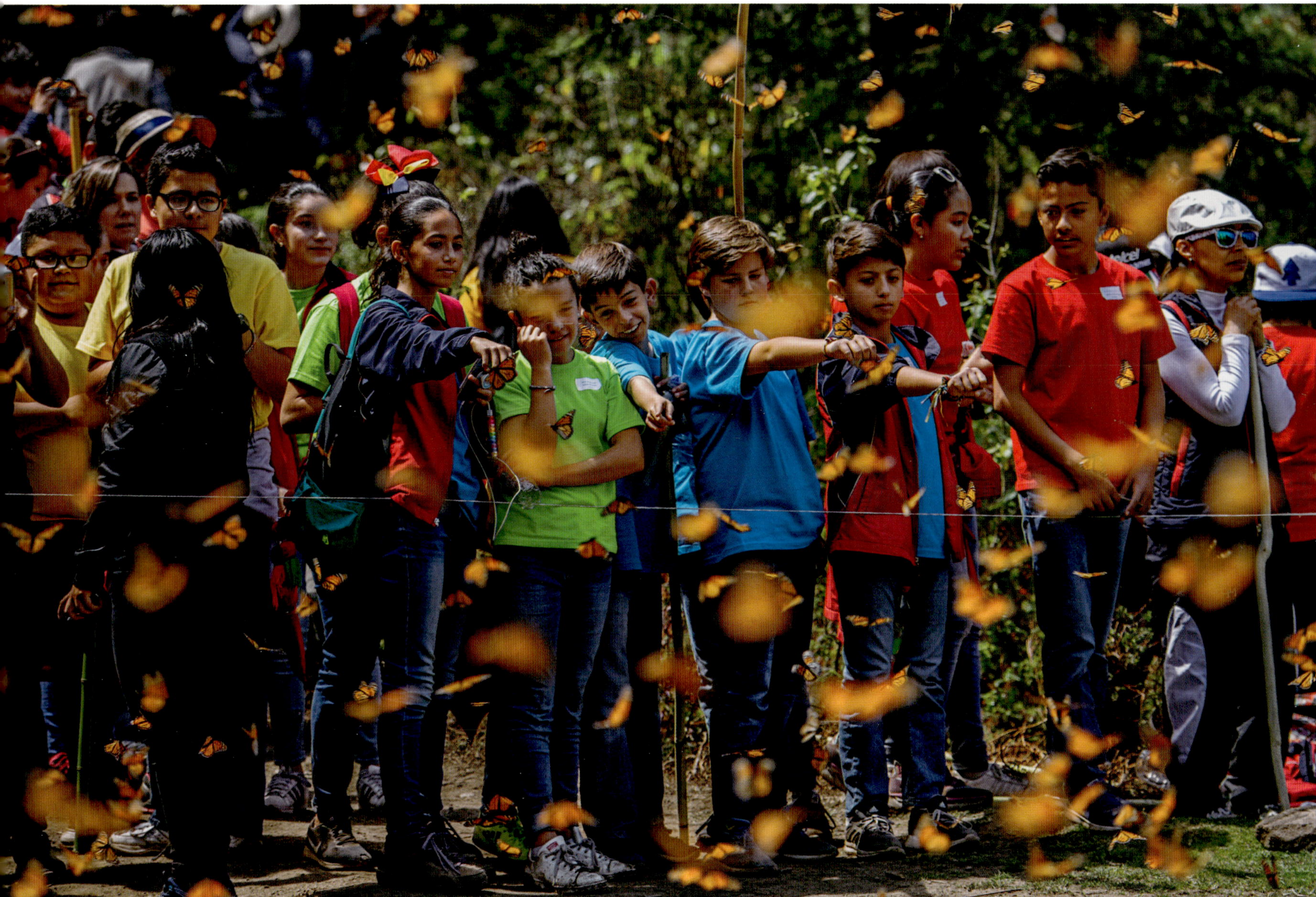

Schoolchildren watch monarch butterflies at midday in El Rosario Sanctuary. Each winter, local students visit the sanctuary to observe the colonies and learn about monarch migration and conservation.

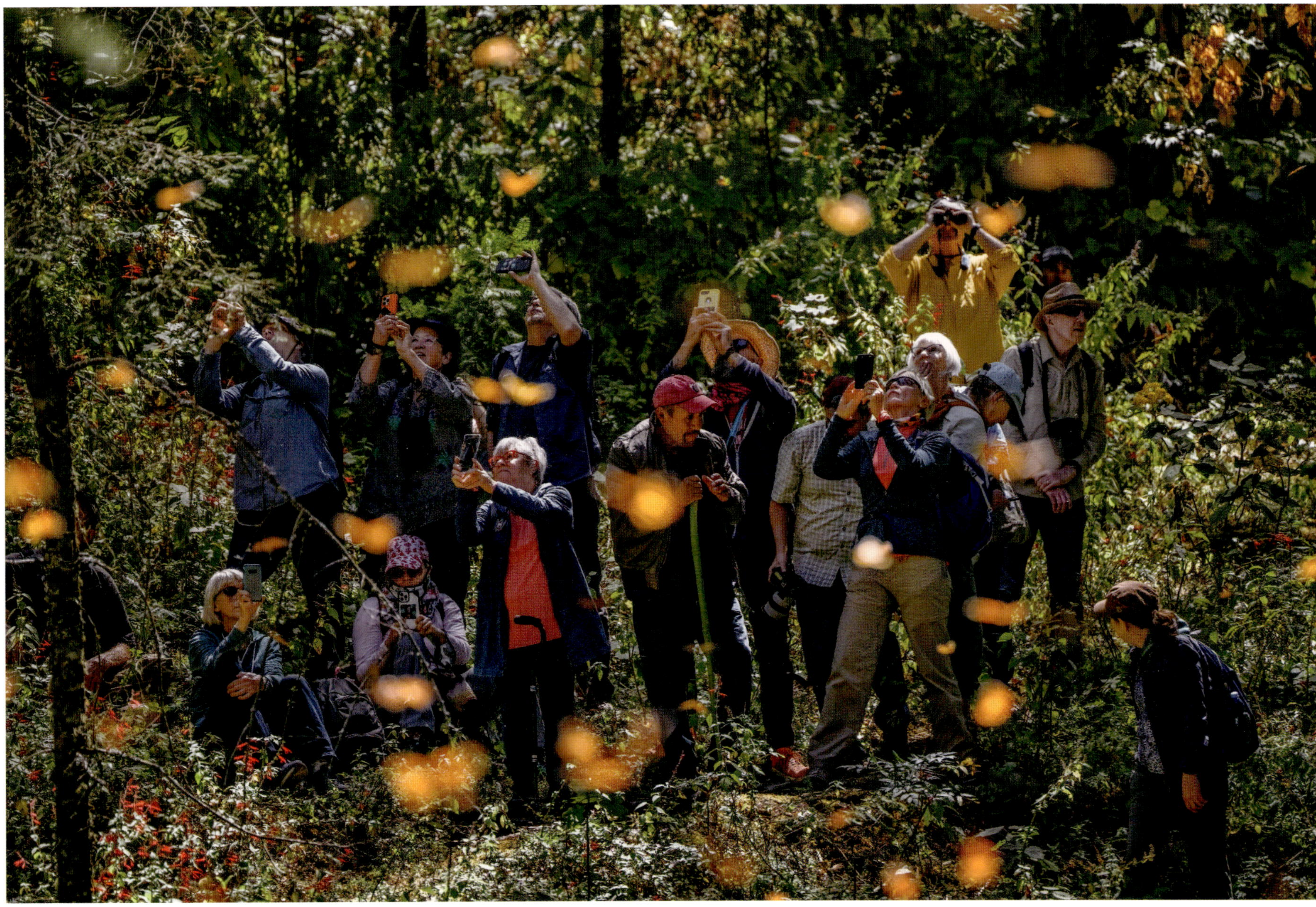

Visitors observe monarchs at El Rosario Sanctuary, where tourism revenue helps local communities restore forests on land once cleared for agriculture or logging, thus promoting conservation while supporting local livelihoods.

PAGES 190–191: Avocado plantations surround the foothills of Cerro Pelón, a volcanic peak within the core zone of Mexico's Monarch Butterfly Biosphere Reserve.

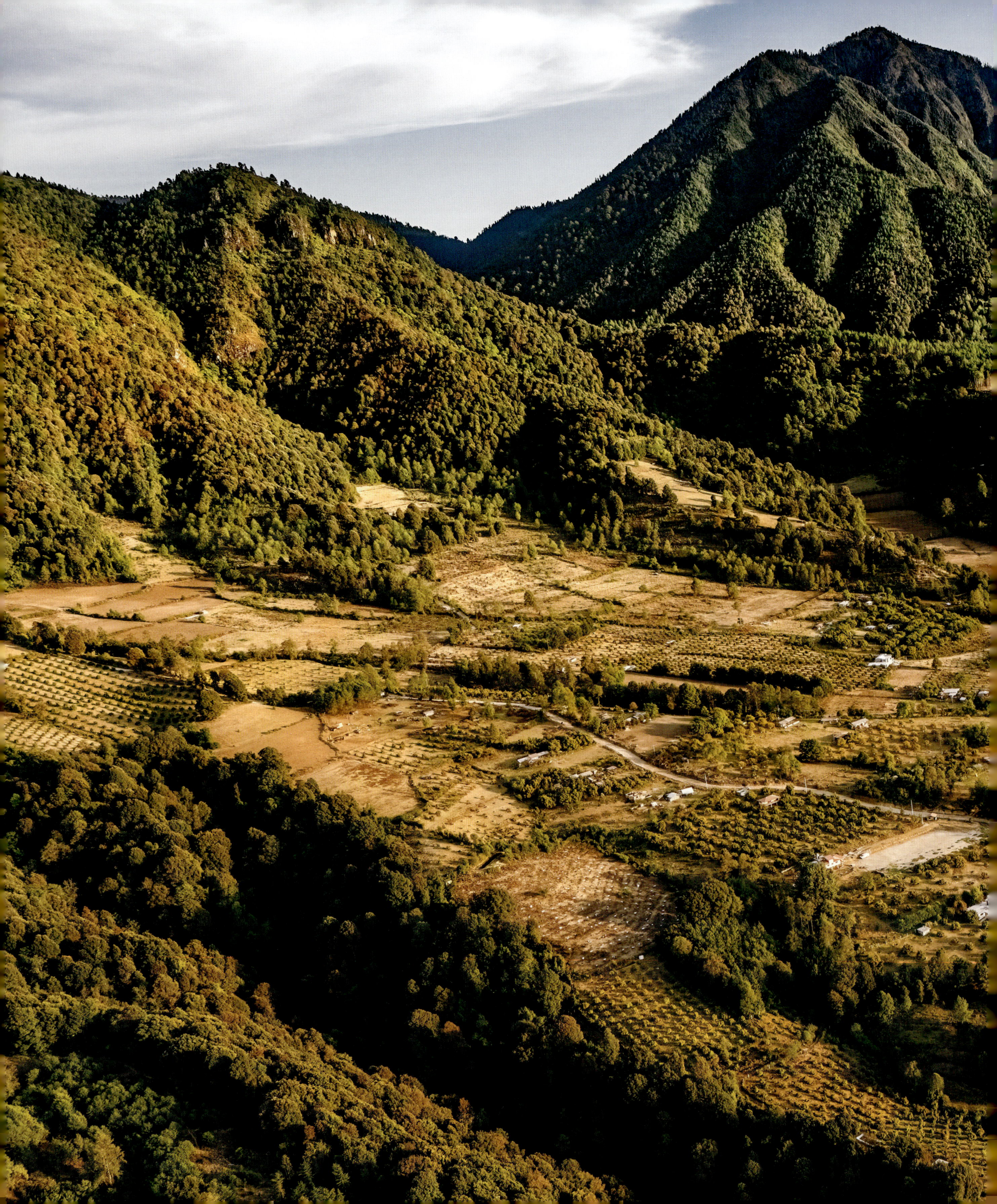

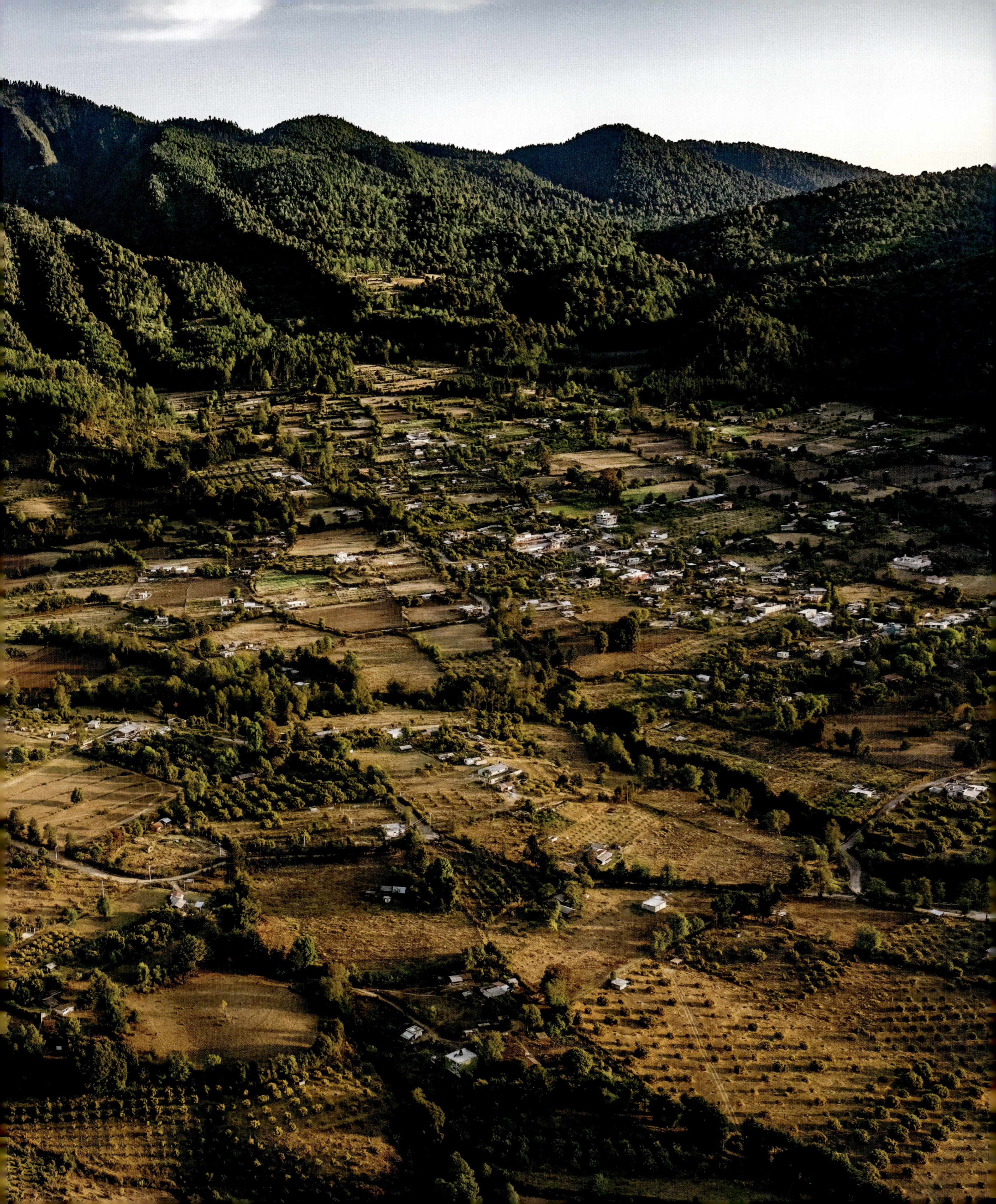

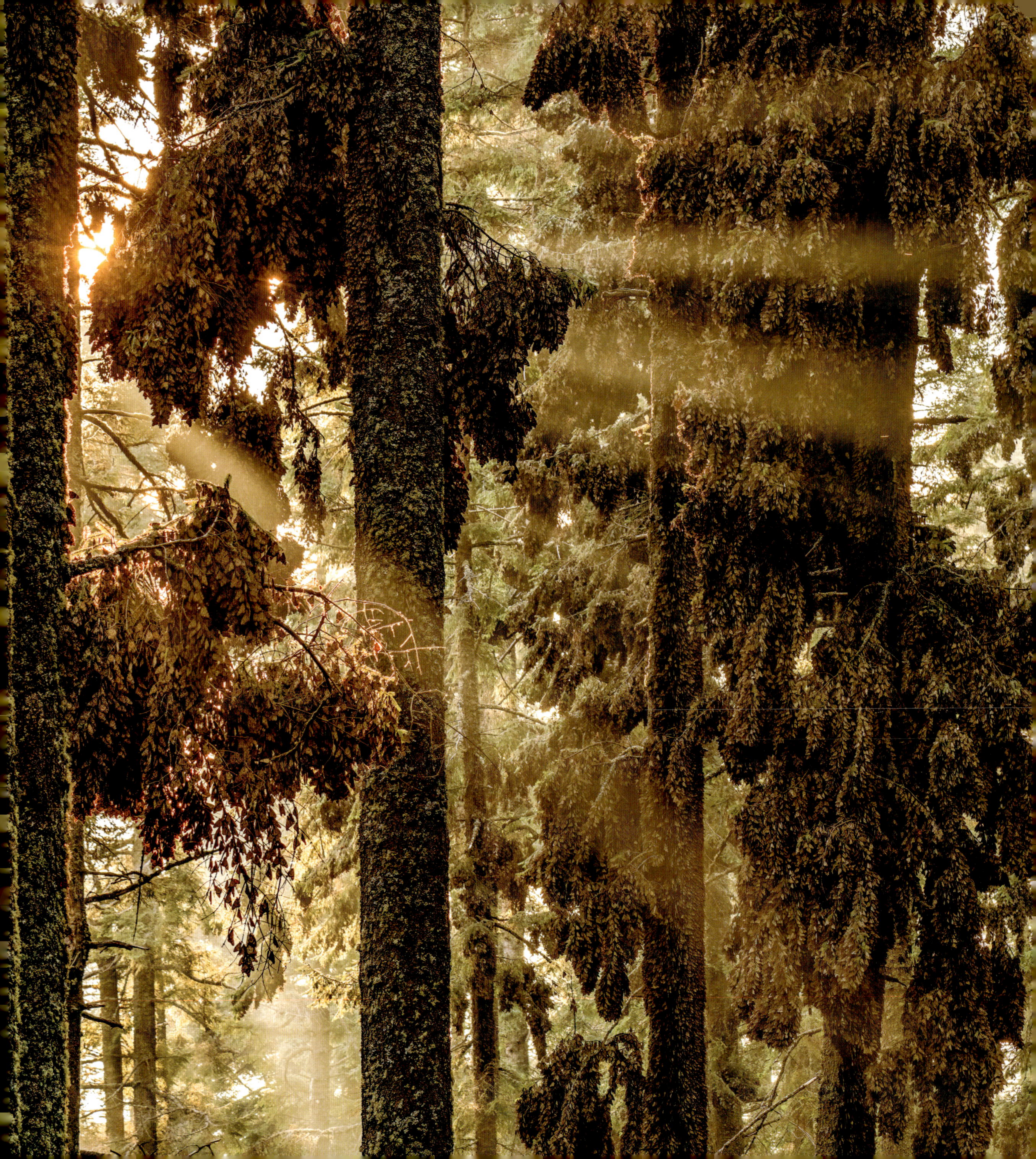

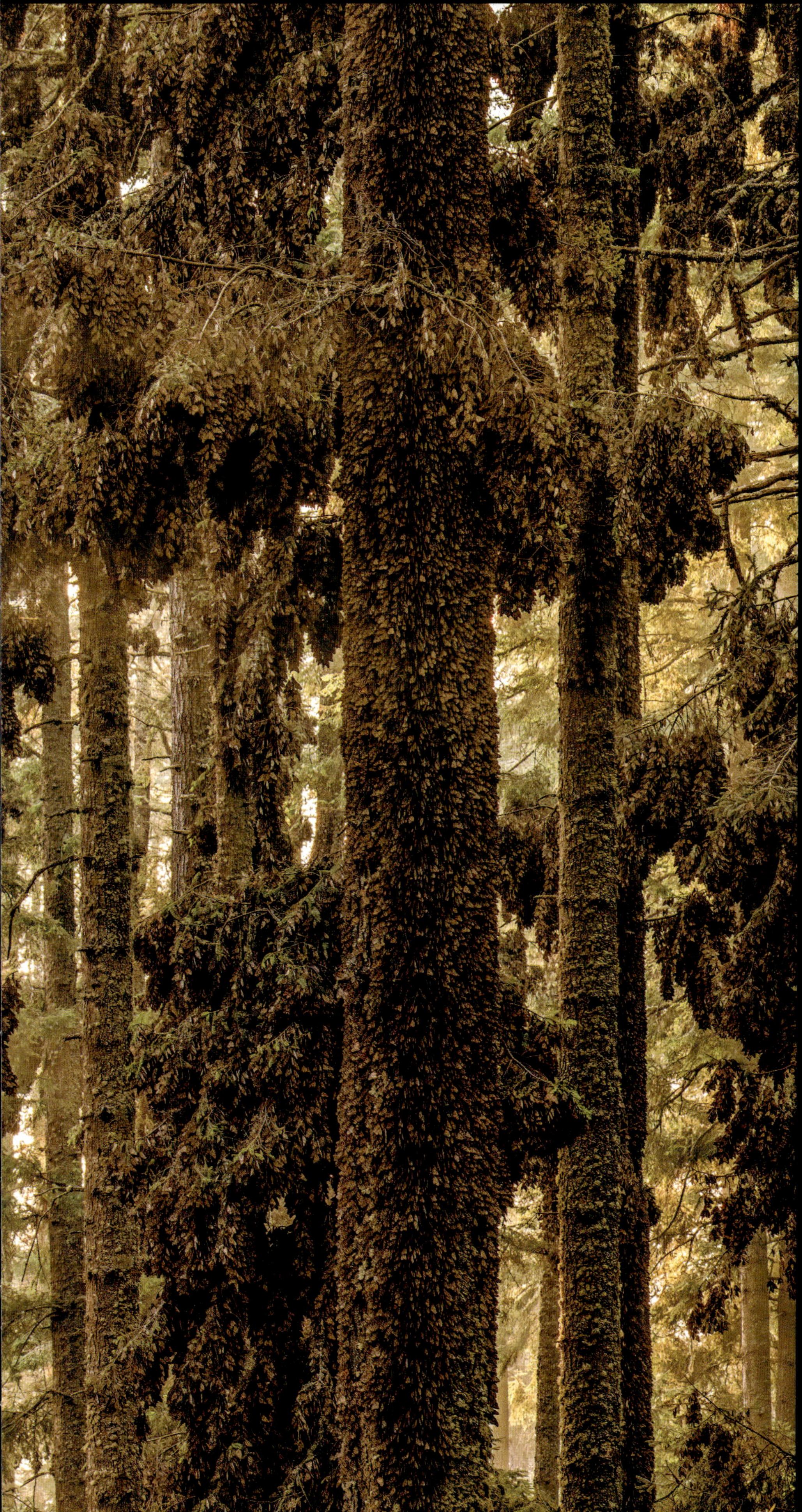

FOREST OF THE MONARCHS

THERE WAS A STRANGE LIGHT that early afternoon, a faint promise that something different might happen. The noon hours had flown by faster than usual, as I tried to get pictures of tourists enjoying the wonderful sight of a blue sky covered in orange butterflies. Around 5:30 p.m., when the sun was low and the visitors had already left, I was ready to go, too. But I noticed something different that made me stay a bit longer. At El Rosario Sanctuary, due to its southwestern exposure and the presence of mountains close by, it is common to have flat, dull light in the afternoons, often overcast because clouds form on the horizon. But that evening, the winter sun, setting quickly, pierced through the gaps of the oyamels. The dust from the visitors was settling and added a dreamy feeling to the scene, like a giant hay-barn window. I found a patch of forest with a nice structure and particularly dense butterfly clusters, and I got ready to shoot the first frames. And then it all happened very quickly. I heard a loud, cracking noise that echoed in the forest, and my guide started whistling and calling to me: "Jaime, come here! A branch covered in monarchs has fallen to the ground!" That was so cool, something I had never seen before. But I was witnessing magic in my viewfinder, and I decided to ignore my guide's call and my itching curiosity. I hastily set up the tripod and the remote trigger before the light faded away, and I grabbed a few more shots. The whole thing was over before it started, the magical scene quickly washed out by gray, hazy light. But I knew I had witnessed something unique and captured an image of the moment. Sometimes you just know. I called out to my guide to find out where he was and walked downhill to the spot where the branch had fallen. It still had many monarchs on it, and I was able to shoot some frames with the satisfaction of knowing that on that particular evening, my job was already accomplished.

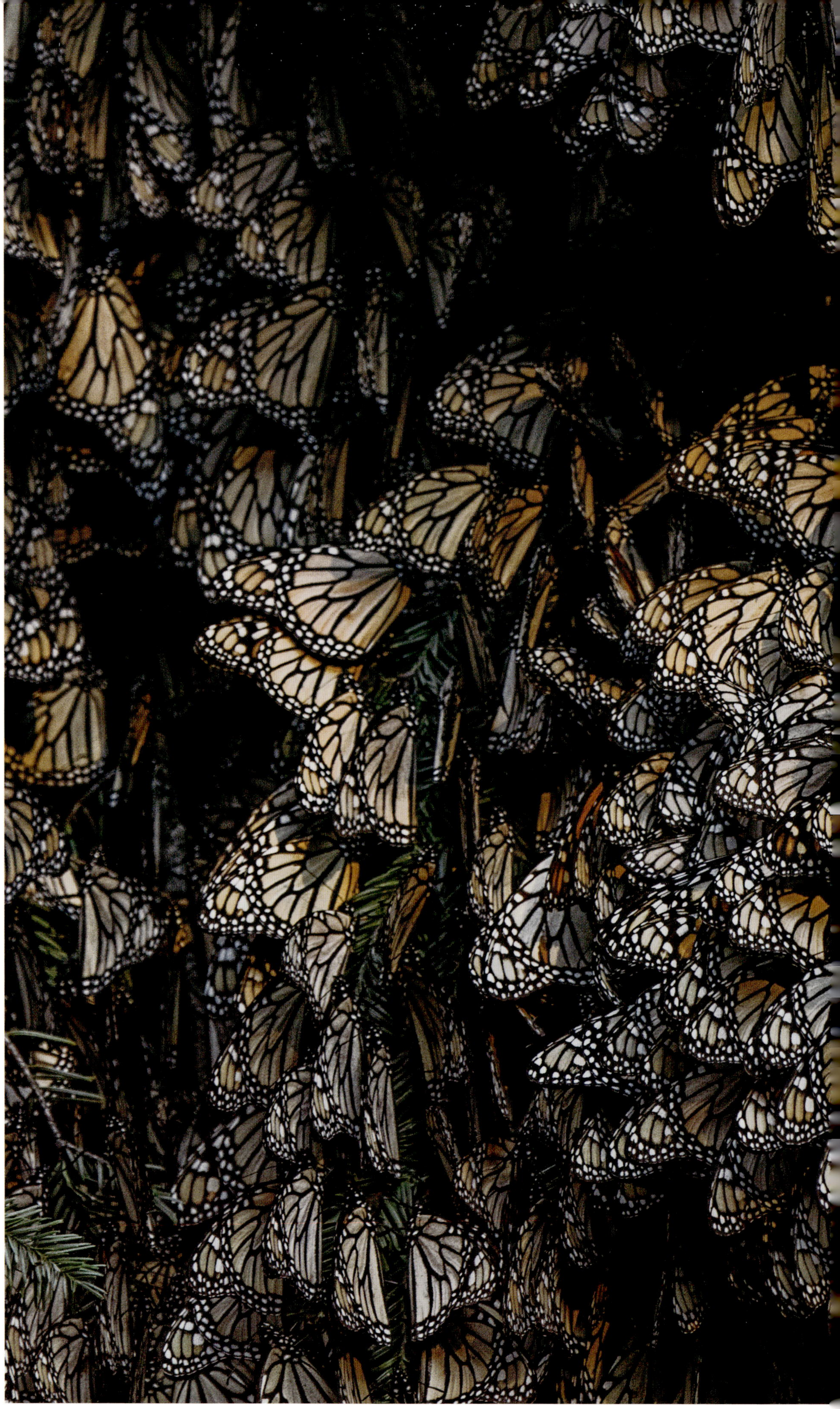

Monarch butterflies cluster in the afternoon light at El Rosario Sanctuary, part of the Monarch Butterfly Biosphere Reserve in Michoacán, Mexico.

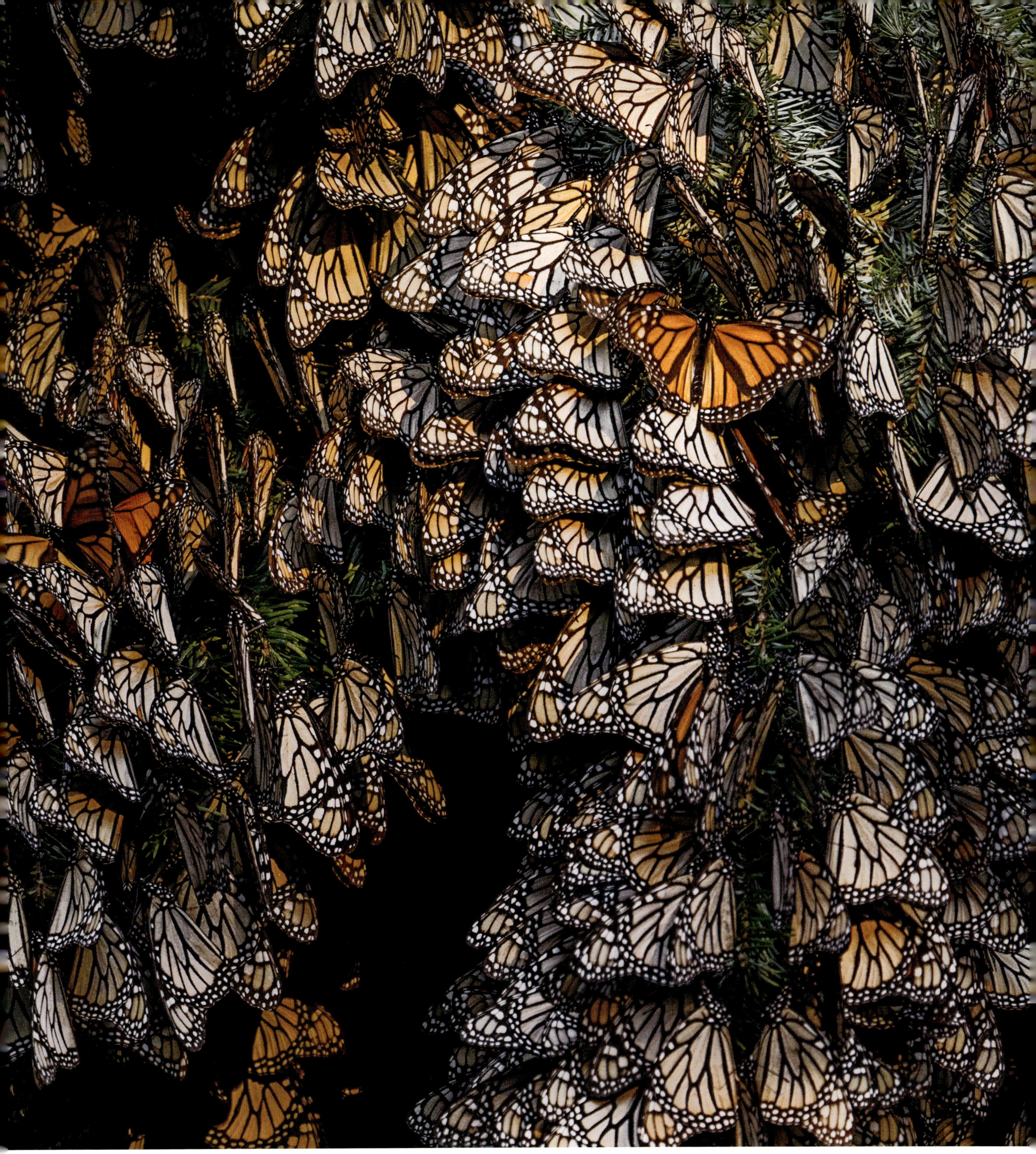

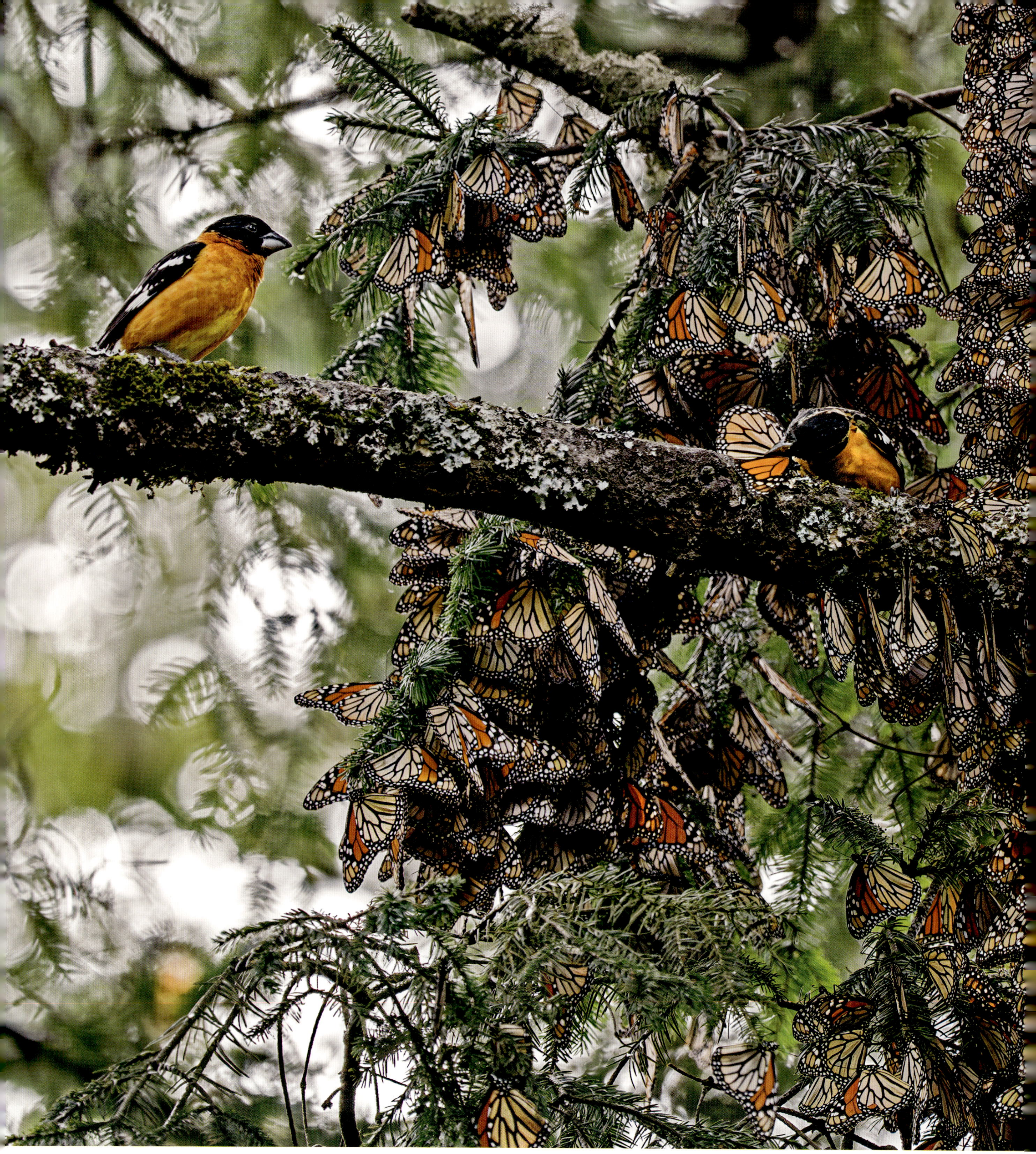

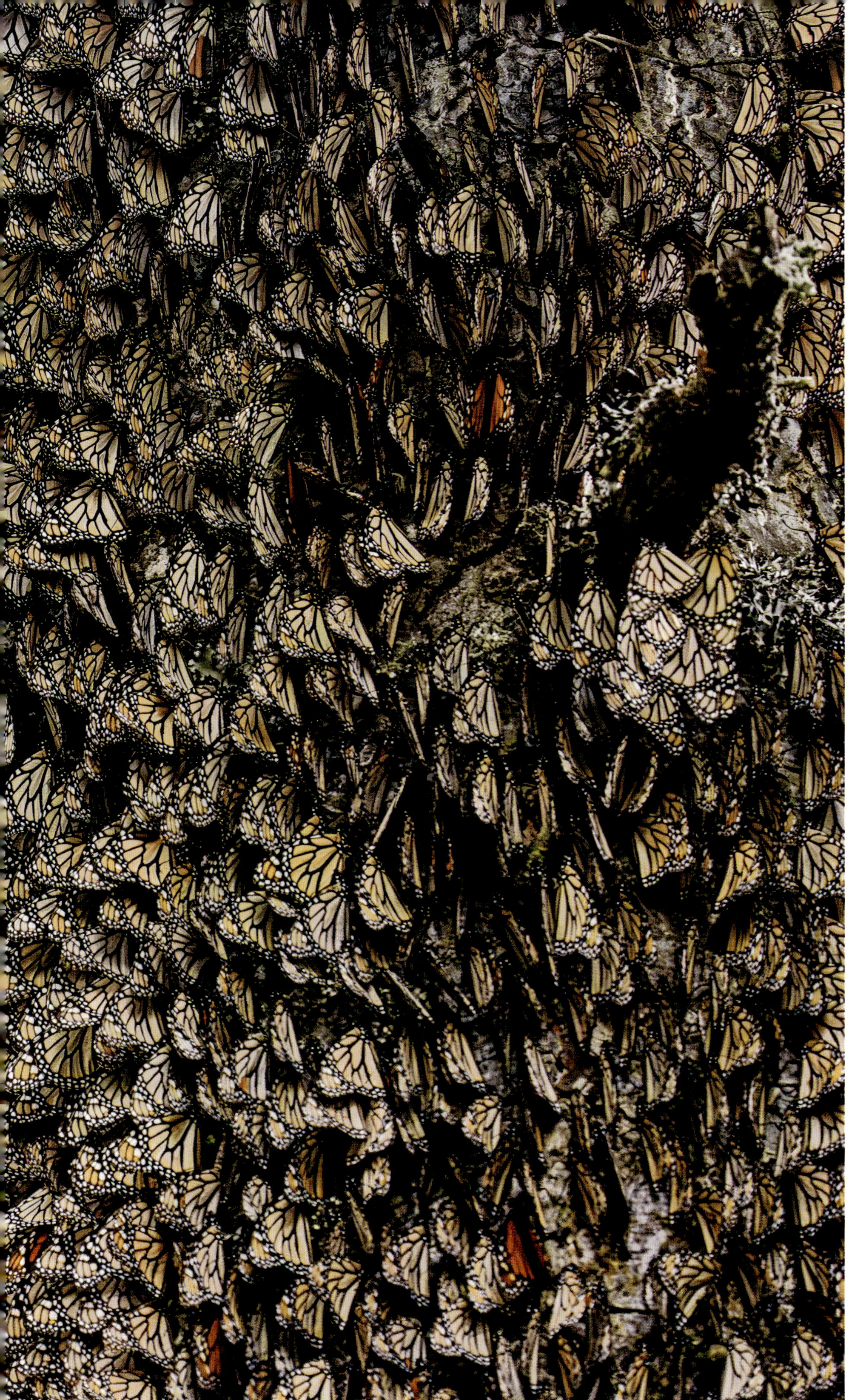

A pair of black-headed grosbeaks, one of the monarch's main predators, feed on butterflies in the early morning at El Rosario Sanctuary. Some bird species in the Mexican sanctuaries have evolved resistance to the toxins that make monarchs poisonous to most predators.

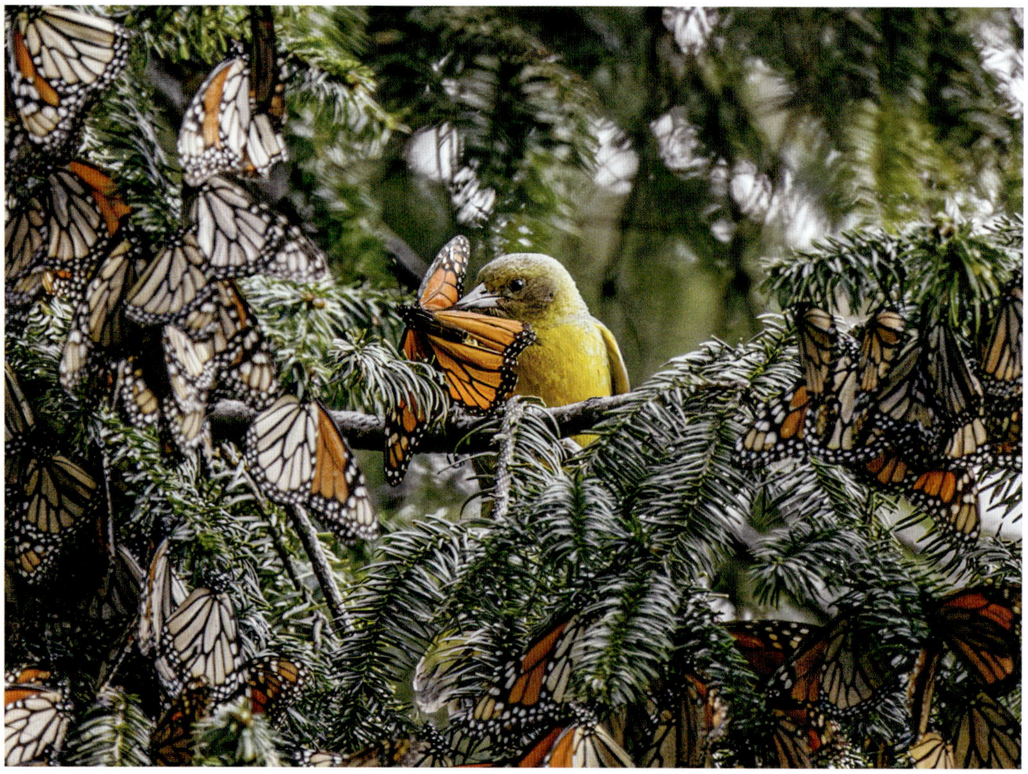

A Scott's oriole feeds on monarchs, selectively consuming only the abdomen, where toxins are less concentrated. This adaptation allows these birds to avoid the effects of cardenolides, the chemicals that make the butterflies poisonous to most predators.

OPPOSITE: A rose-breasted grosbeak—a migratory bird species that is rare in the Michoacán mountains where butterflies gather—feeds on monarchs in the early morning at El Rosario Sanctuary.

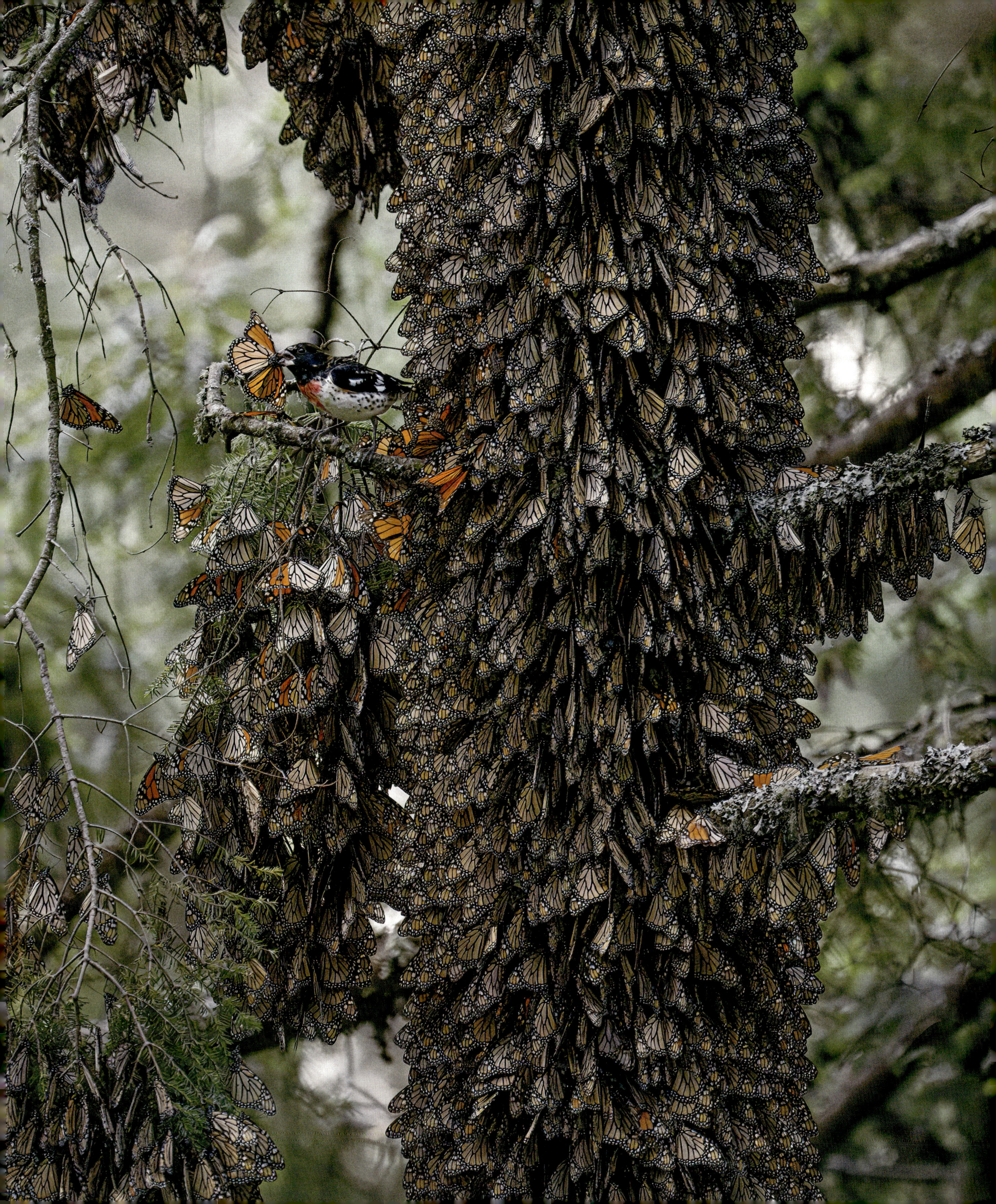

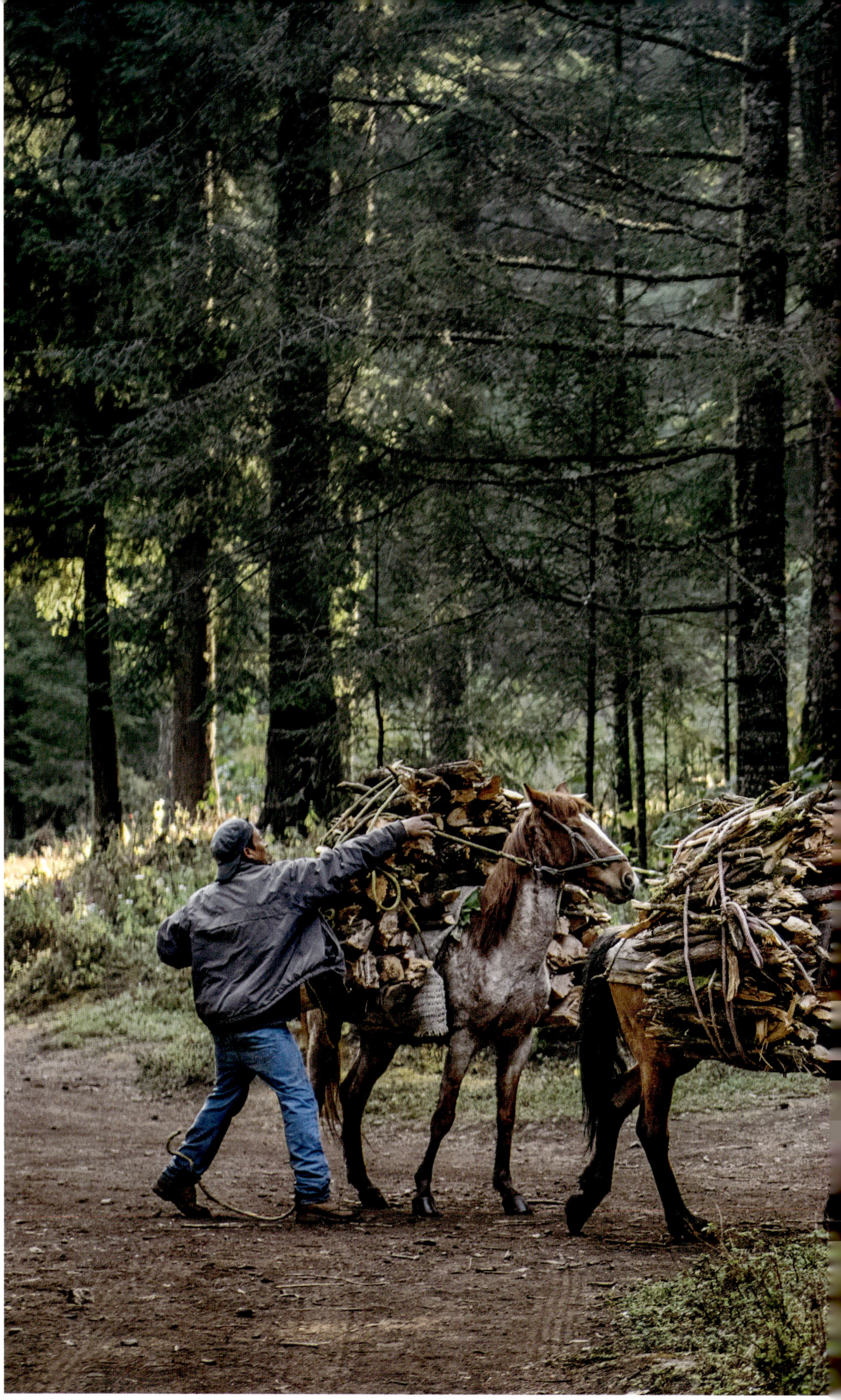

A man collects fallen wood from the forests of El Rosario Sanctuary in the Monarch Butterfly Biosphere Reserve, Michoacán, Mexico. Firewood is an essential source of energy for many local households, but removing so much from the ground can gradually affect soil health and forest structure.

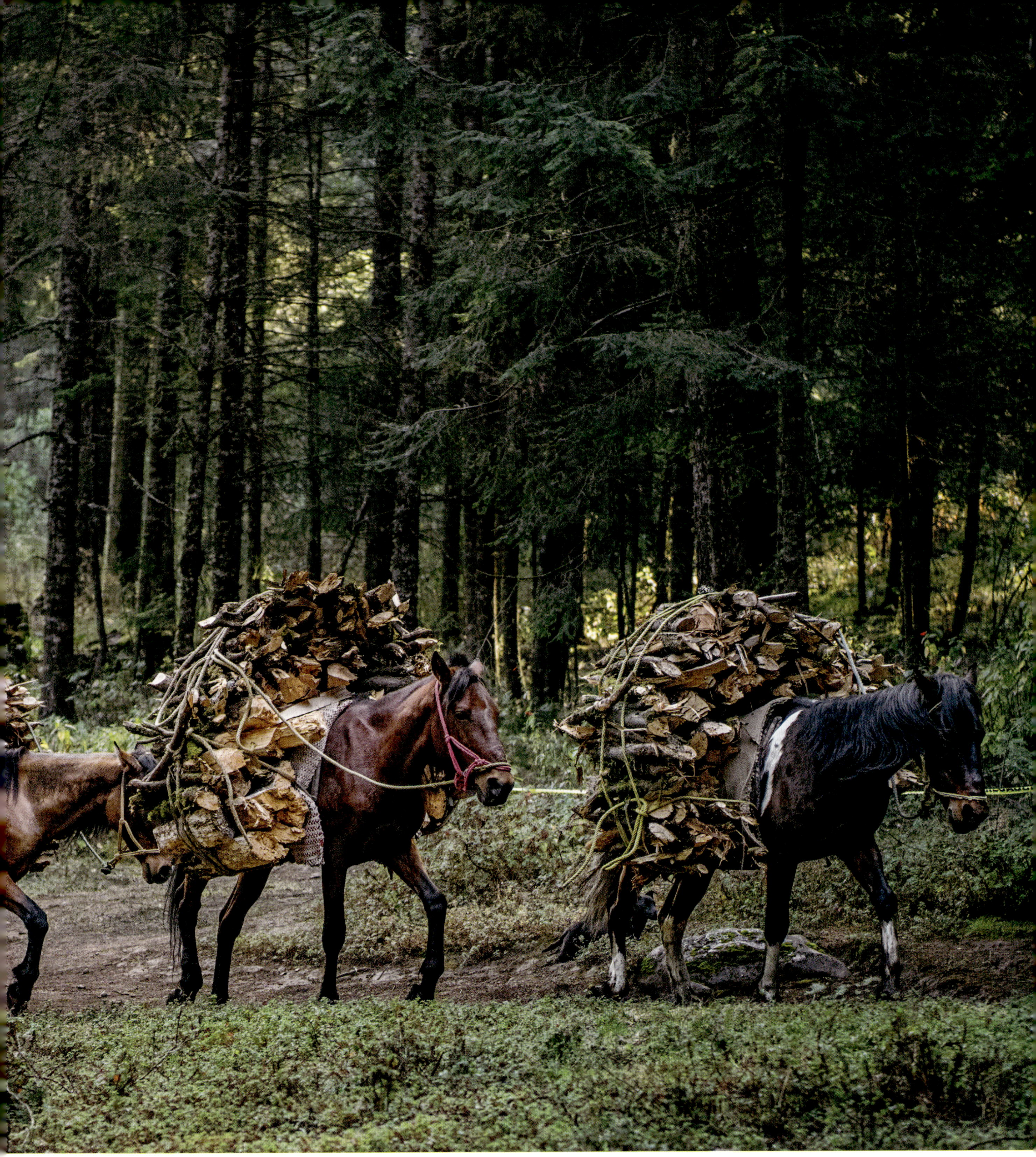

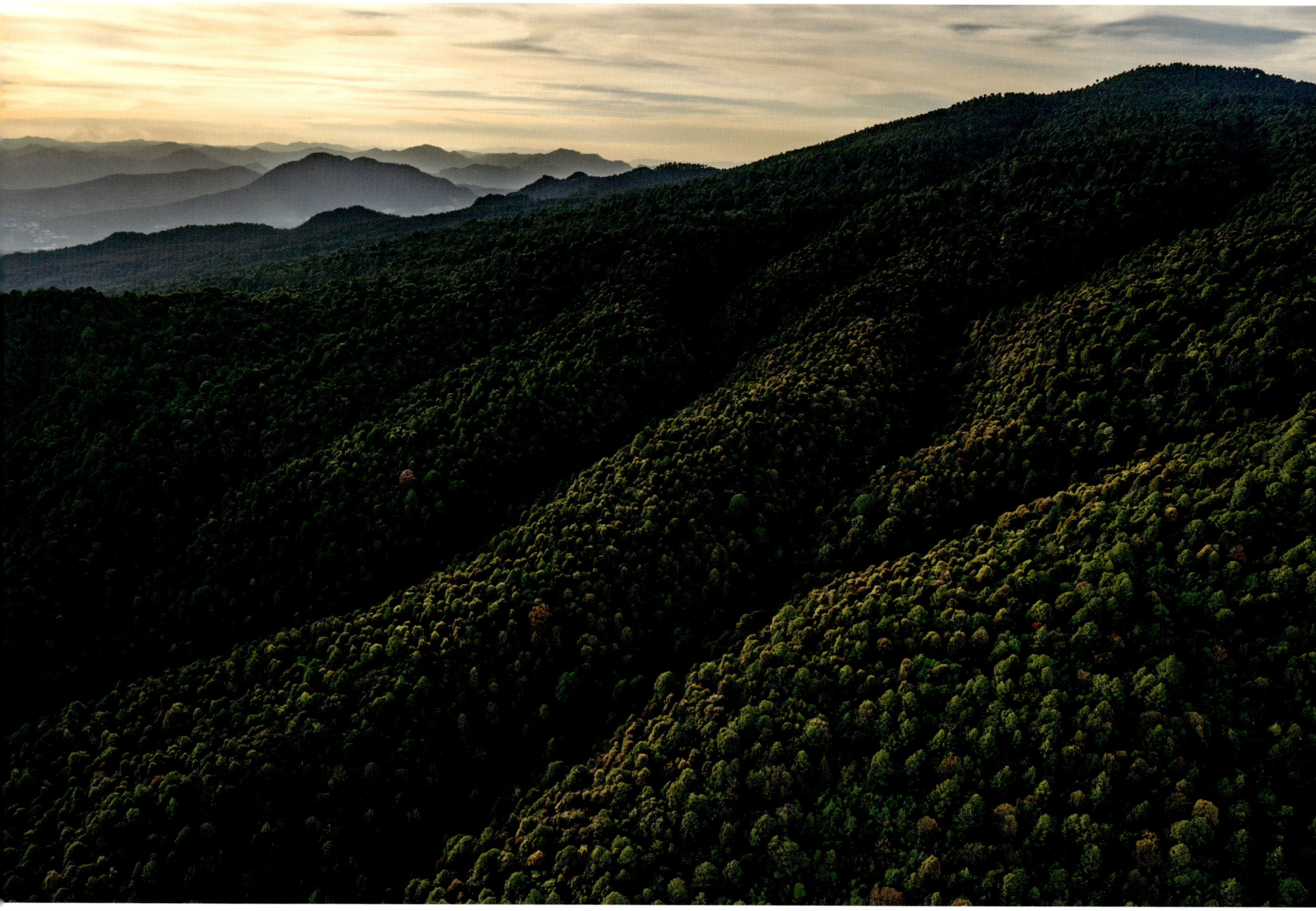

The overwintering colonies of monarch butterflies are found on only a dozen isolated mountaintops covered with oyamel fir forest in the Trans-Mexican Volcanic Belt. The forest canopy protects the butterflies from rain, wind, and temperature extremes, creating the stable microclimate they need to survive the winter.

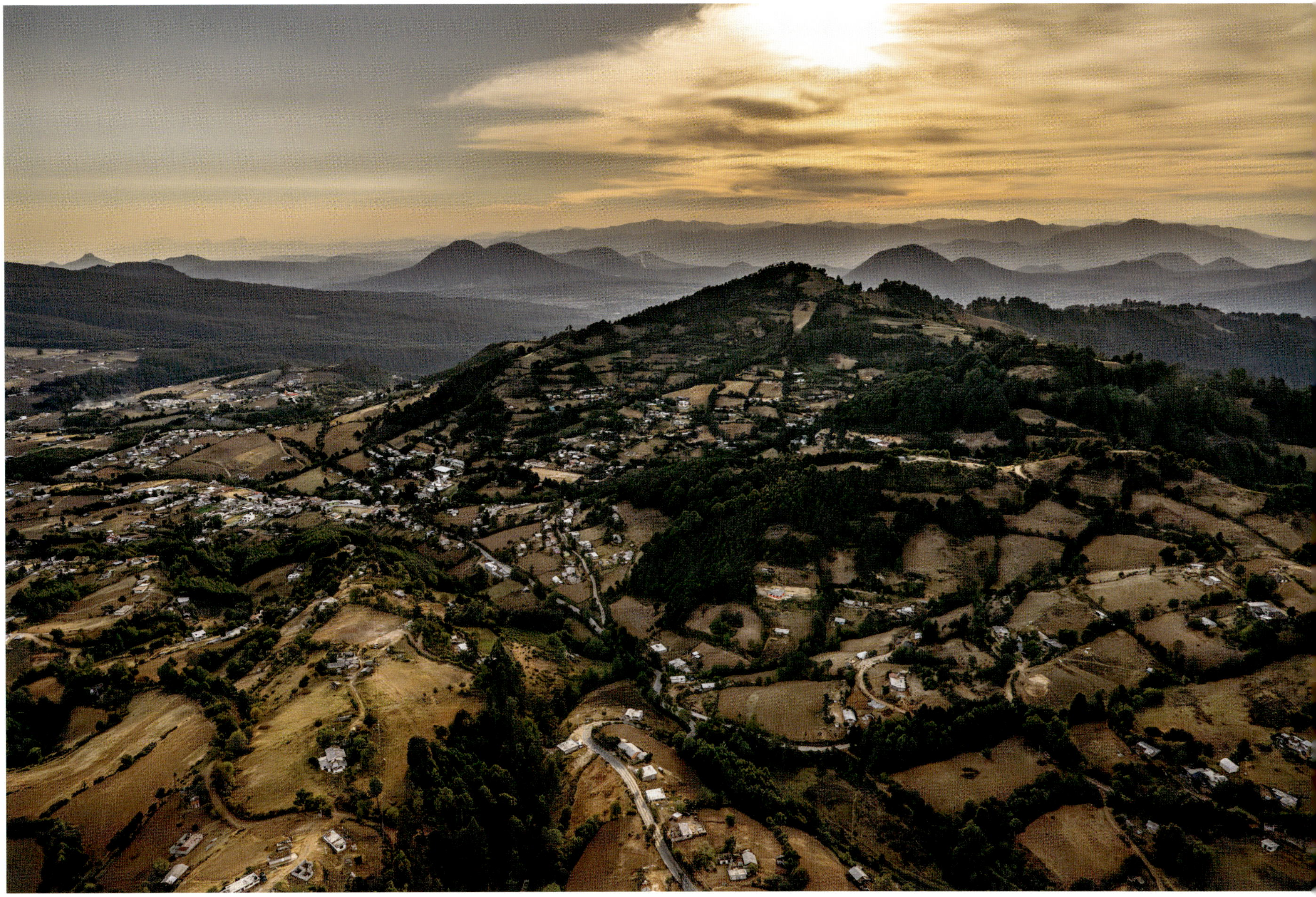

A patchwork of farms and pastures spreads across the foothills near Angangueo, on the edge of the Monarch Butterfly Biosphere Reserve in Michoacán, Mexico. Over generations, small-scale agriculture has shaped this landscape, supporting livelihoods but creating challenges for conserving the native forests on which monarchs depend.

PAGES 204–205: Students at the local Instituto Americano Leonardo da Vinci marvel as Moisés Acosta—founder of the Papalotzin Environmental Education Center, formerly located on the outskirts of the Monarch Butterfly Biosphere Reserve—shows the different stages of the monarch butterfly's life.

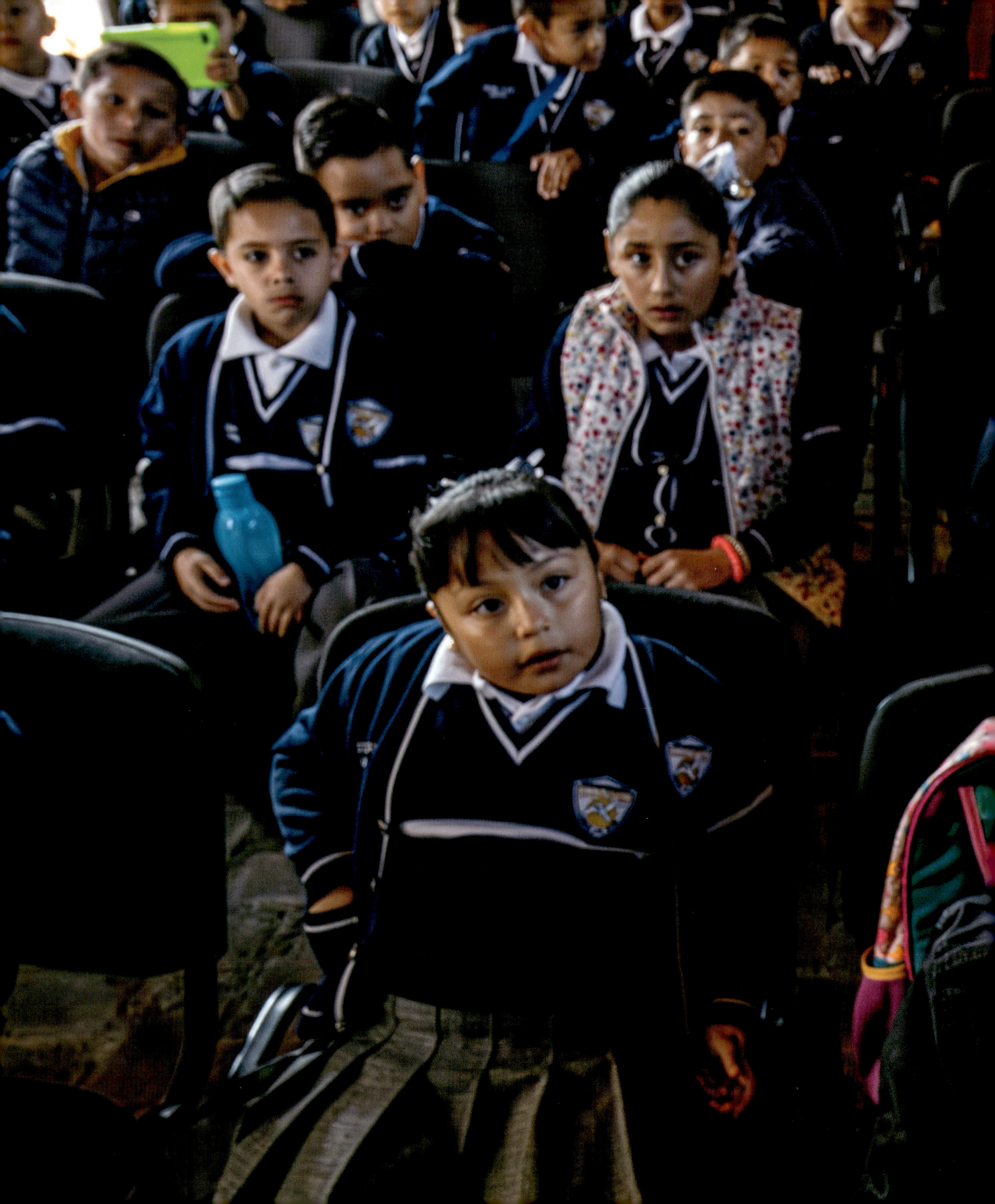

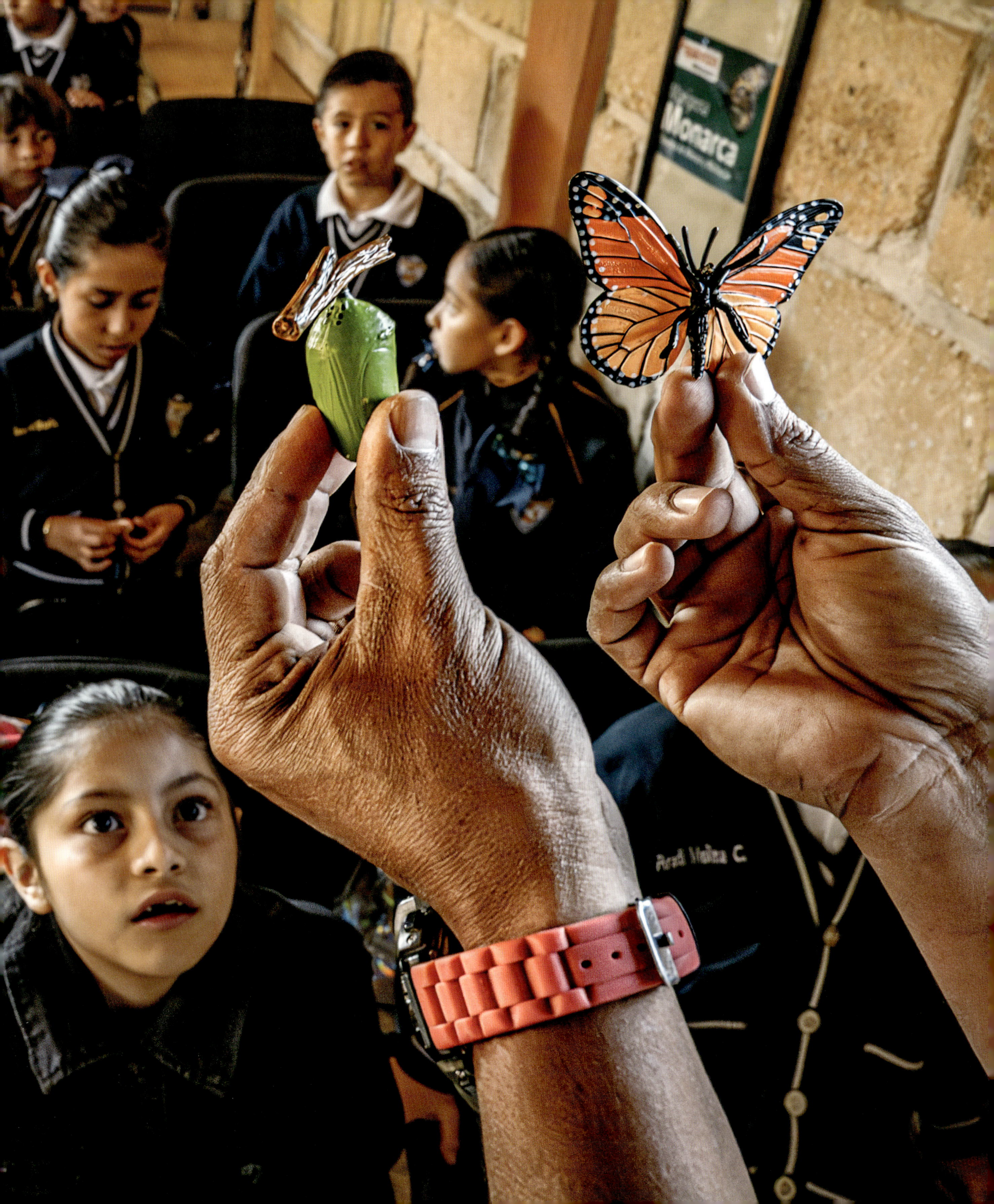
Monarca

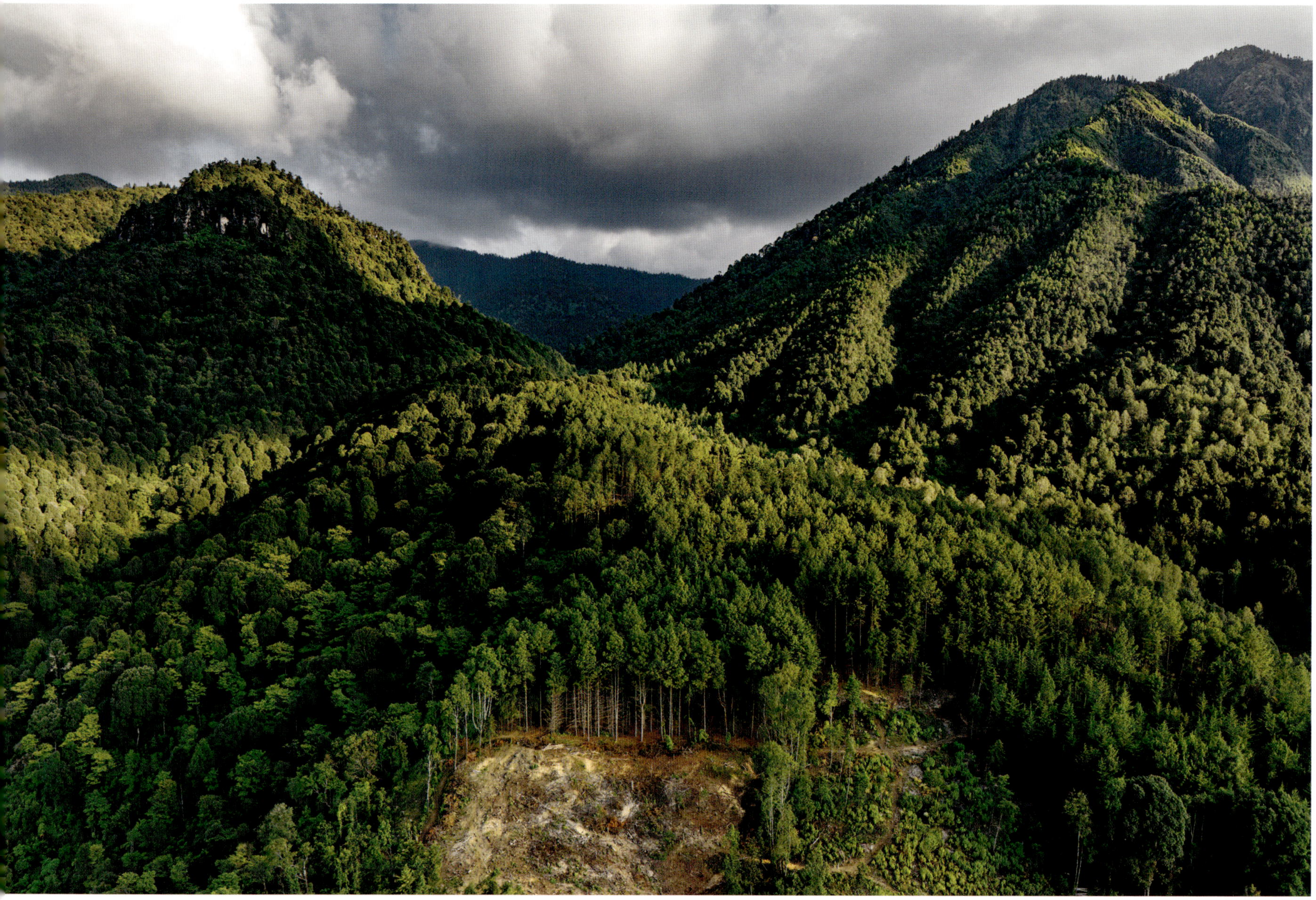

A patch of forest recently cleared close to the buffer zone of Mexico's Monarch Butterfly Biosphere Reserve. Here, native oyamel fir forest coexists with forest plantations, where tree cutting and replanting represent both ongoing pressure on the reserve and efforts toward sustainable land management.

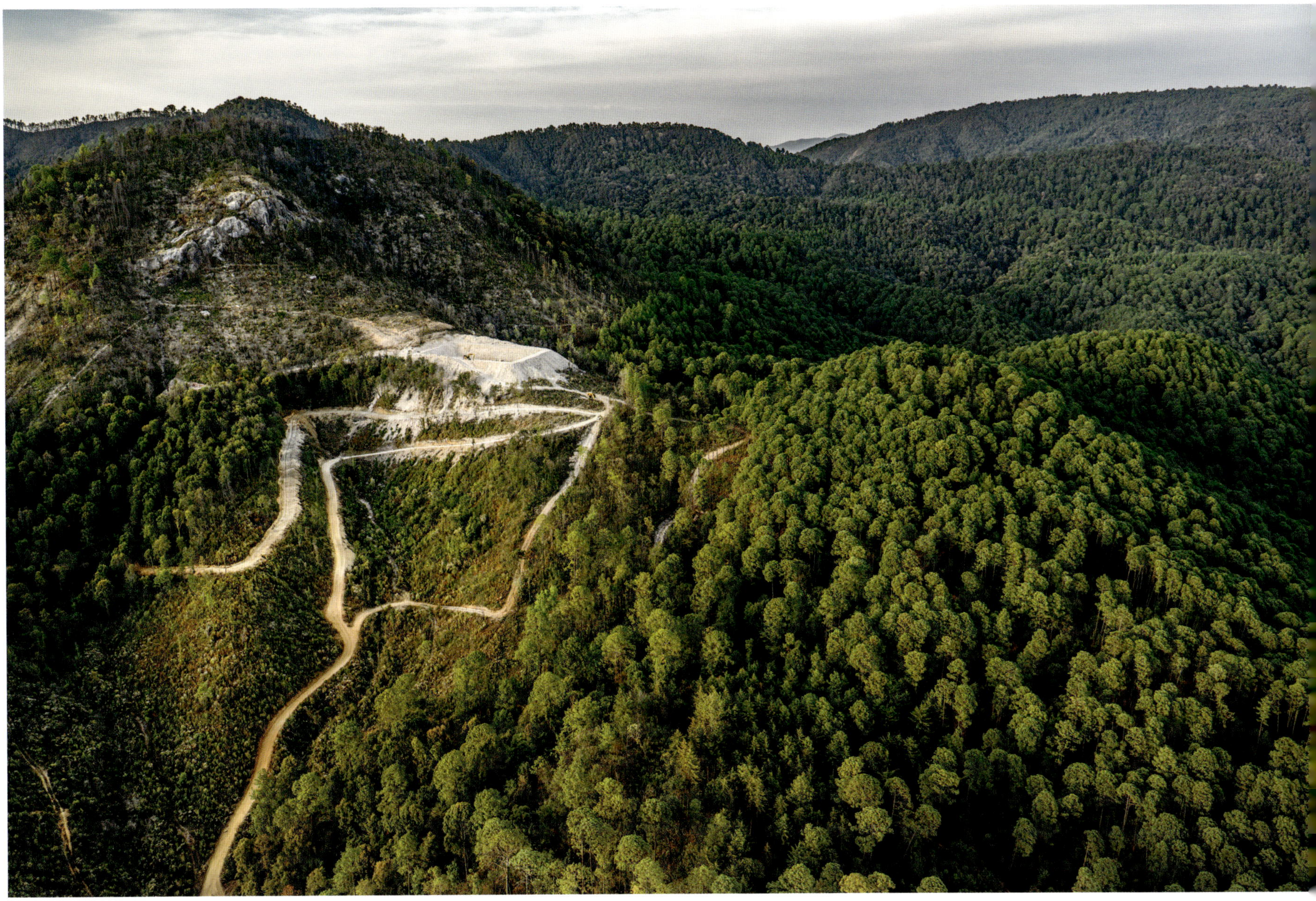

Land clearing and quarry activity near Aporo, on the northern edge of the Monarch Butterfly Biosphere Reserve. Small extraction sites like this contribute to habitat fragmentation and erosion in the buffer zone surrounding the forests where monarch butterflies overwinter.

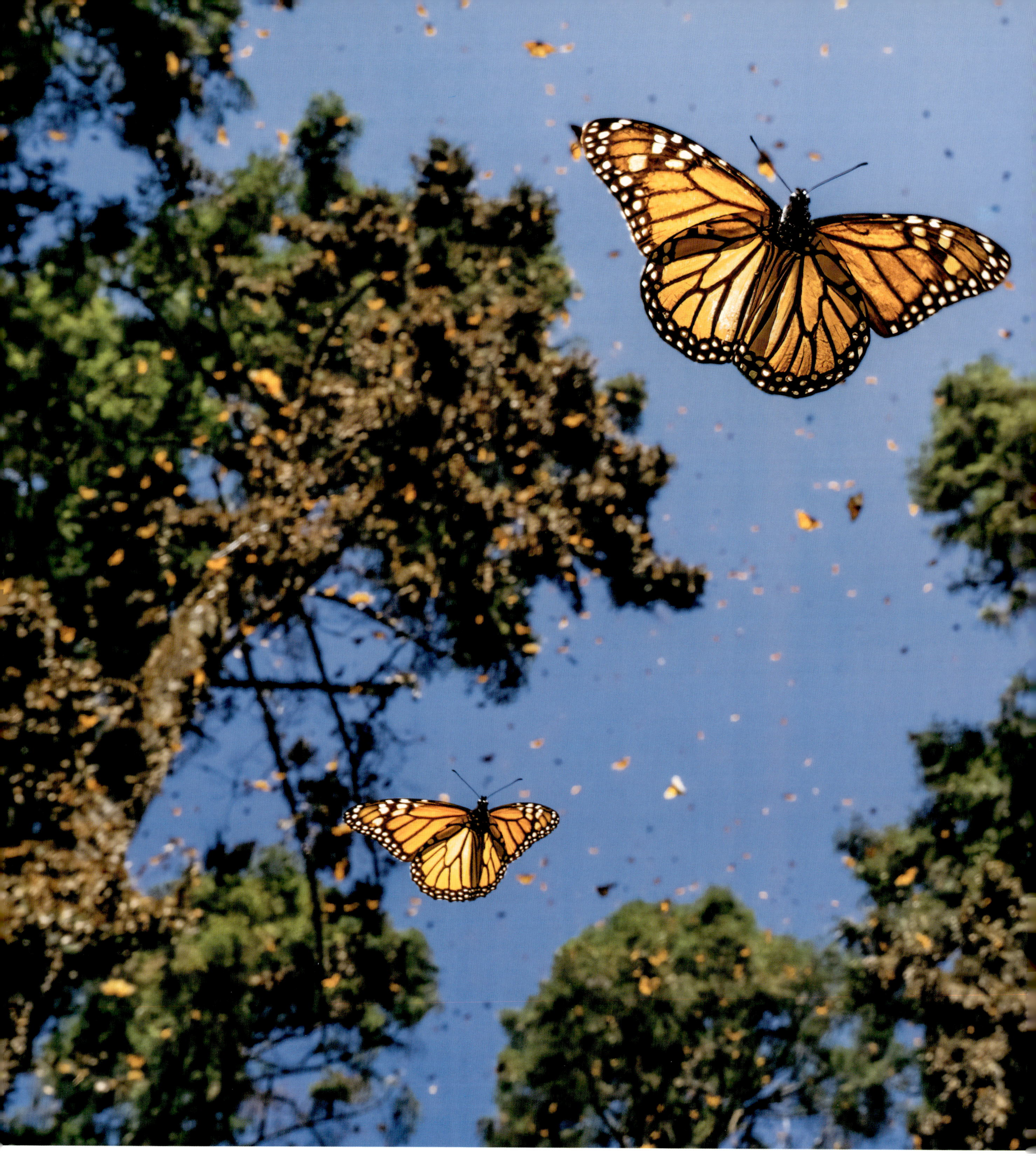

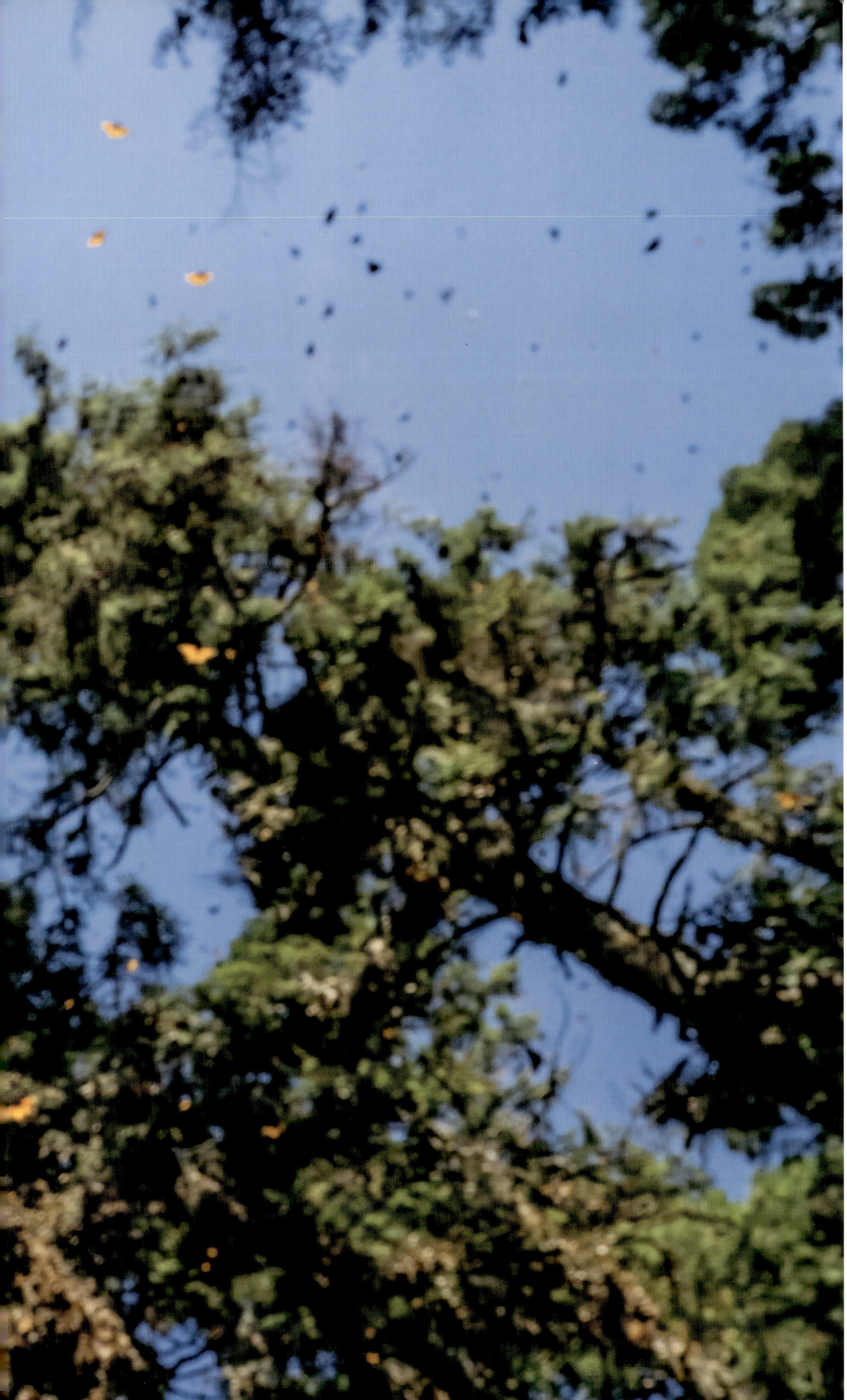

On warm days, thousands of monarchs become active in El Rosario Sanctuary, gliding down from the trees to drink water from streams and moist soil on the forest floor.

Martín Cruz Piña, Eduardo Rendón Salinas, and Genaro Mondragón Contreras (left to right) work with local community leaders to monitor monarch colonies at Ejido San Pablo Malacatepec, Michoacán, Mexico. WWF-Mexico began these surveys in the 1990s to track colony size and forest conditions, providing essential data for managing and protecting the Monarch Butterfly Biosphere Reserve.

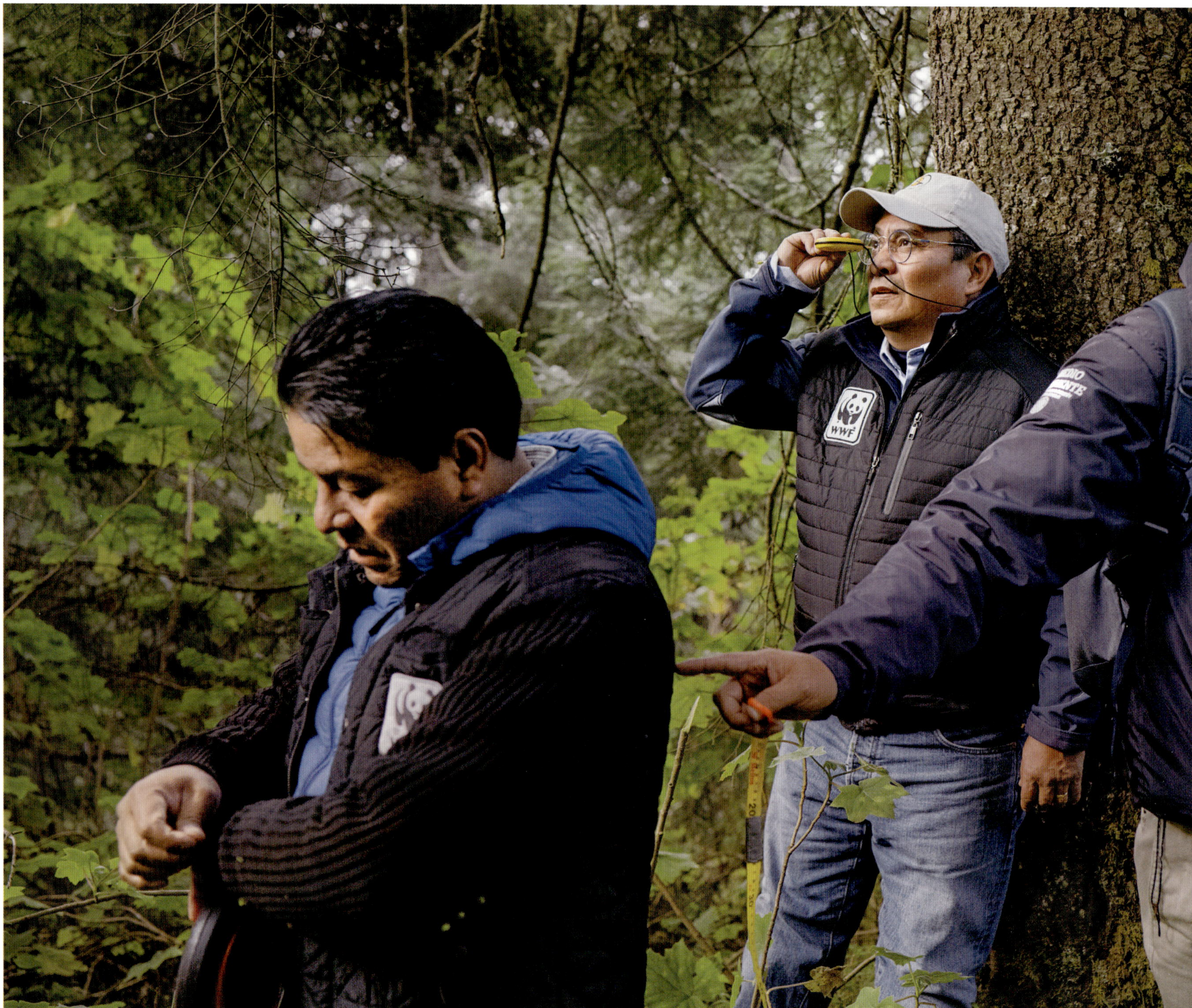

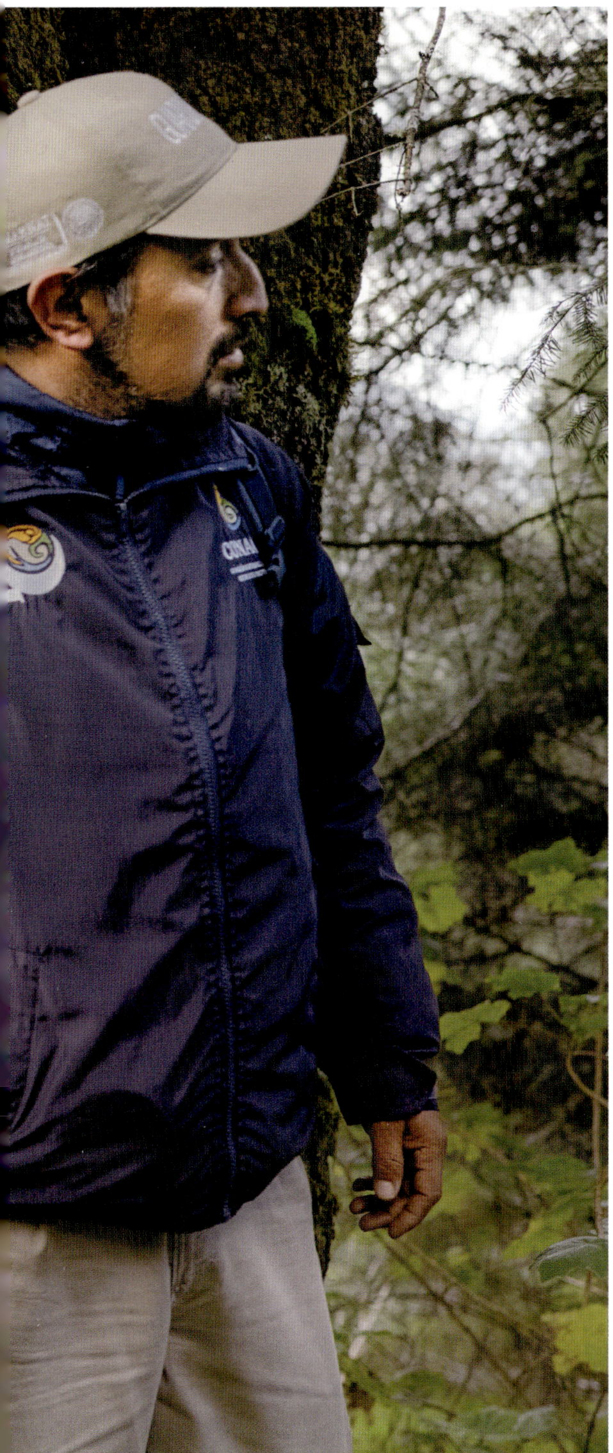

FIELD NOTES FROM

EDUARDO RENDÓN SALINAS

TO TALK ABOUT HOW I DECIDED to dedicate my life to preserving the monarch butterfly, I will share a story. It was October 1993 when Alfonso Alonso invited me to join Lincoln Brower's team for a monarch butterfly research project in Mexico during the November 1993 to March 1994 overwintering season. I gladly accepted, since I had listened to Professor Héctor Pérez talk enthusiastically about the monarchs during my university entomology classes.

In November, we began visiting the monarch sanctuaries on the border between Michoacán and Estado de México. One of our objectives was to measure the size of the overwintering colonies of the eastern monarch butterfly, which migrates to Mexico each year from Canada and the United States. To measure the area occupied by each colony, we established a polygon enclosing the forest areas where the butterflies congregate, using a method originally designed by William Calvert. This was the beginning of the systematic monitoring that provides an indirect measure of the monarch butterfly population density. To date, we have data from 33 continuous years.

The first sanctuary where we started surveying that season was Cerro Altamirano. Don Braulio, our local community guide, took Alfonso Alonso, Eneida Montesinos Patiño, and me up the mountain to the colony. After work, we decided to take a rest. Exhausted, I lay down on the ground at the edge of the butterflies' colony and was asleep before I knew it.

The first thing I saw upon waking was an image I will never forget: Thousands of monarchs were now covering the trees under which I was lying, and I felt deeply protected by them. That is when I decided to make conservation of the monarch butterfly my purpose in life.

We lived, camping out, in Sierra Chincua, for four and half months. Our rustic field laboratory was in the city of Zitácuaro, Michoacán, where every two weeks we analyzed the butterflies we had collected from all the colonies. Alfonso, Eneida, and I spent many nights in the lab using microscopes, scales, and other equipment to study thousands of butterflies. Our goal was to determine the butterflies' size, physical condition, and energy, comparing them when they arrived with when they left at the end of the winter as they started their migration back north.

In 1993, we joined with Eligio García Serrano, a biologist and later a National Geographic Explorer, to form an extended team. At the time, he was working for the Monarch Butterfly Special Biosphere Reserve, a smaller predecessor of the current protected area. From 1994 to 2003, Eligio continued monitoring the monarch colonies in coordination with us. Three key decrees were made by the Mexican government to protect the monarch butterfly, in 1980, 1986, and 2000, when the Monarch Butterfly Biosphere Reserve (MBBR) was created, protecting 139,019 acres. The main overwintering colonies are in the core zone (33,493 acres) of the MBBR, and no logging is permitted in this area. Sustainable projects are allowed in the buffer zone of 105,526 acres surrounding the nucleus.

Mexico's monarch region is a vast area covering almost 2.5 million acres, including the MBBR and other protected areas where monarch butterflies also establish colonies. Not only are the forests of this region the sites selected naturally by the migratory monarch butterfly, but they also form the hydrological basins of the Cutzamala System, which supplies water to more than six million people in Mexico City's metropolitan area.

In March 1994, I returned to Mexico City to conclude my undergraduate work. But as soon as I finished in August, I went back to Sierra Chincua, where I spent the next two years. During this time, I recorded daily temperatures in the forests, studied the forests' regeneration, and carried out reforestation experiments. In August 1996, I began my graduate studies, and I became a doctoral candidate in 1998. I finished all doctoral credits in 2002. Then I began my family, who are the light of my life: my wife, Mary, and my daughters, Fer and Normita, to whom I dedicate this essay, as well as to the memory of my mom.

In early 2003, I coordinated the design of forest monitoring in the MBBR core zone for World Wildlife Fund (WWF). The goal was to quantify annual changes in forest cover as a scientific baseline for the Monarch Fund. Administered by the Mexican Nature Conservation Fund, the Monarch Fund is an economic incentives strategy for the Indigenous communities and ejidos, or farm communities—the two main types of social land tenure in Mexico, together representing all the people who own the forests inside the MBBR. Currently these owners receive additional payments from the National Forestry Commission for hydrological environmental services—financial support that has been crucial for MBBR conservation. Indeed, illegal logging is almost absent in the core zone of the reserve, and the forests are significantly recovering.

In the middle of 2004, after serving for a year as the deputy director of the Monarch Butterfly Biosphere Reserve, I was invited again to work with World Wildlife Fund, and I put together the first team focused on monarch butterfly conservation with Pablo Ángeles Hernández, who was responsible for forest restoration. We were later joined by Guadalupe Gómez Gutiérrez (administration management), César Ávalos Tello (forest management), Caterine Forero Díaz (sustainable development), and, more recently, Gabriela Sánchez Carmona (administrative assistant) and Adriana Valera Bermejo (pollinator coordinator). I still proudly belong to this extraordinary team.

We have been monitoring the monarch butterfly colonies and the forest in coordination with the National Commission of Natural Protected Areas (CONANP), and the results are fundamental to conservation actions. For example, we have established reforestation programs. From 2005 to 2024, we reforested 53,165 acres of forest in the monarch region in collaboration with local communities. Of the 22.5 million trees we planted, 21.1 million had been propagated in community nurseries established in 2007 by World Wildlife Fund. Today, we are continuing to support plant production in coordination with Martín Cruz Piña and Alejandra Hinojosa Mendoza of Danaidas, A.C., our local partners.

Our accomplishments are the sum of the work of many people representing federal, state, and local governments as well as private, nonprofit, and academic institutions interested in preserving the monarch butterfly. But the most important agents are the local people who are committed to conserving and sustainably managing their forests to benefit the monarchs, providing ecosystem services that benefit us all.

Eduardo Rendón Salinas is the lead monarch butterfly scientist at WWF-Mexico. He has worked in the Monarch Butterfly Biosphere Reserve in Mexico for more than 30 years.

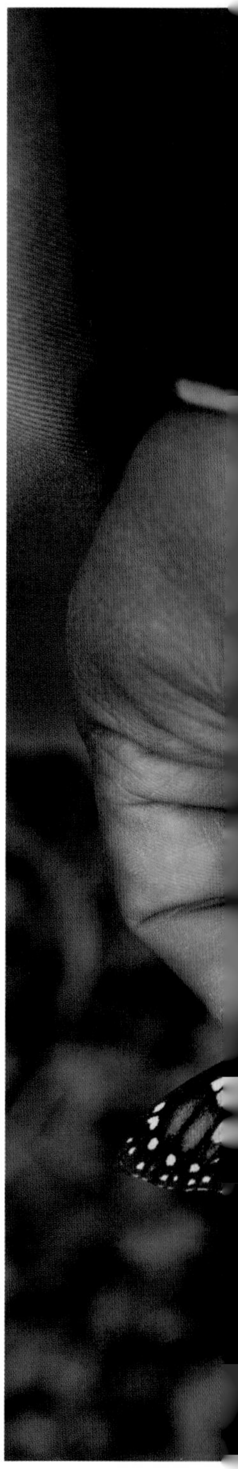

Eduardo Rendón Salinas from WWF-Mexico shows the physical differences between a male (left) and a female (right) monarch butterfly to the monitoring team at Ejido San Pablo Malacatepec. Identifying the monarchs' sex helps researchers better understand colony composition and population dynamics. Males have thinner veins in their wings and a small black spot on each hindwing.

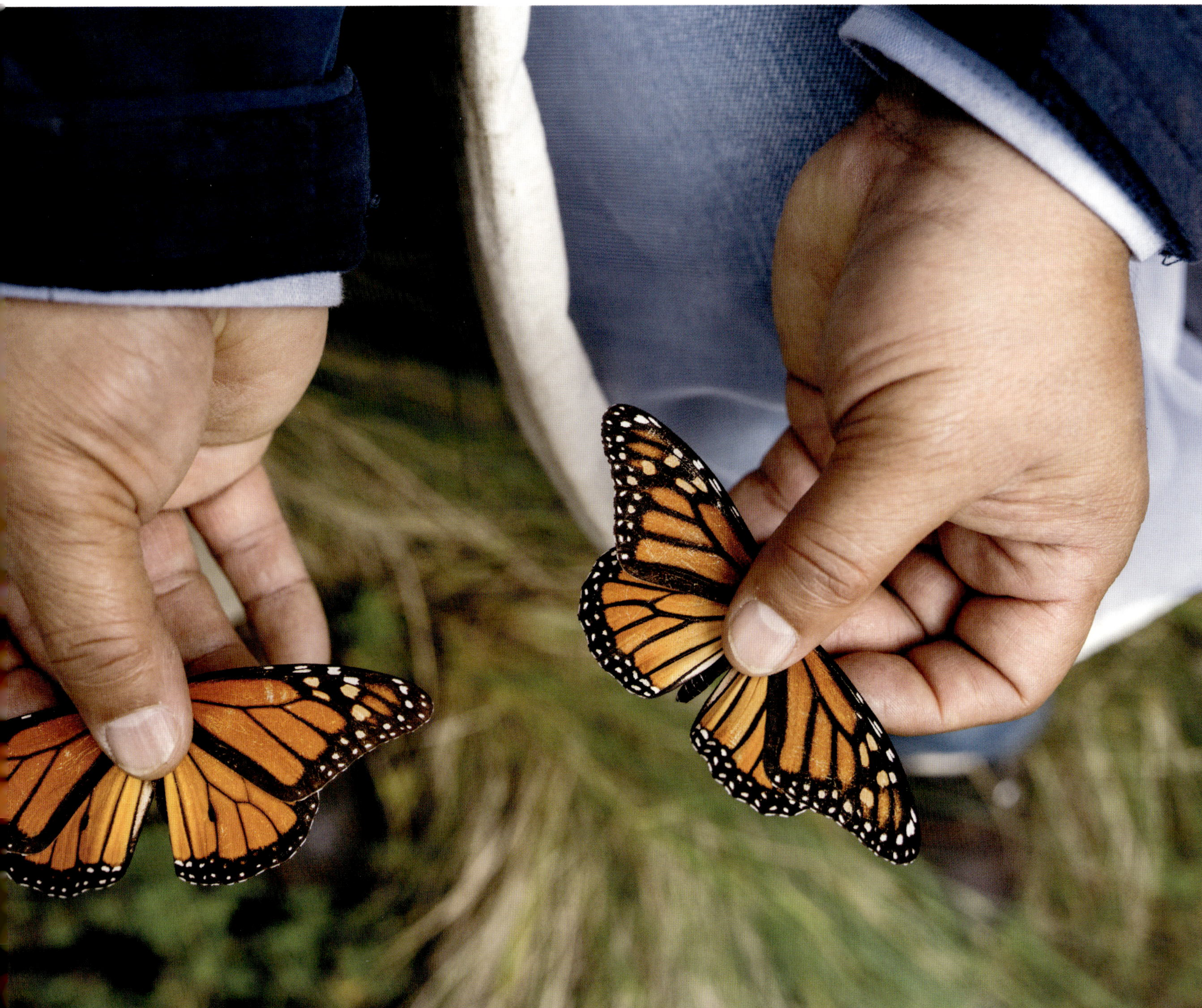

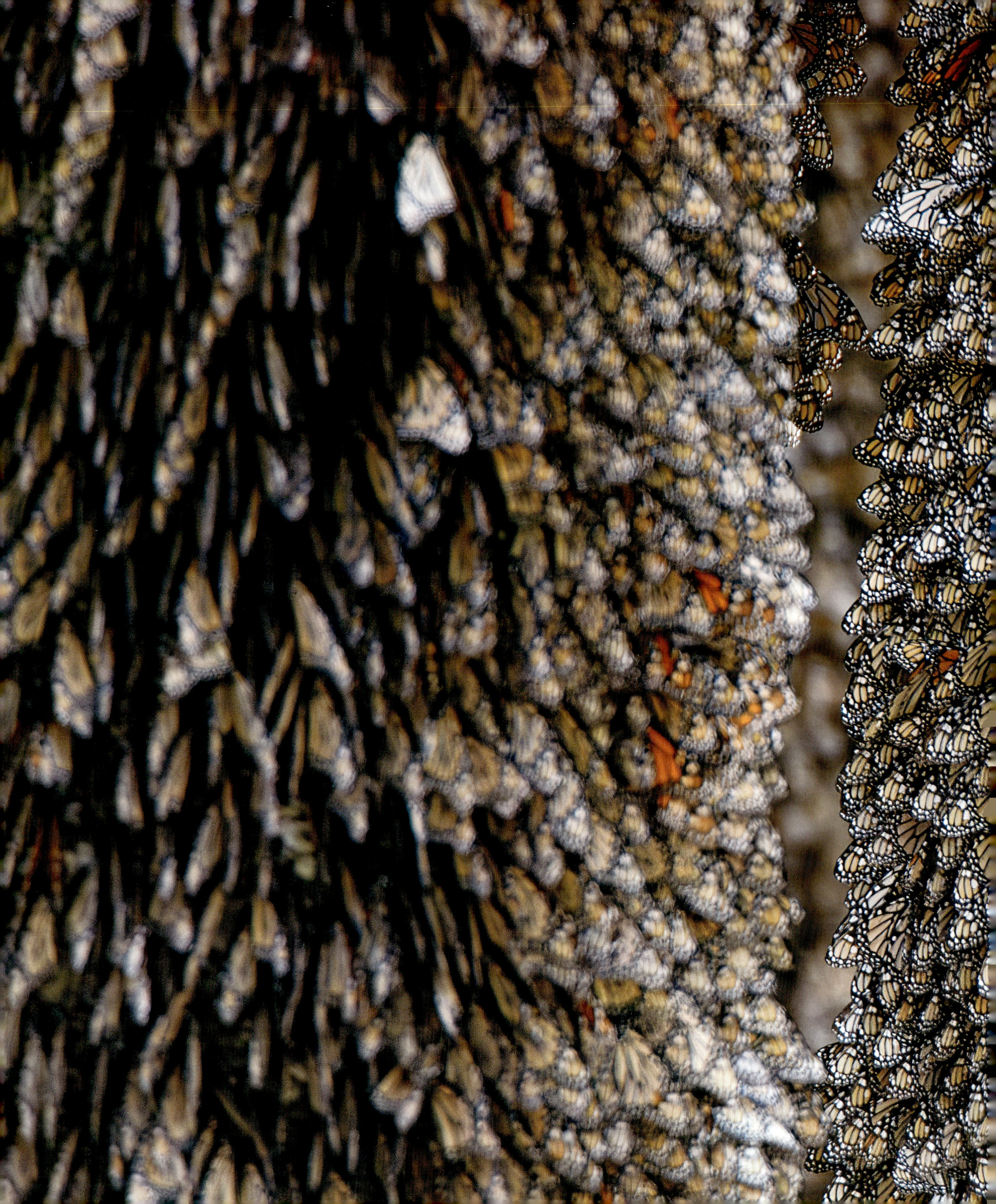

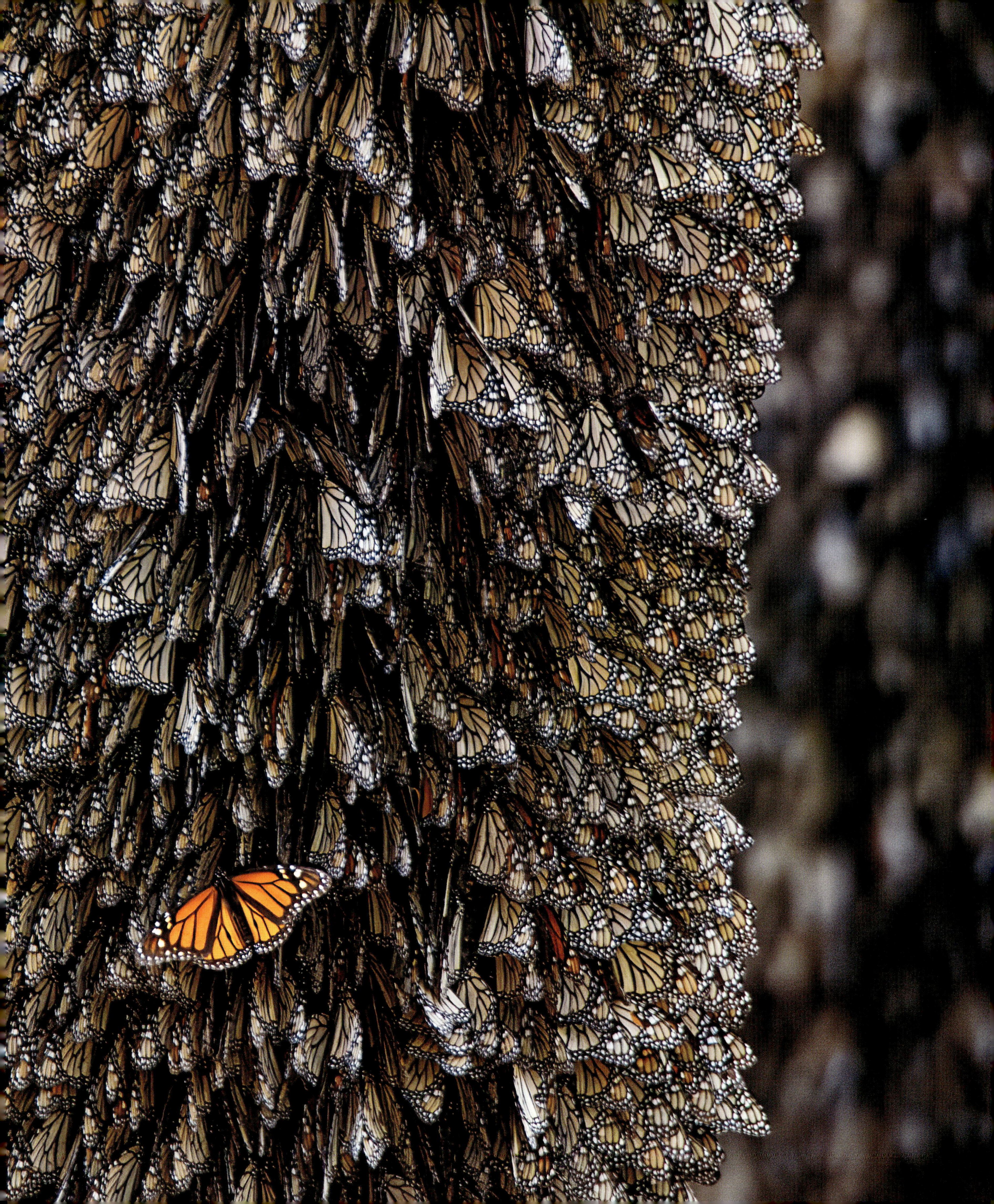

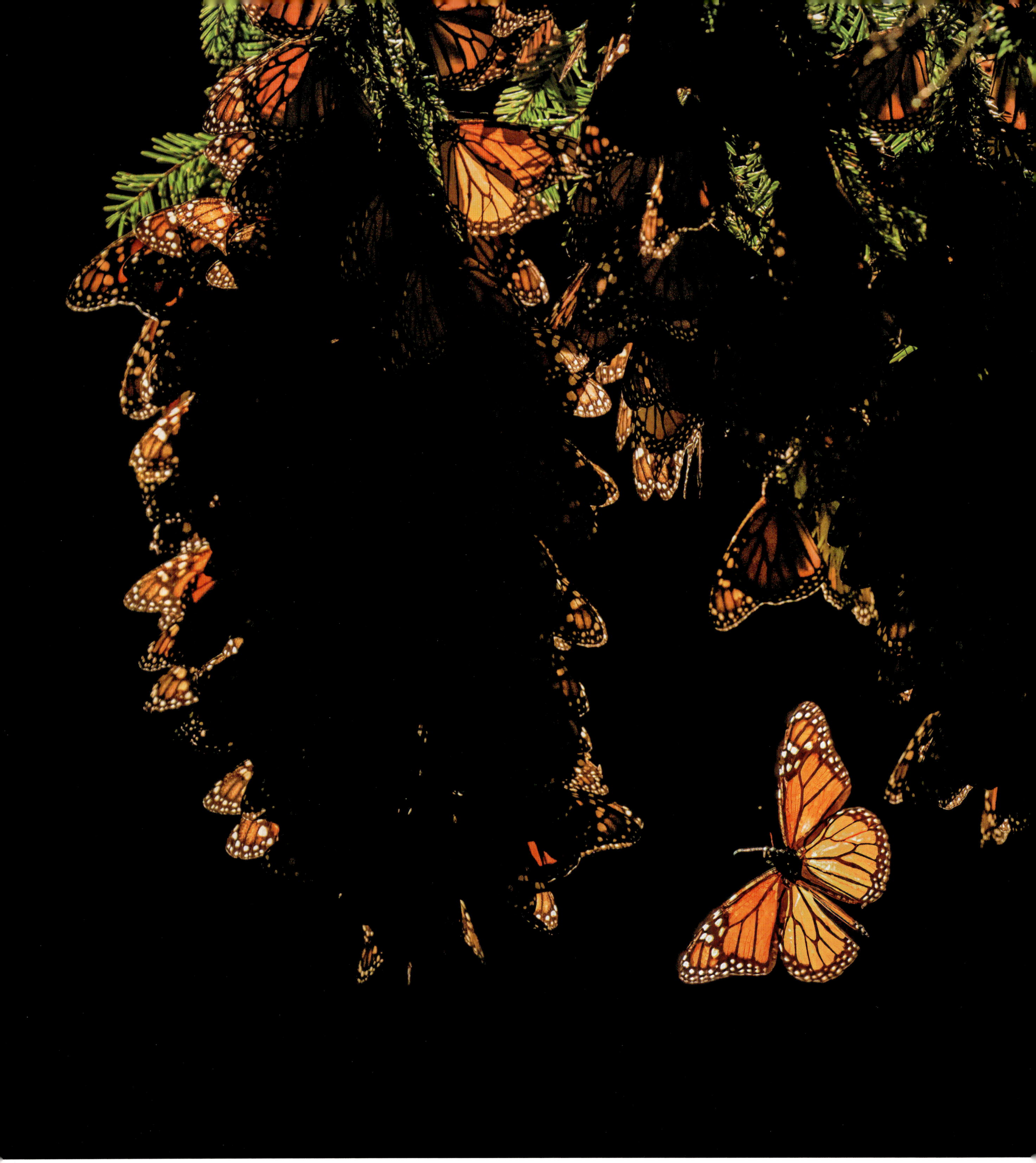

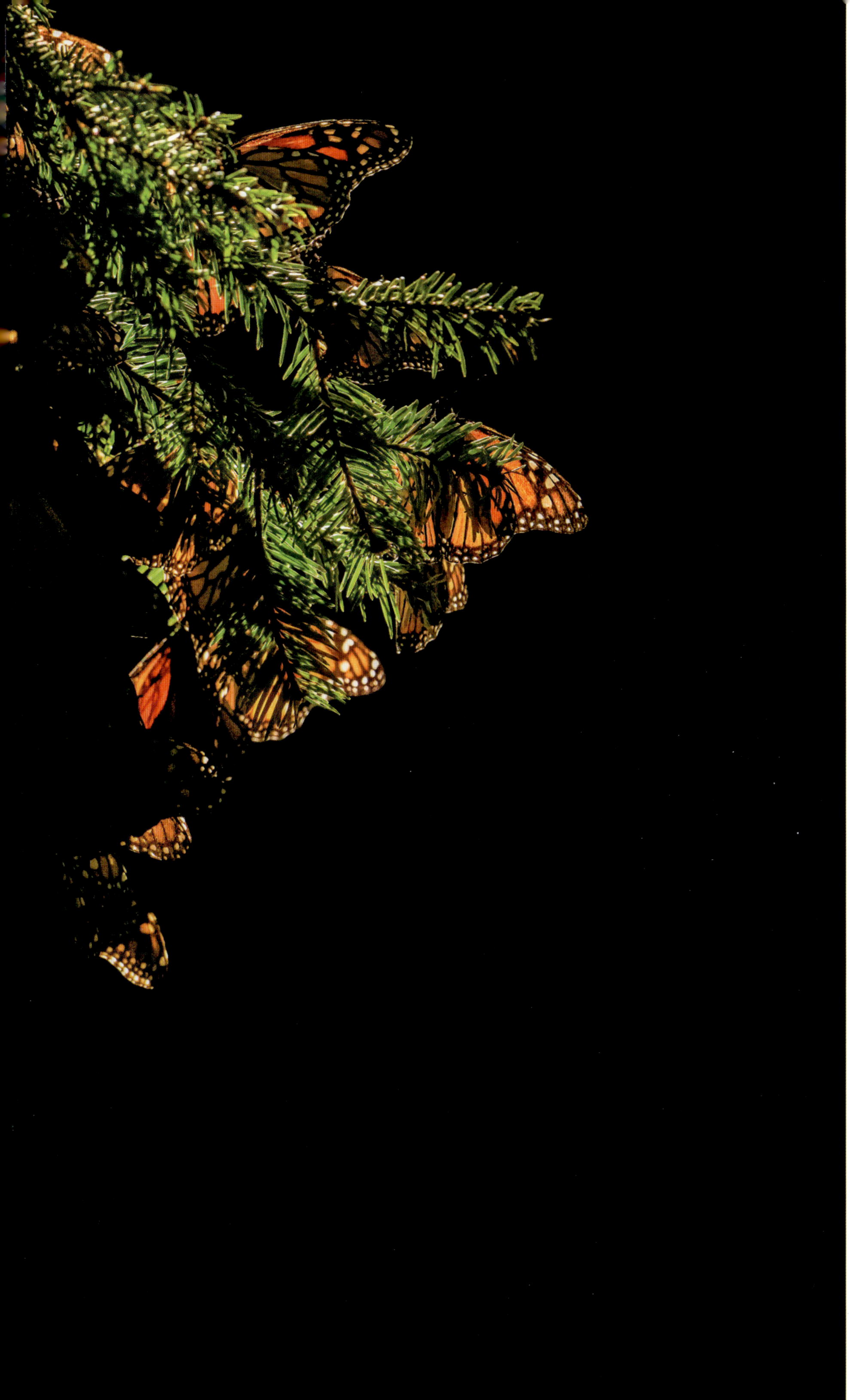

As sunlight filters through the forest canopy, monarch butterflies stir and reposition within the colony at El Rosario Sanctuary. Their response to temperature and light synchronizes with the forest's daily rhythm, shaping their activity throughout the winter months.

PAGES 214–215: Monarch butterflies densely cluster on oyamel fir trunks in Sierra Chincua Sanctuary. Their collective body heat helps retain warmth and moisture through the cold nights, while the trees act as natural heat reservoirs, preventing the colonies from freezing.

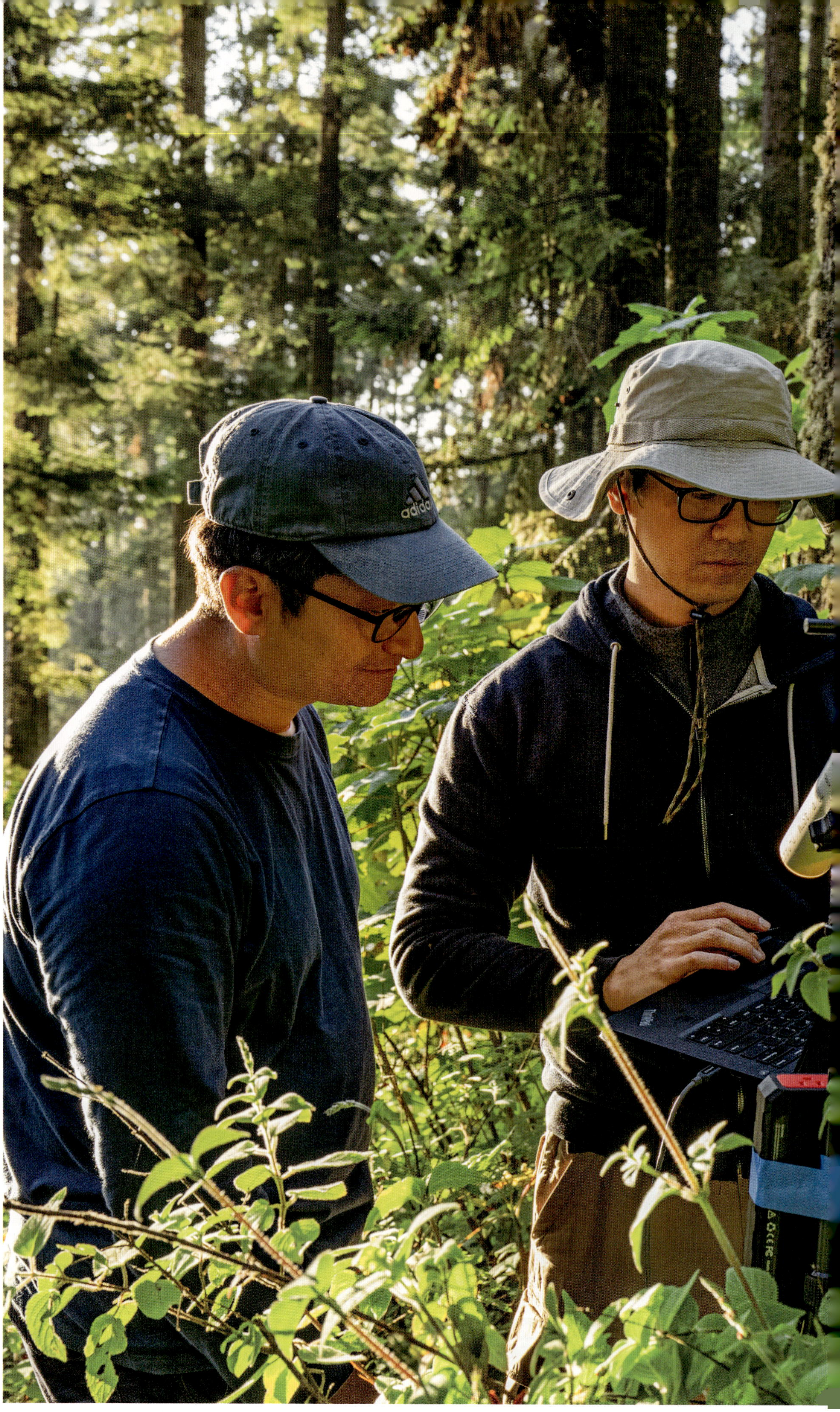

D. André Green II (far right) and his team from the University of Michigan (left to right: Inhee Lee, Hun-Seok Kim, and David Blaauw) deploy antennas in El Rosario Sanctuary, Michoacán, Mexico. They are searching for signals from monarch butterflies previously tagged in Texas with miniature sensors—data that will be analyzed with AI to reconstruct their migration routes.

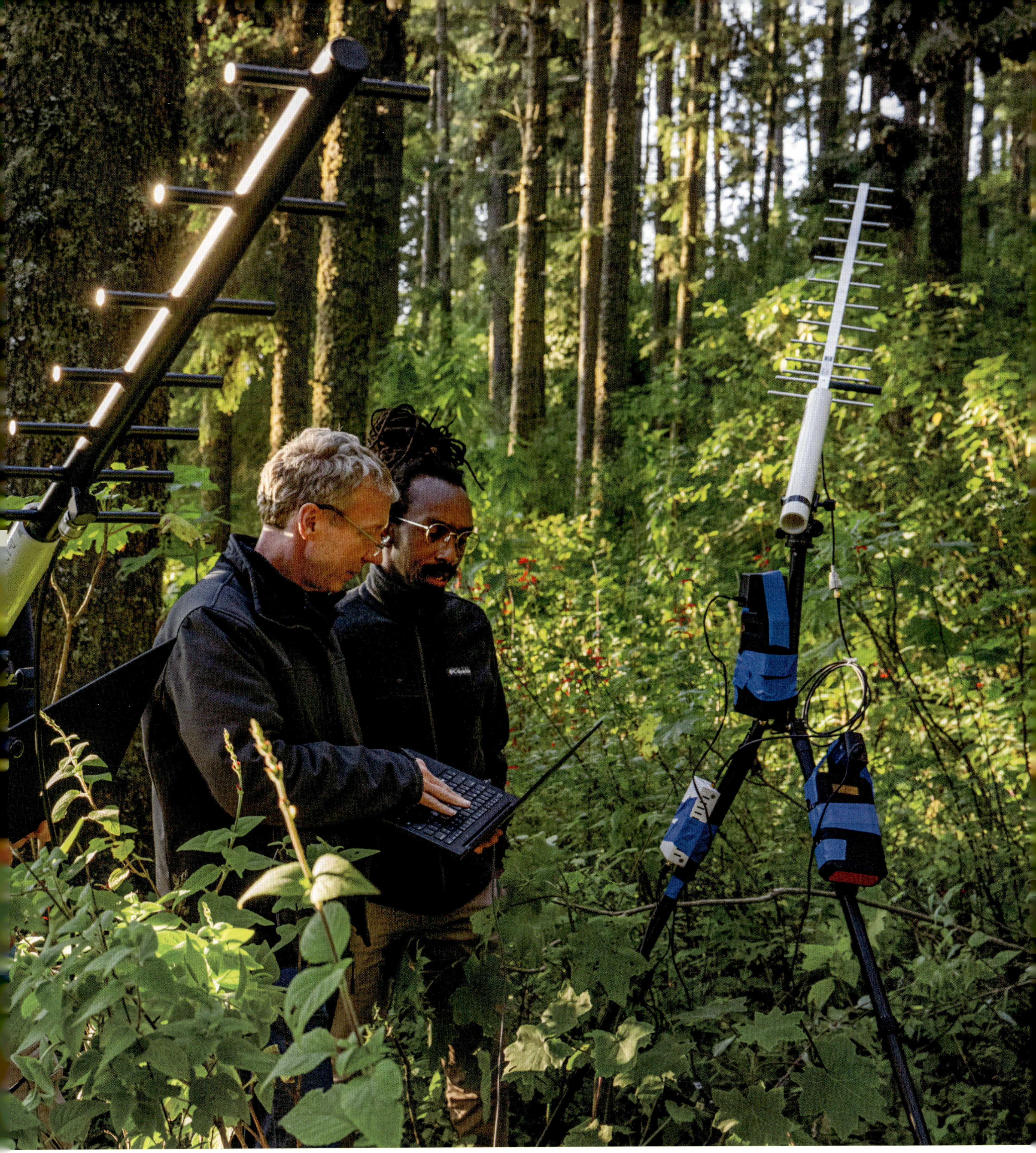

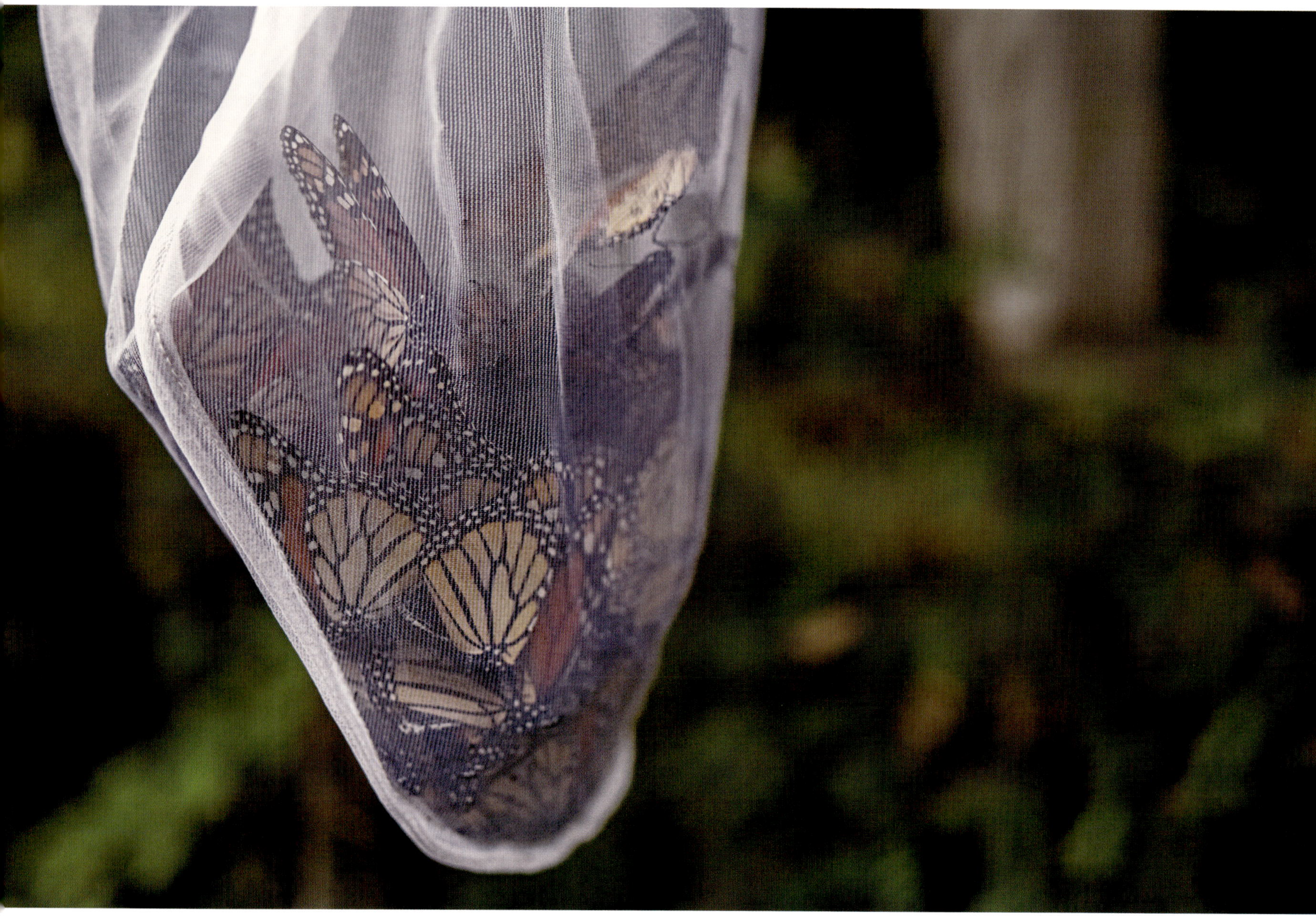

In Mexico, monarch butterflies are briefly captured for scientific study to analyze lipid reserves and other indicators of their health and readiness for the long journey back north.

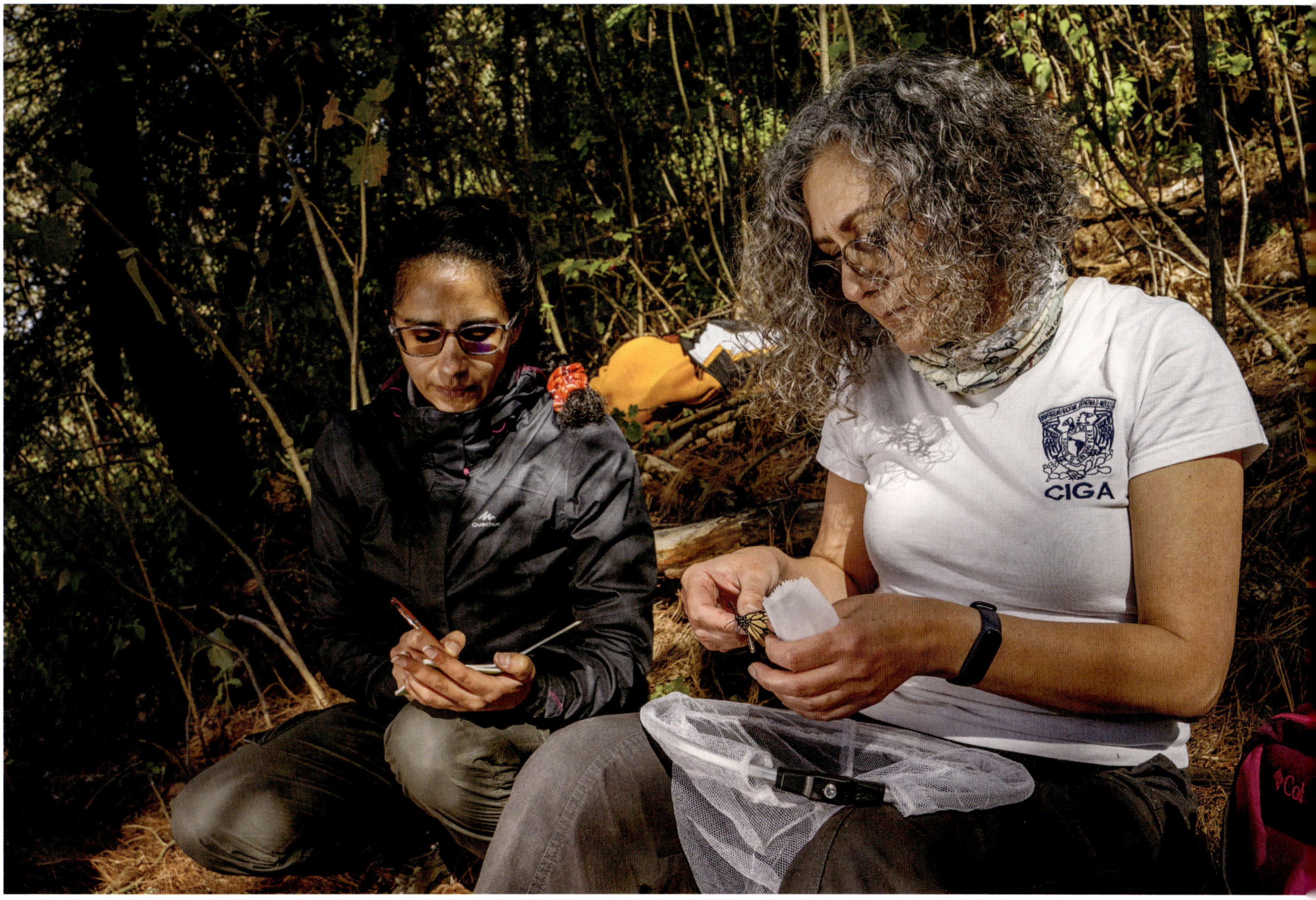

Isabel Ramírez (right), a scientist from the Center for Research in Environmental Geography at the National Autonomous University of Mexico, and her assistant, Viviana Andrade, study monarch butterflies in Sierra Chincua Sanctuary. Ramírez is part of an international project examining lipid content and other physiological indicators to learn how the butterflies manage their energy reserves throughout the winter and during their remarkable migration.

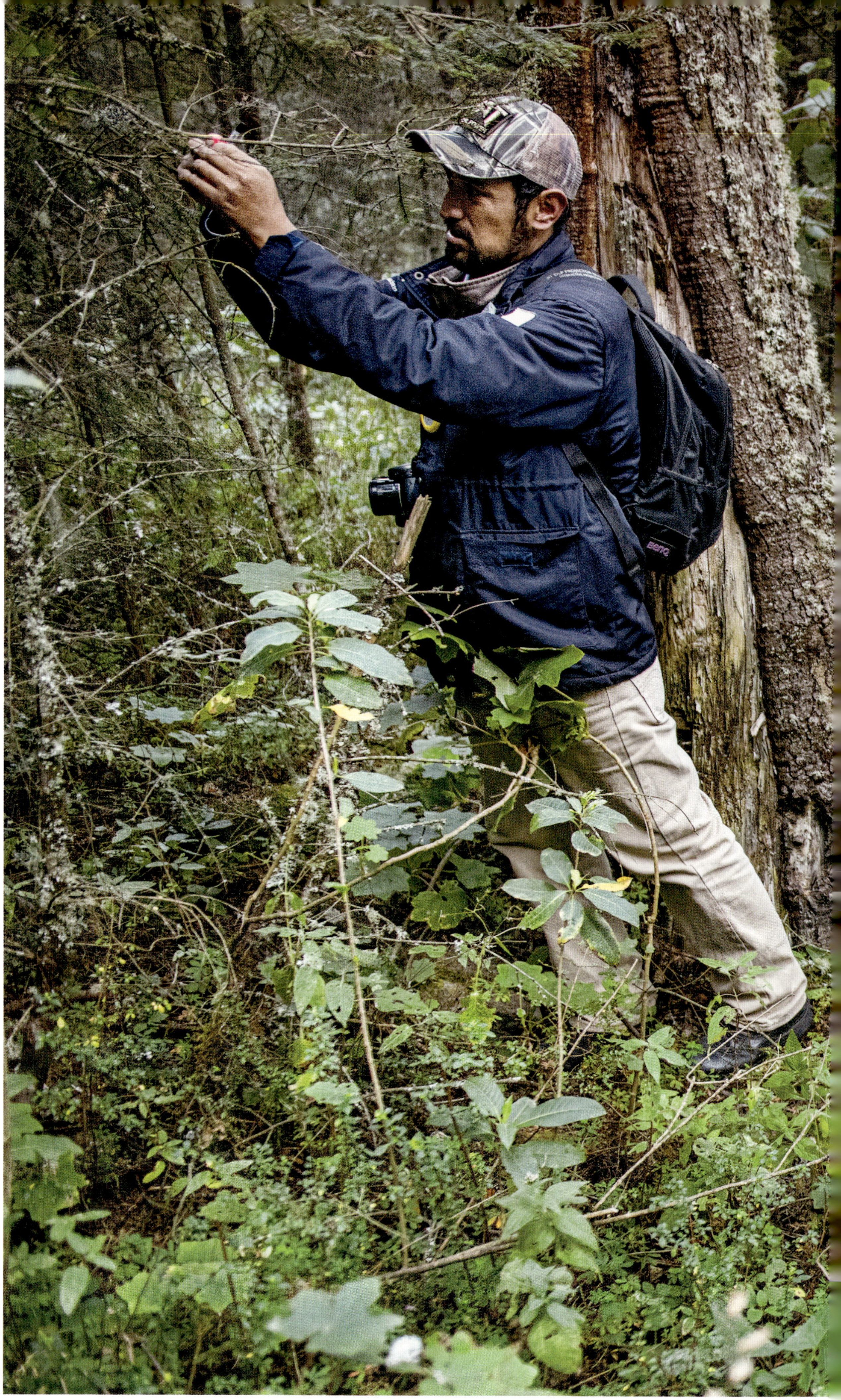

From left to right, Genaro Mondragón Contreras, Israel González Tellez, and Angélica Reyes Ruiz monitor monarch colonies in the Monarch Butterfly Biosphere Reserve. This joint effort by WWF-Mexico, the National Commission of Natural Protected Areas (CONANP), and local communities has proven essential for tracking population size and habitat conditions each winter.

PAGES 224–225: At dusk in El Rosario Sanctuary, monarch butterflies cluster densely on the branches of oyamel fir trees to rest. As temperatures drop, they remain still through the night, conserving energy until the forest warms again the following morning.

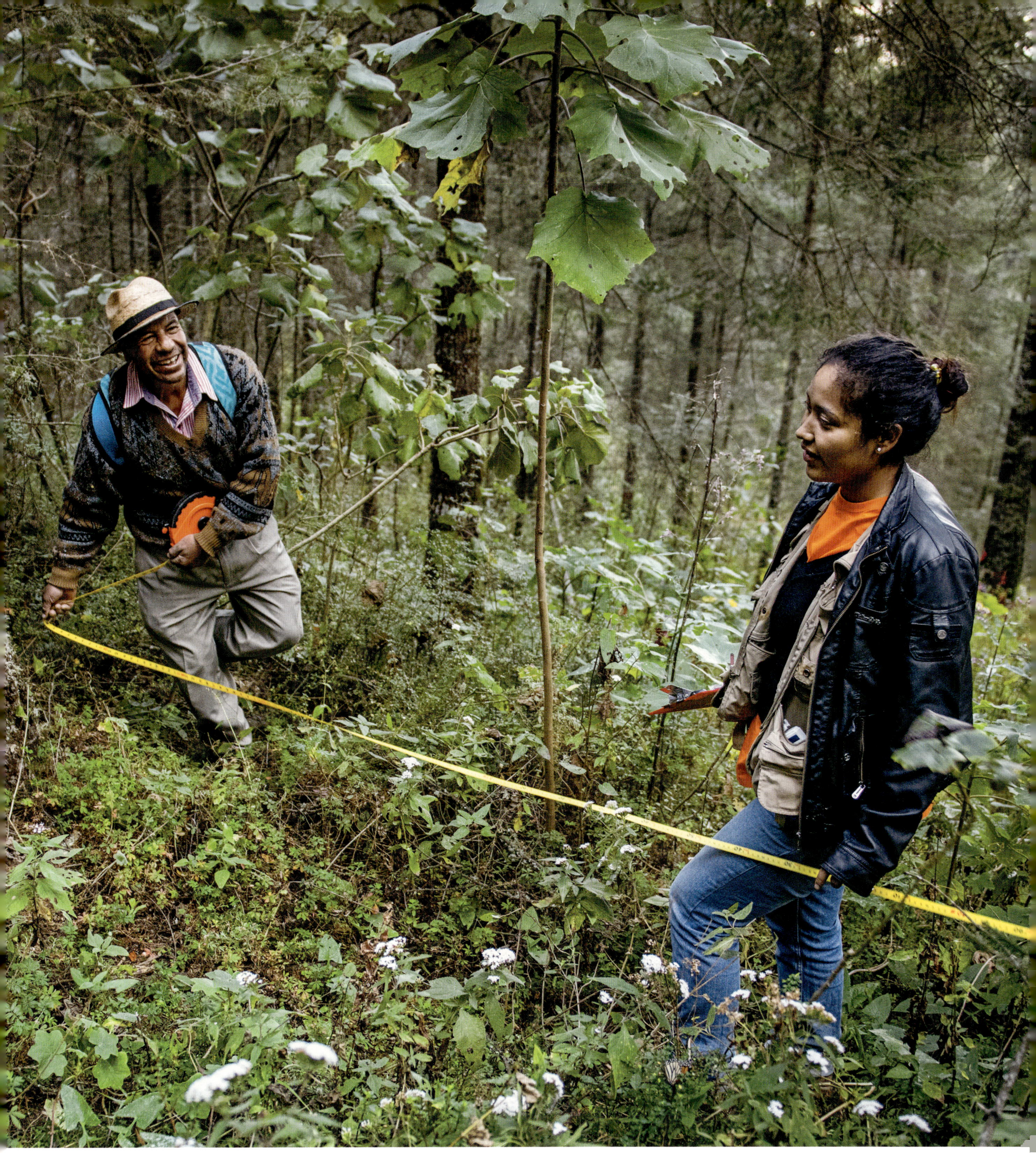

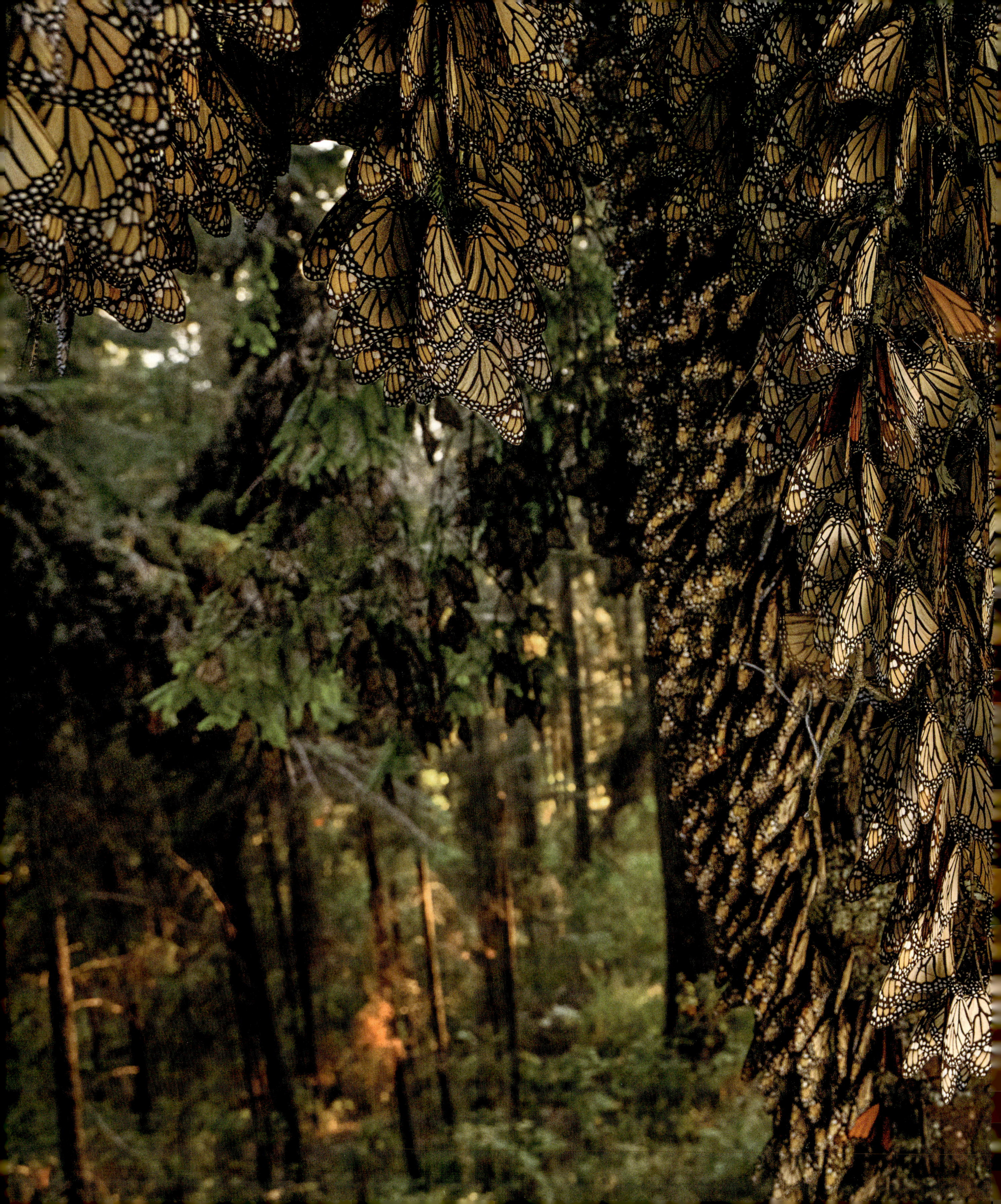

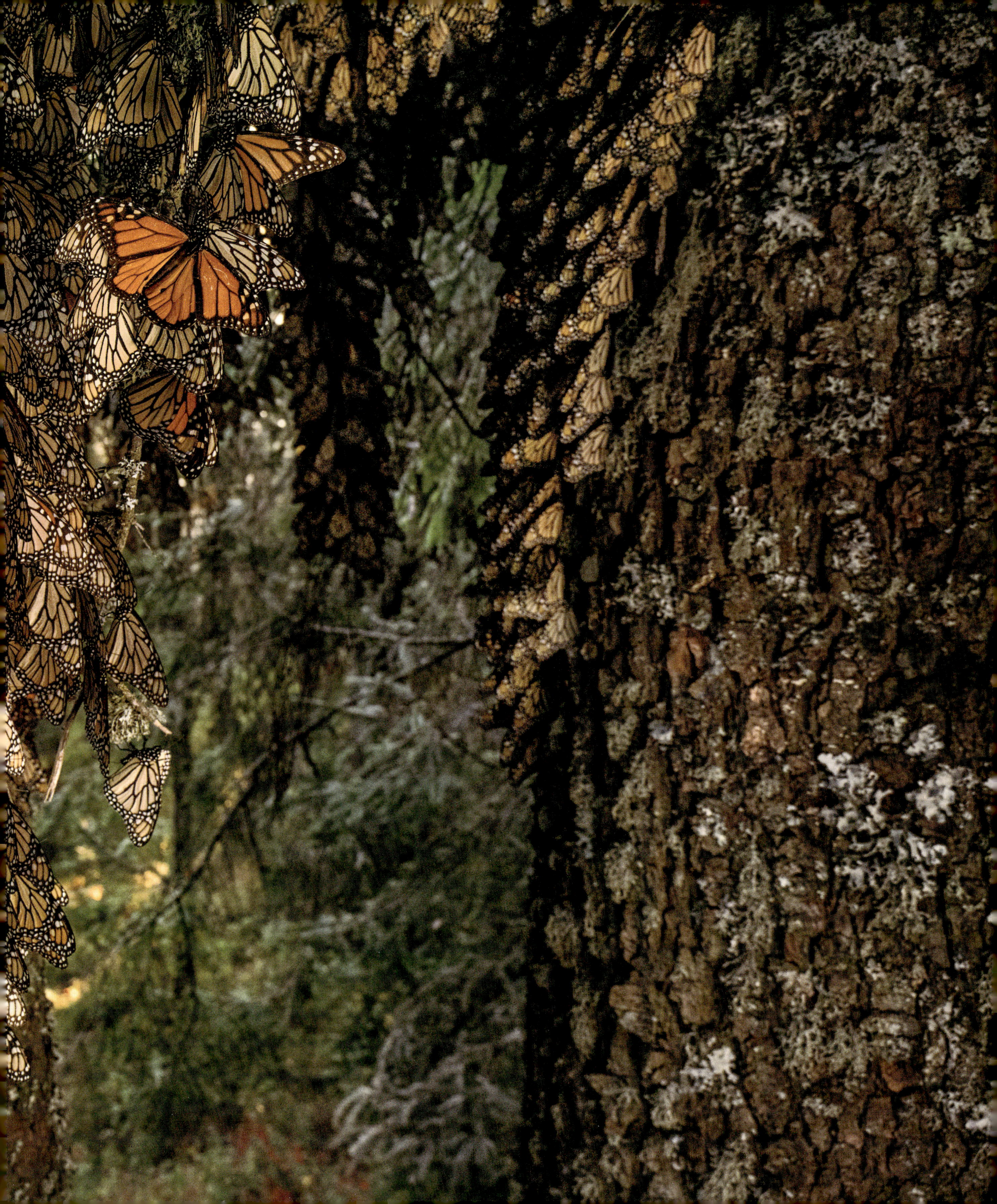

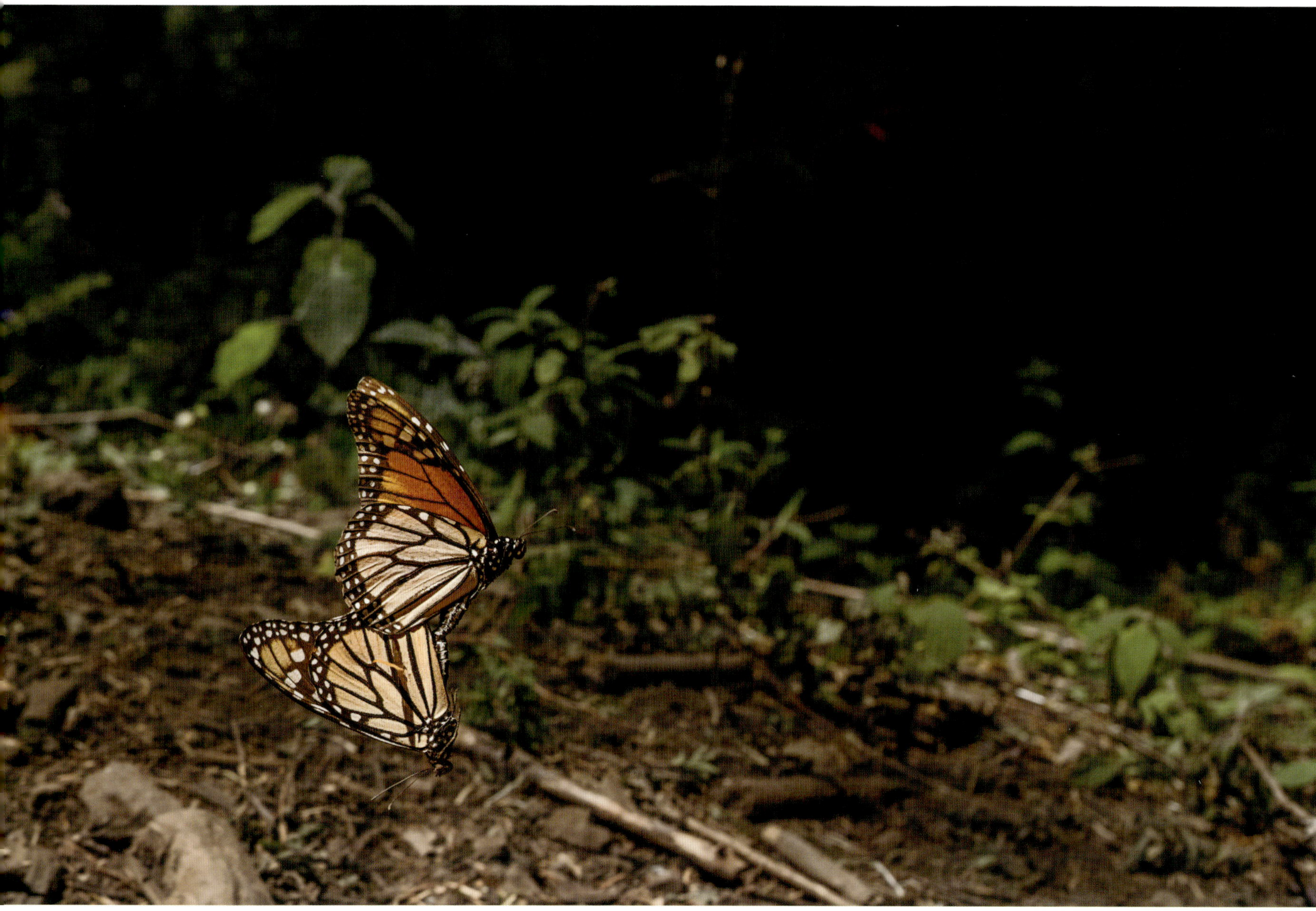

Aerial mating at El Rosario Sanctuary: The male grabs the female midair, often dragging her to the ground before taking off again. This behavior, part of monarchs' mating strategy, may allow females to select stronger partners capable of sustained flight.

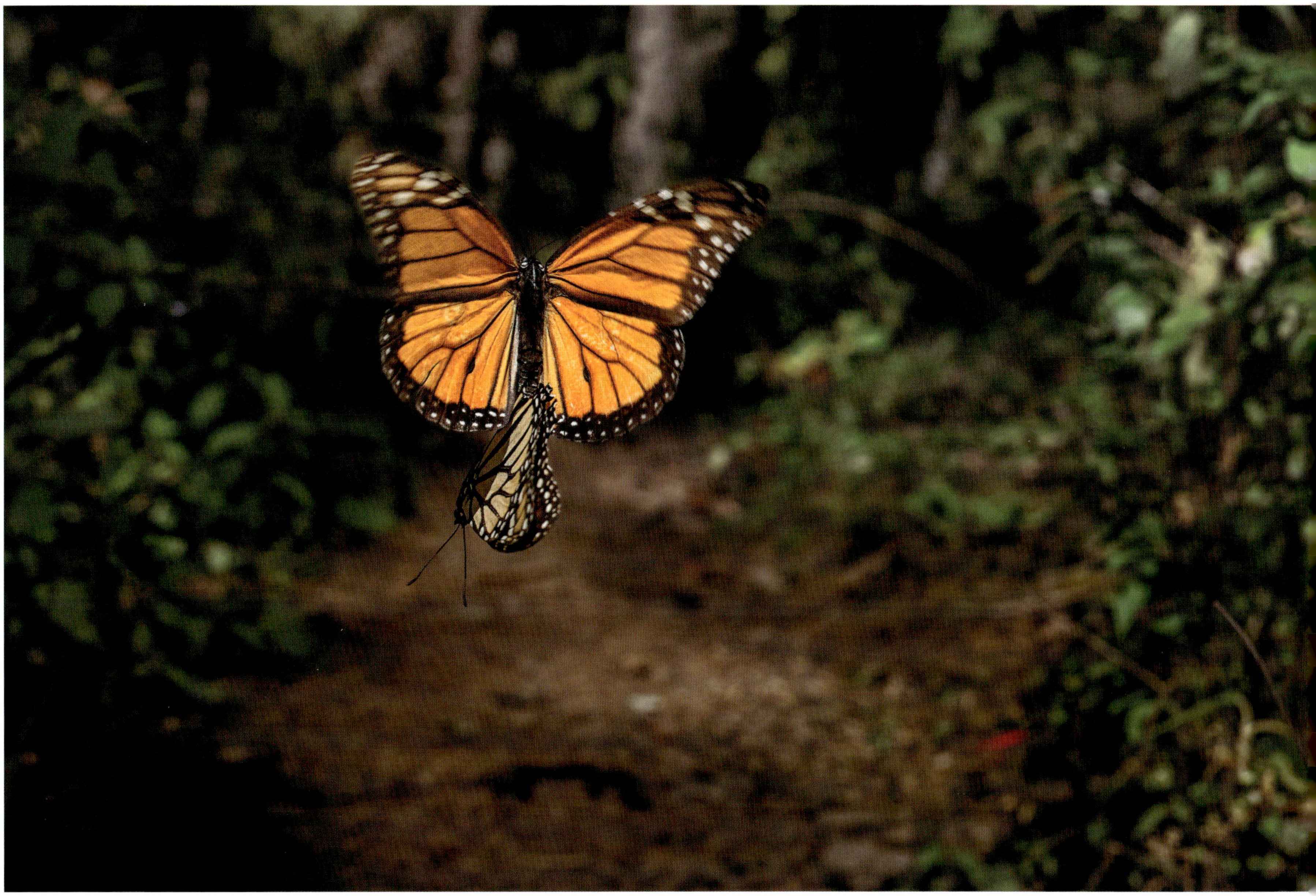

A male monarch drags a female through the undergrowth in El Rosario Sanctuary. The mating process can last up to 16 hours, consuming much of their remaining energy.

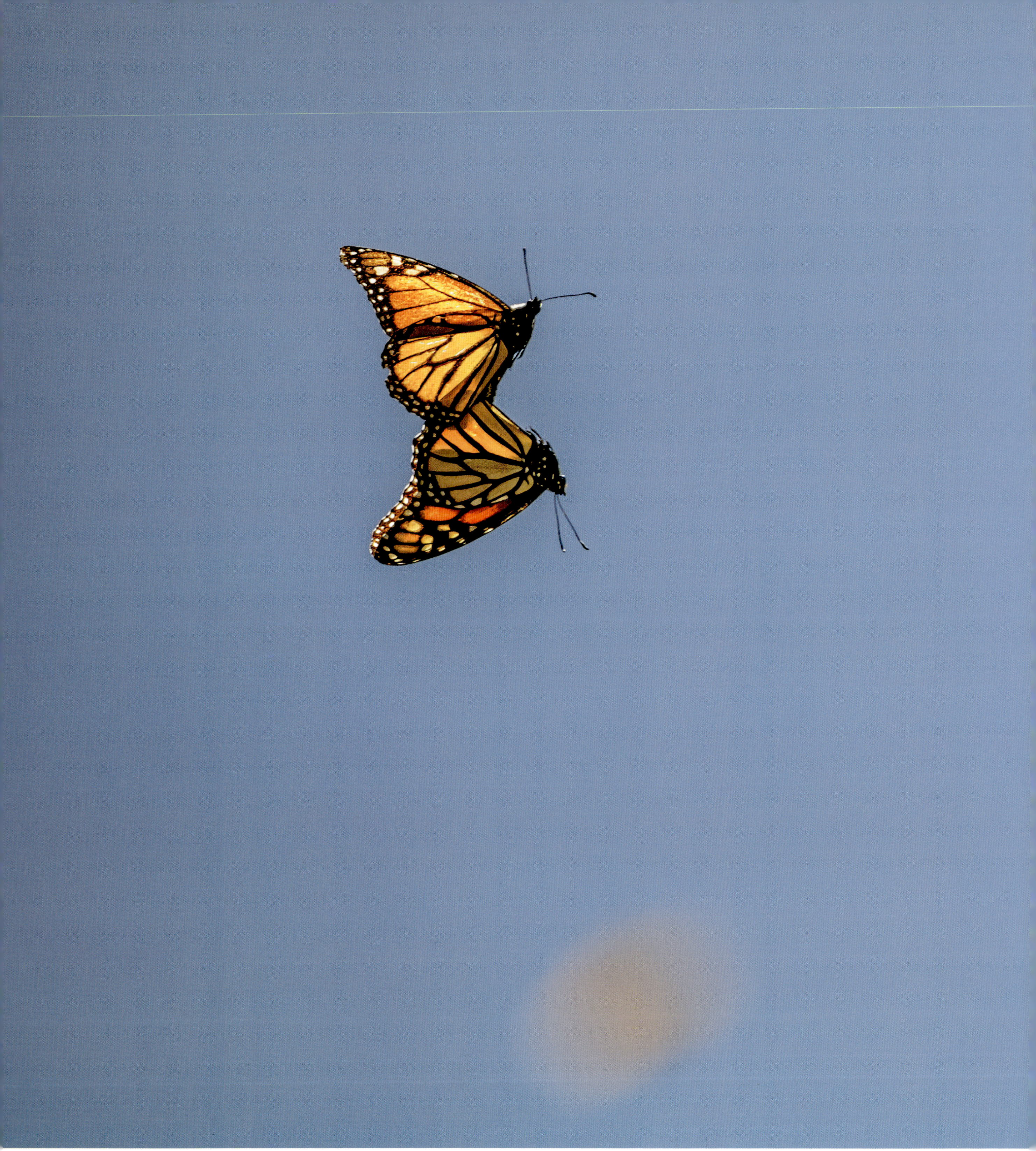

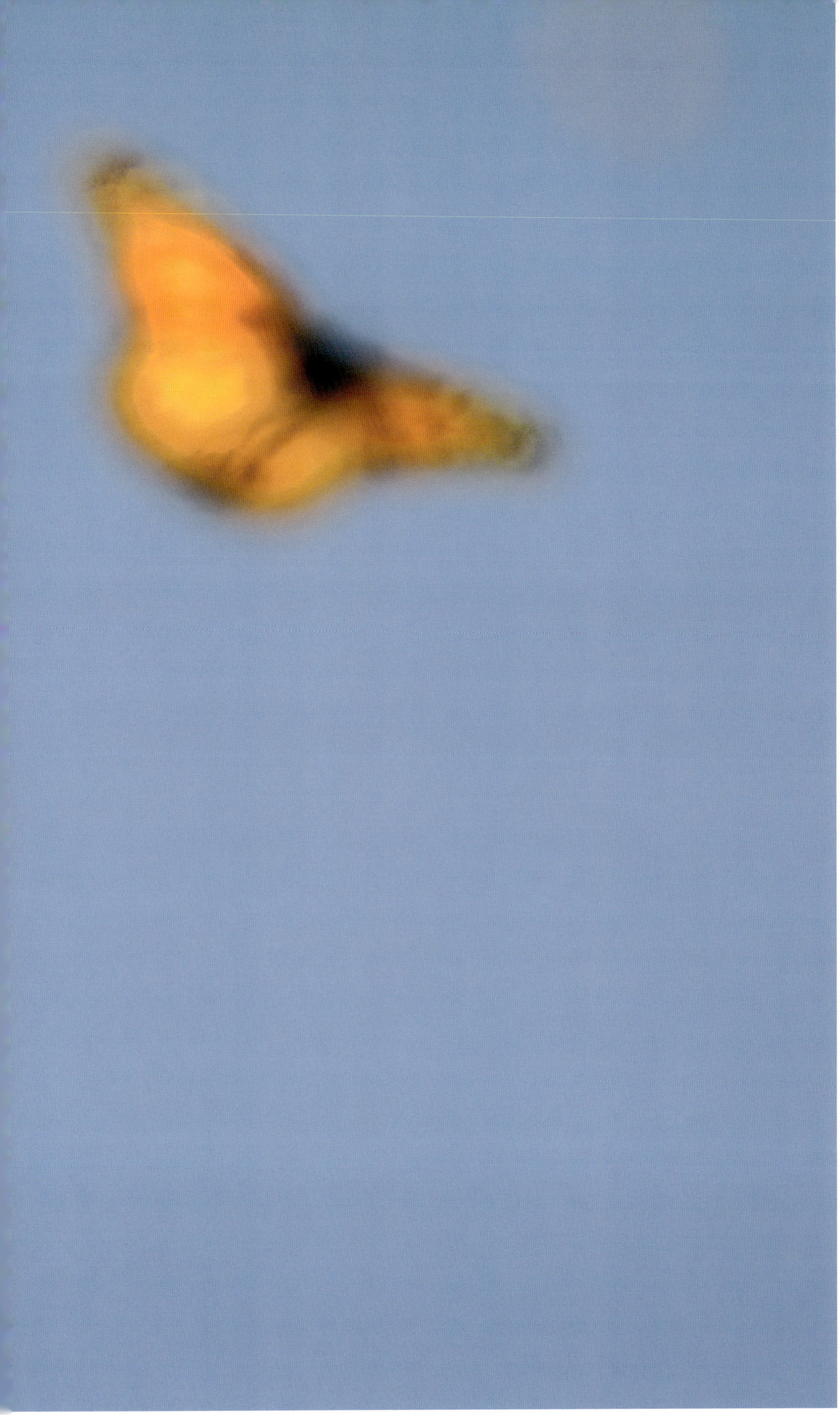

Monarch butterflies mate in flight against the blue sky of El Rosario Sanctuary. The overwintering colonies, once dormant, come alive as reproductive activity begins—an annual transition marking the start of their multigenerational migration north across the continent.

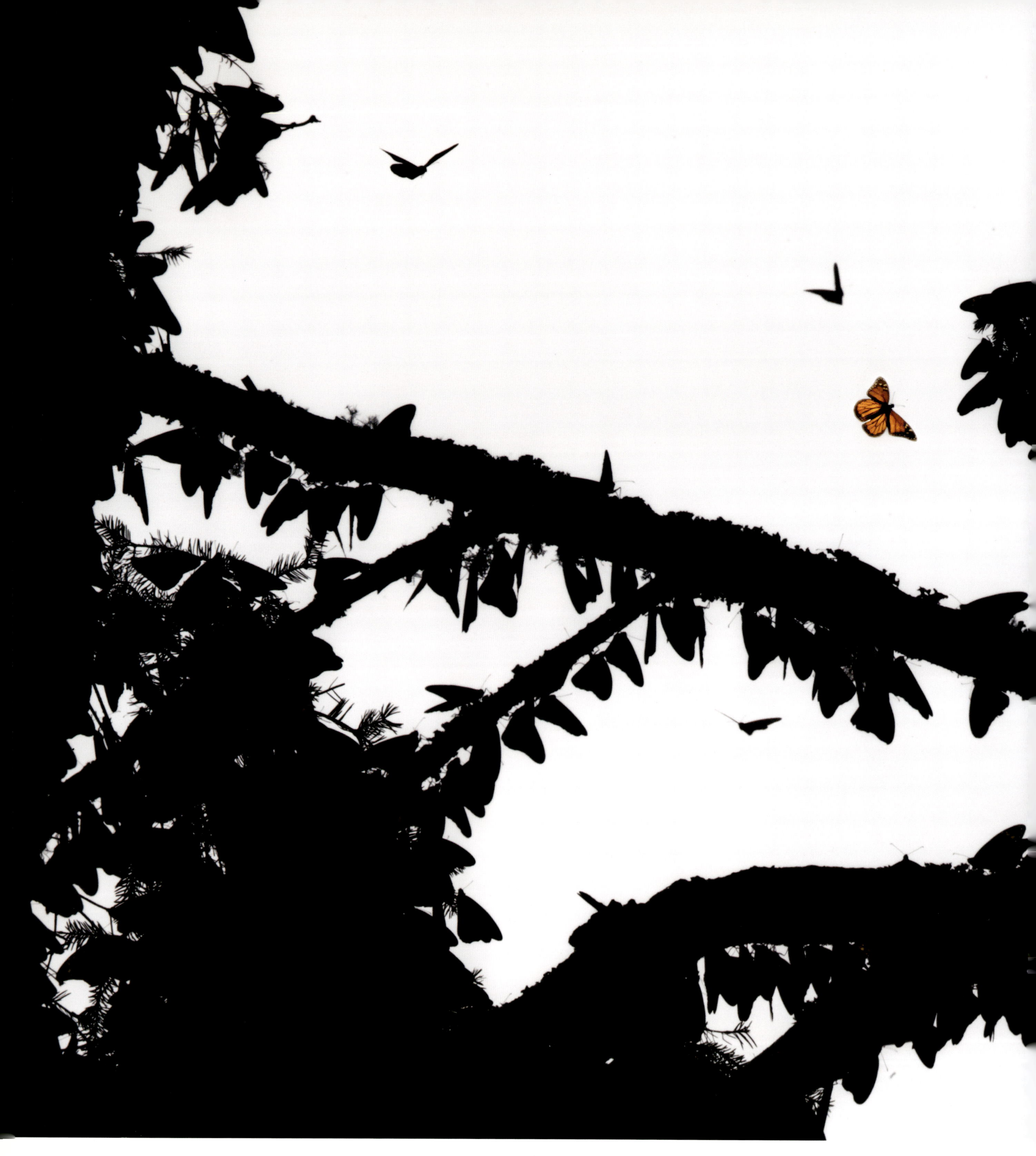

A single monarch butterfly breaks from its roost, illuminated by a ray of sun while others remain in shadow. As the colony warms, the butterflies become active in one of nature's most synchronized spectacles.

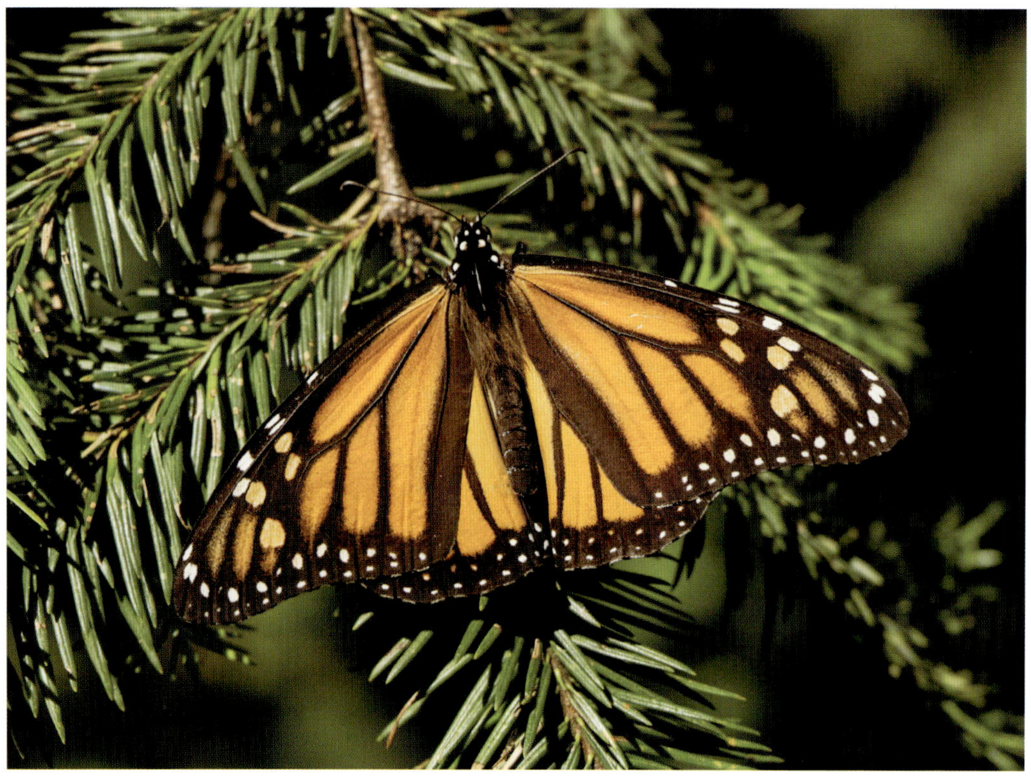

In the oyamel fir forests of Michoacán, a monarch butterfly basks in the morning sun. To be able to fly, its body needs to reach a temperature of about 55°F.

OPPOSITE: Two backlit butterflies take off from an oyamel fir tree where they roosted overnight. As the morning sun warms the forest, the monarch colony gradually becomes more active.

PAGES 234–235: When conditions are right, the sky above El Rosario Sanctuary fills with monarch butterflies. At their peak abundance, they can momentarily dim the sunlight.

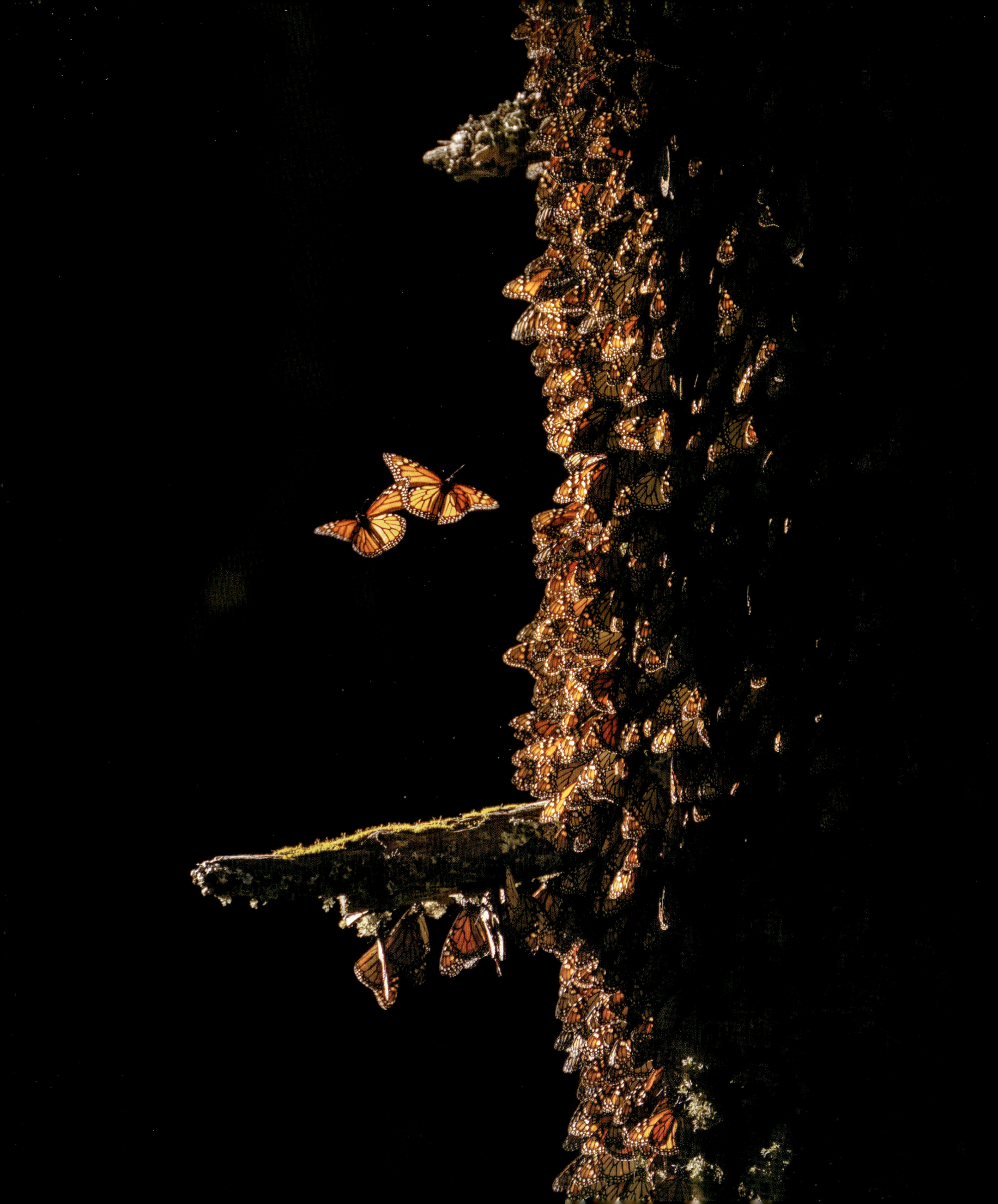

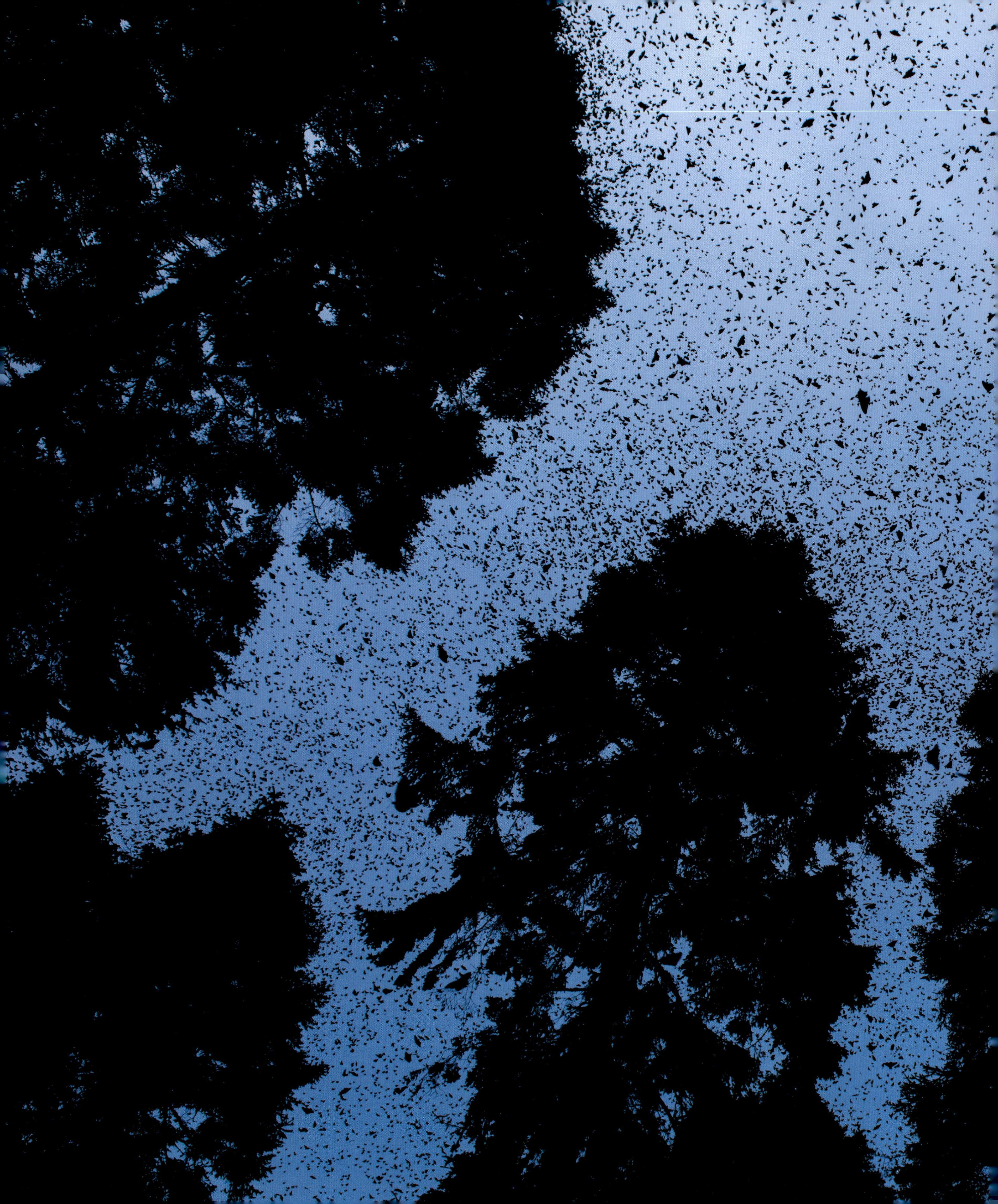

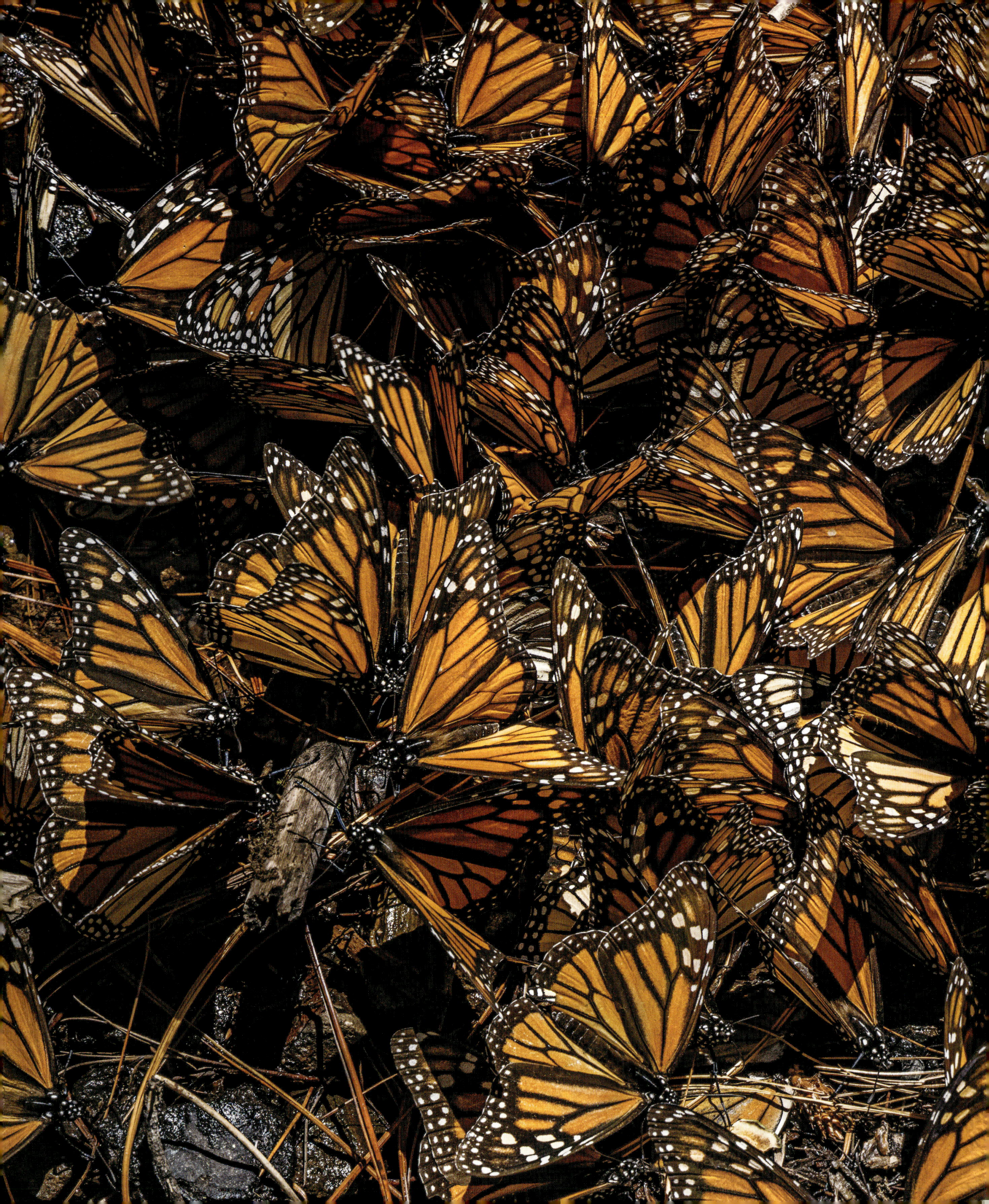

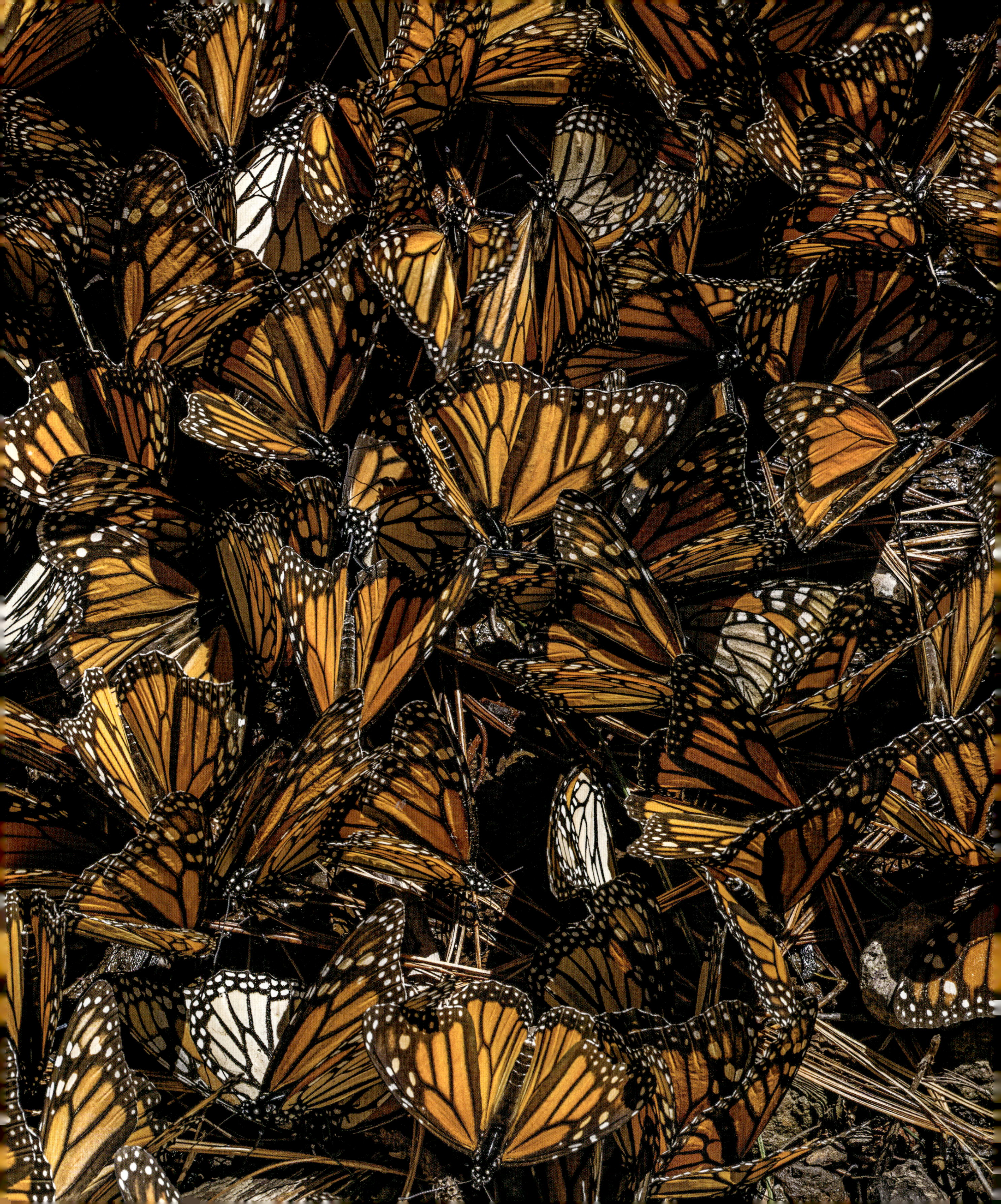

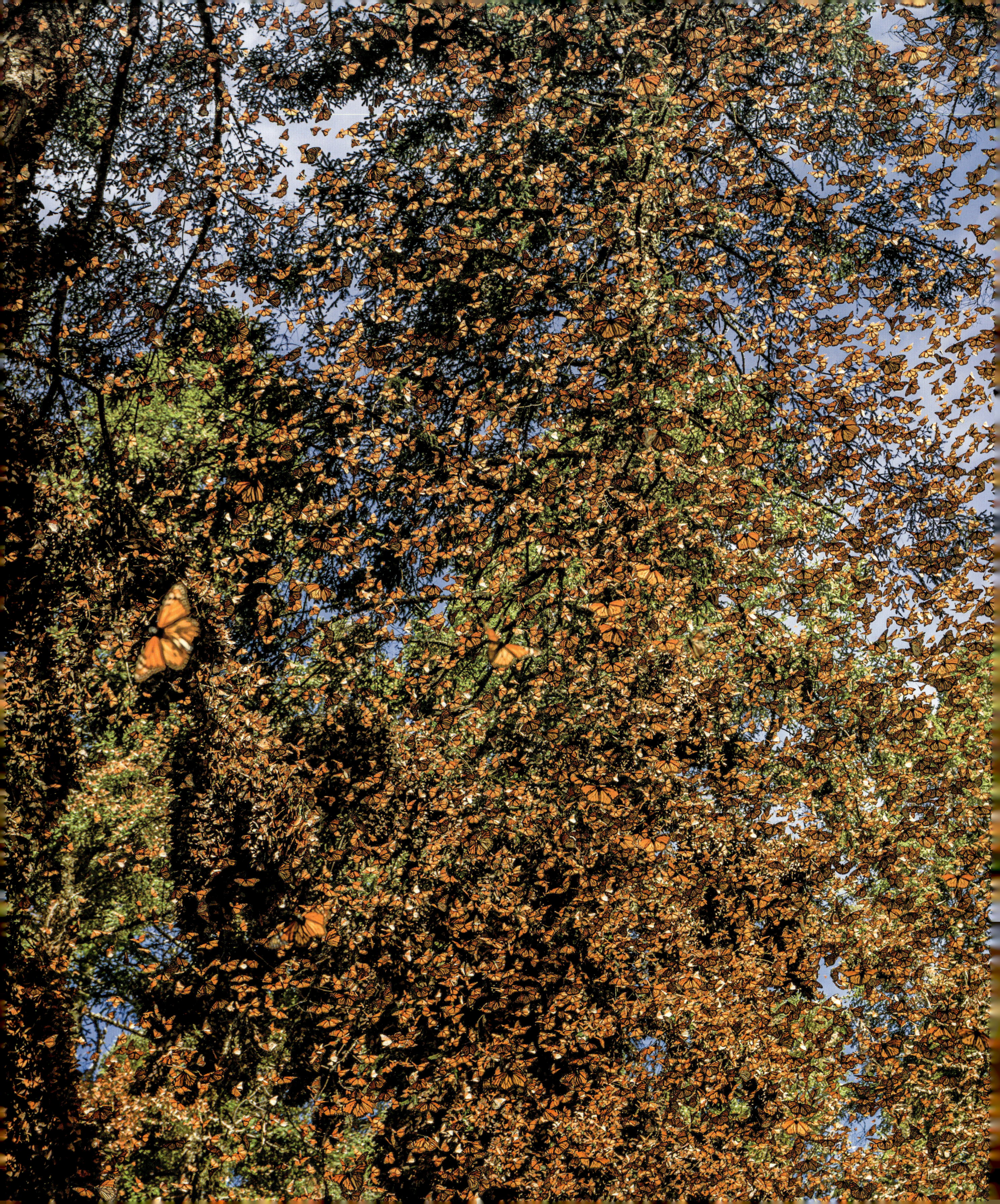

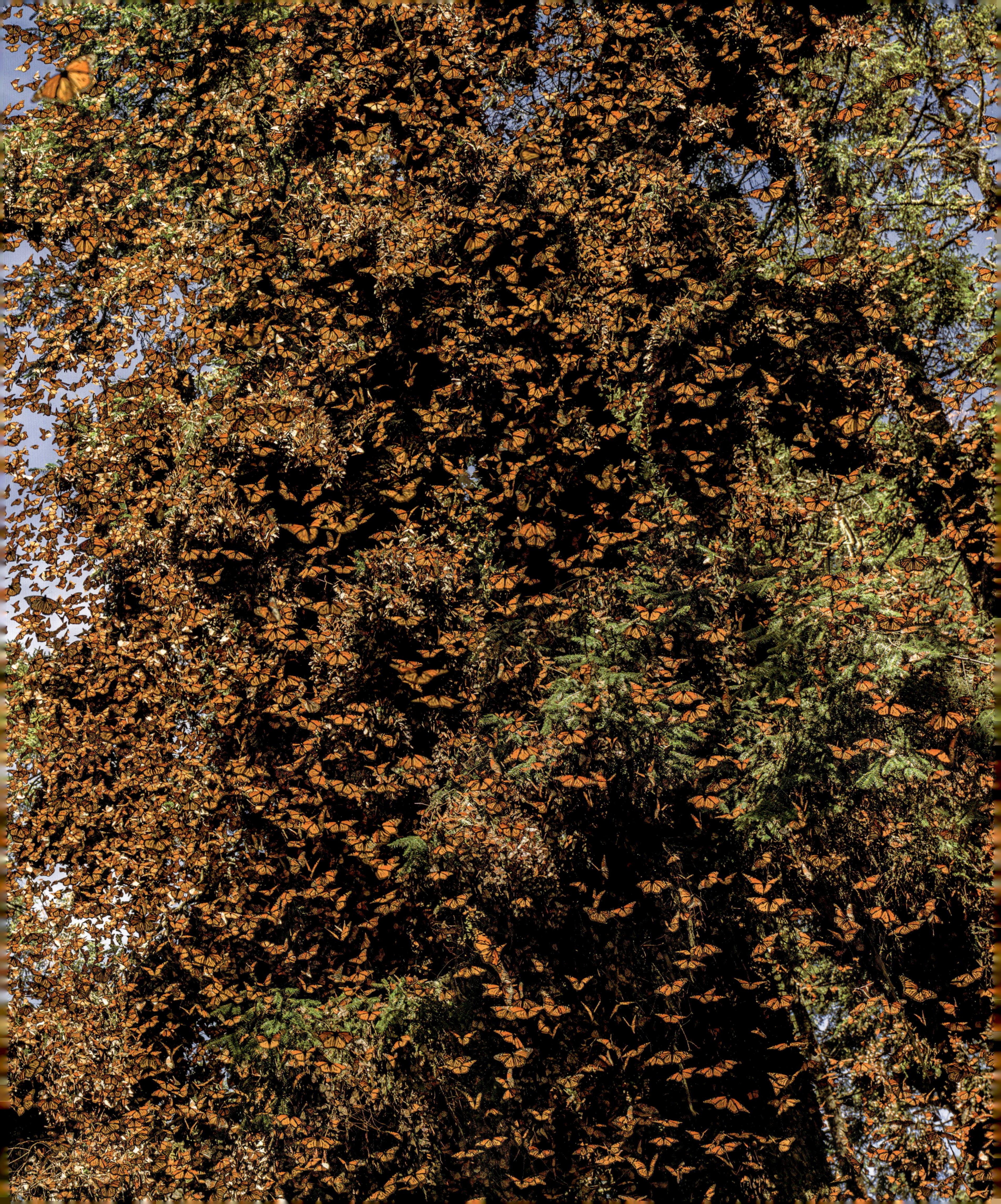

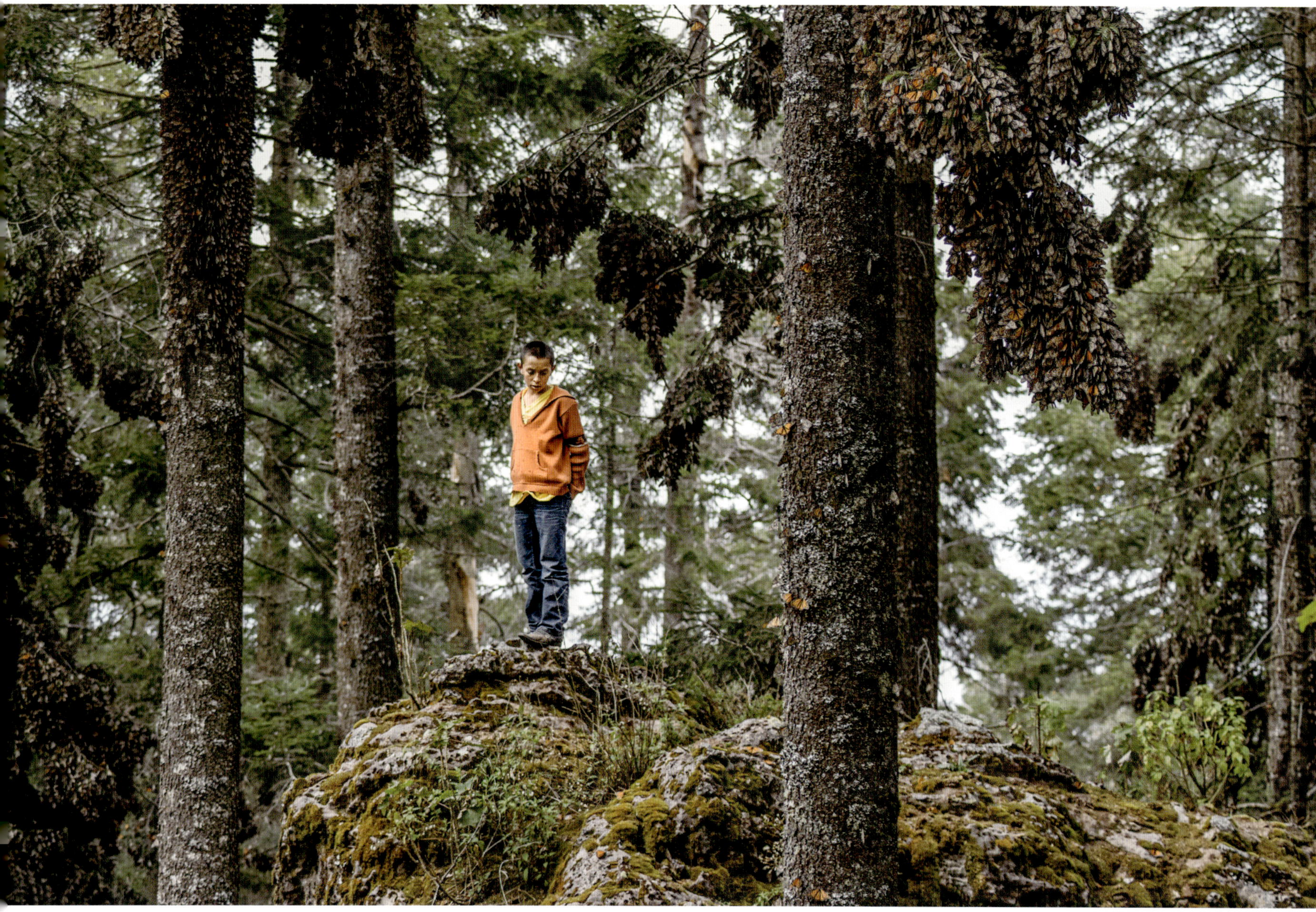

A boy from a nearby community stands among monarch clusters in Sierra Chincua Sanctuary, Michoacán, Mexico. Many local families participate in tourism during the monarch season, with younger members often helping as guides.

PAGES 236–237: Monarch butterflies gather by the thousands at a small stream at El Rosario Sanctuary. During the warmest hours, they fly down from the trees to drink water.

PAGES 238–239: Thousands of monarch butterflies take flight almost simultaneously at midday in El Rosario Sanctuary. Once their body temperature rises above 55°F, clusters suddenly erupt as the butterflies lift off together.

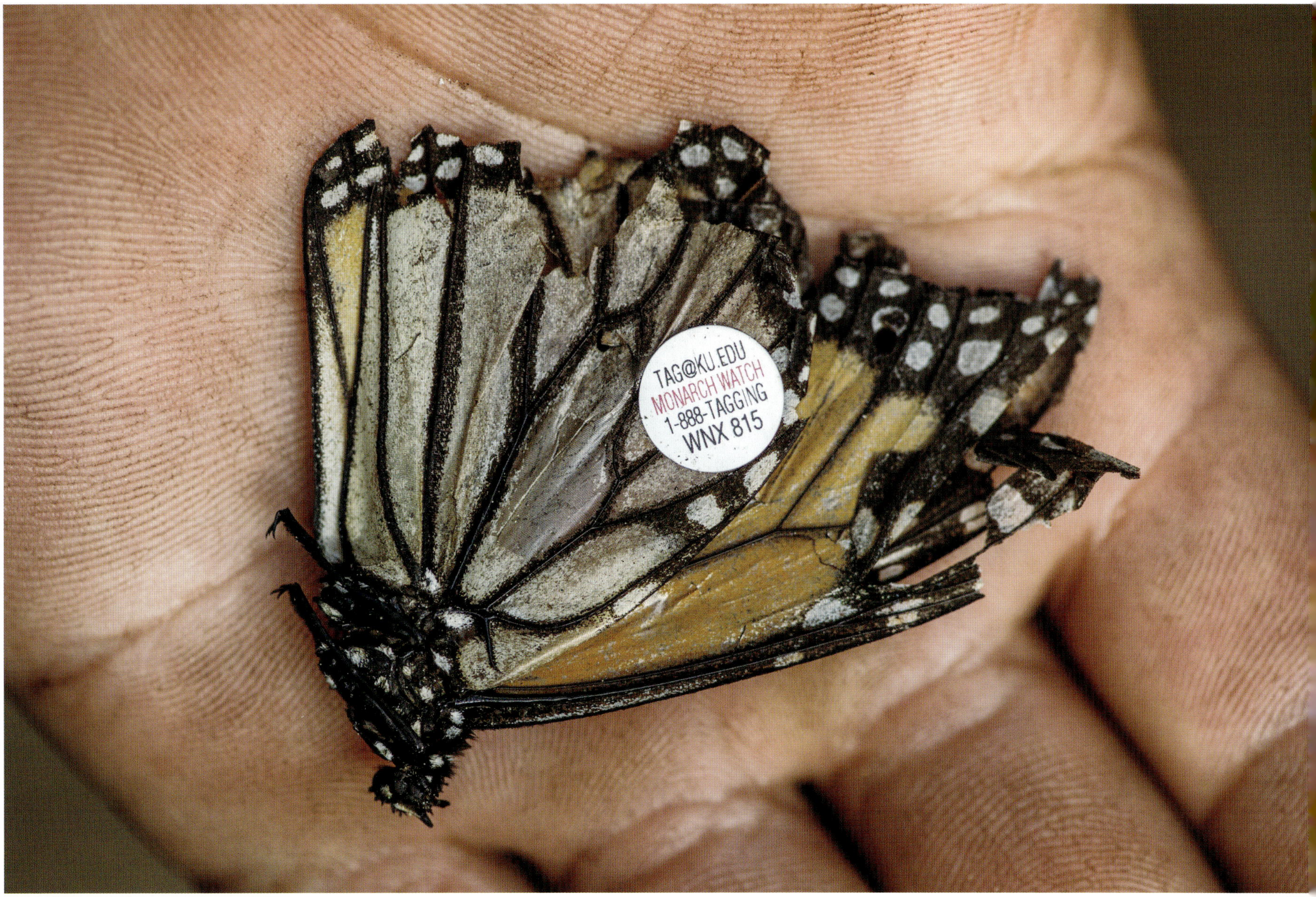

A guide in Sierra Chincua Sanctuary holds a tagged monarch butterfly found dead on the forest floor. Local residents help recover the tags, which contribute valuable data to track migration routes across North America.

PAGES 242–243: Sunlight filters through branches covered with dense clusters of monarch butterflies in El Rosario Sanctuary.

PAGES 244–245: Monarch butterflies swirl among oyamel fir tree branches. Monarchs are a symbol of North America and one of the best indicators of the continent's environmental health.

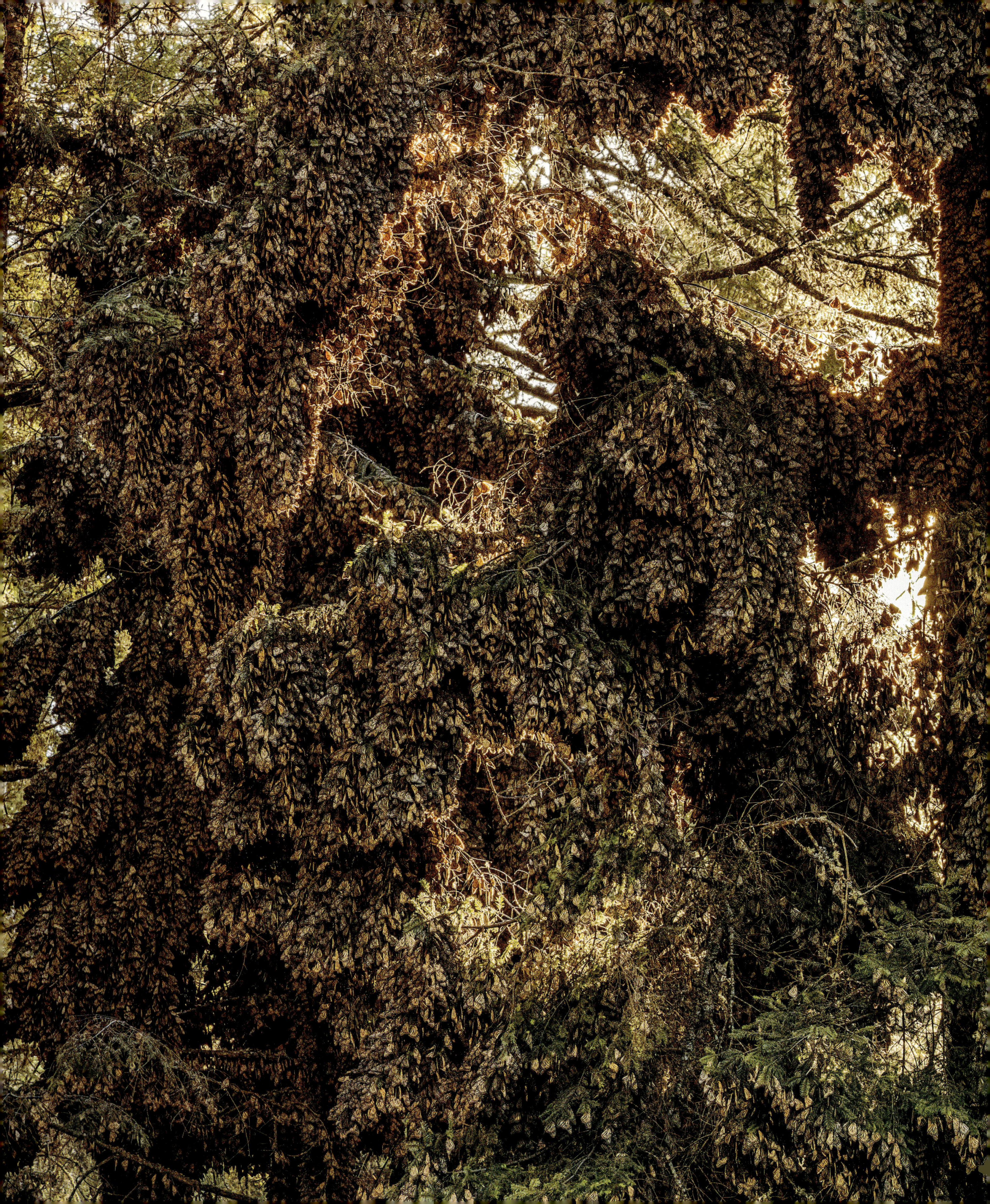

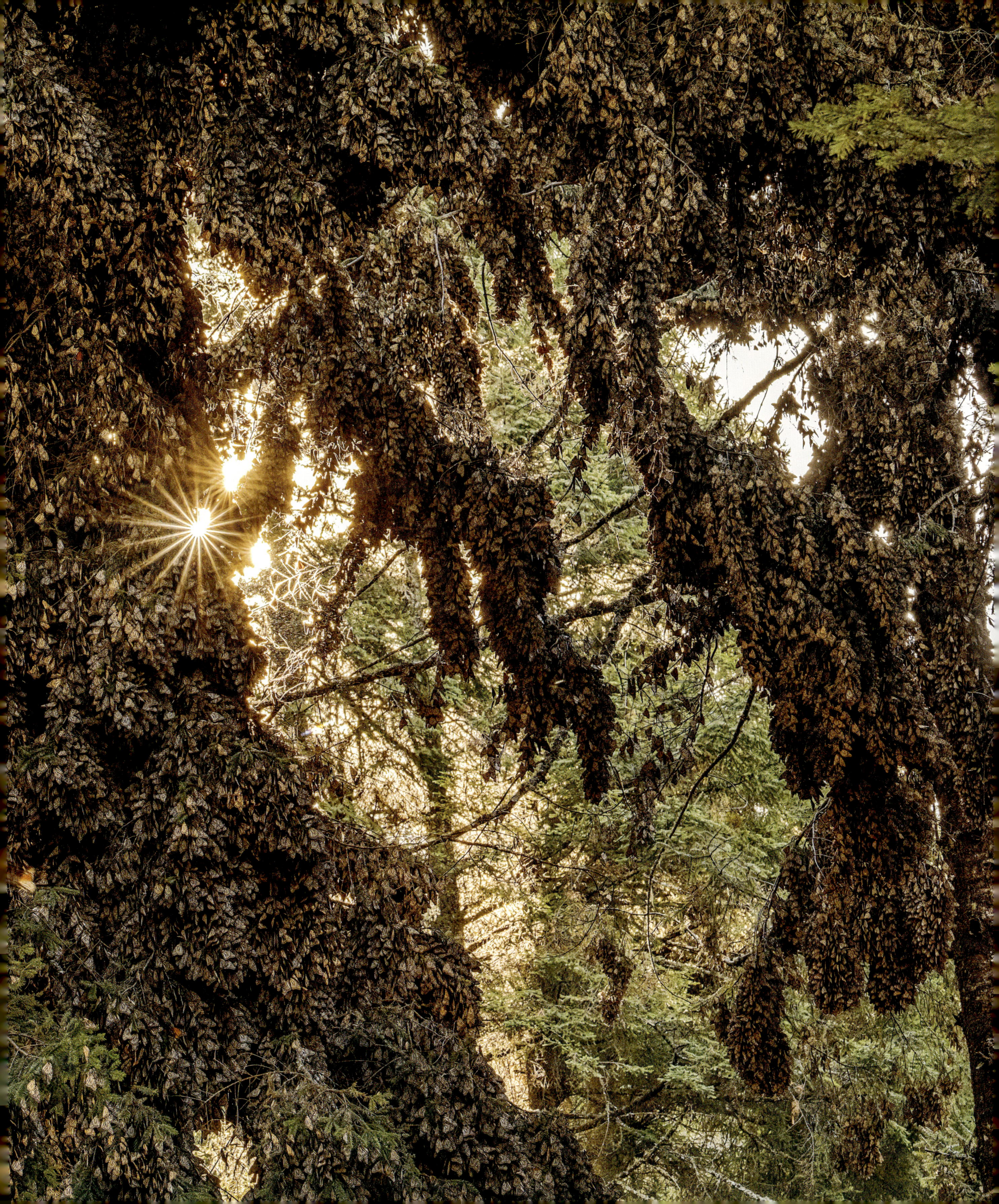

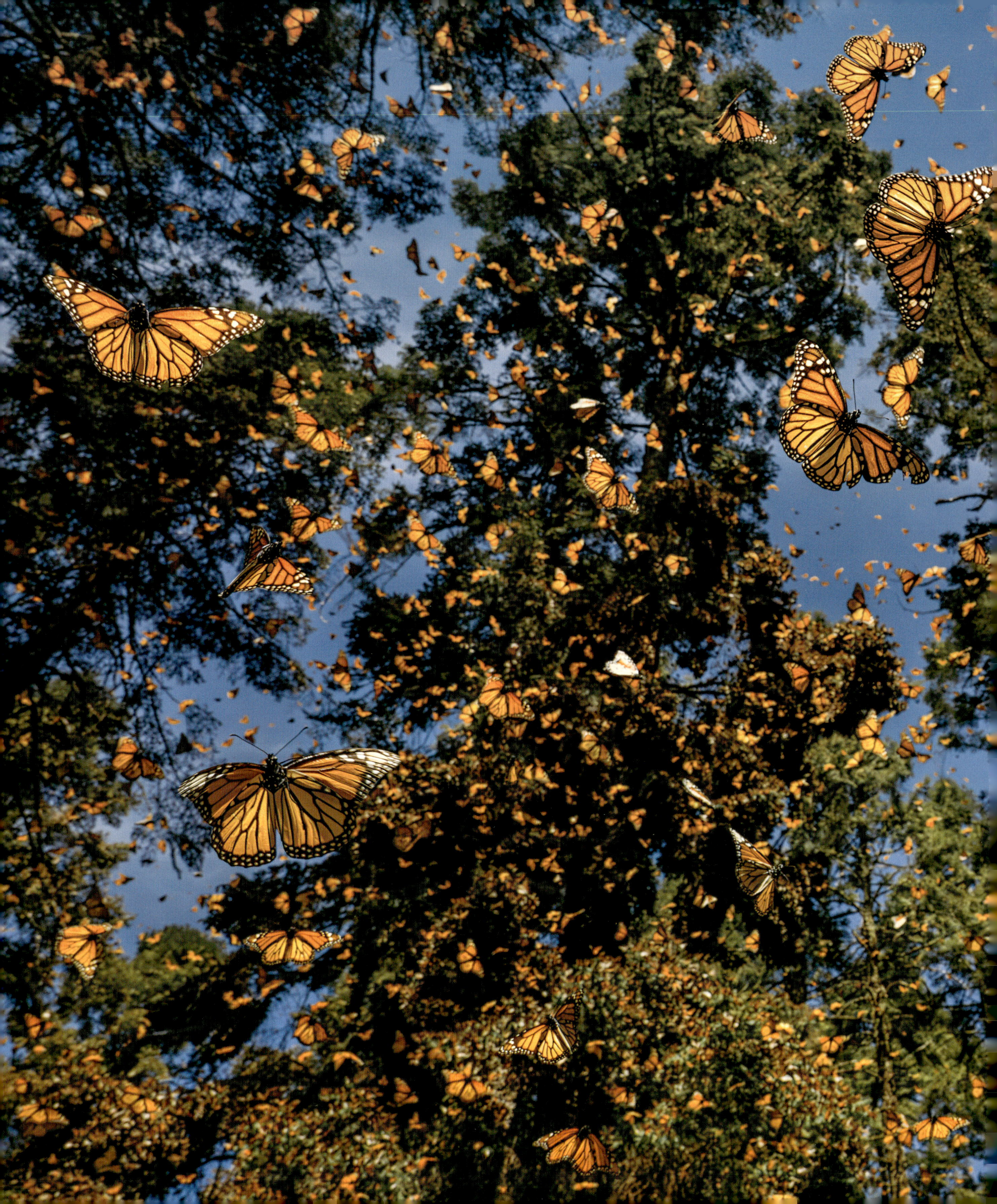

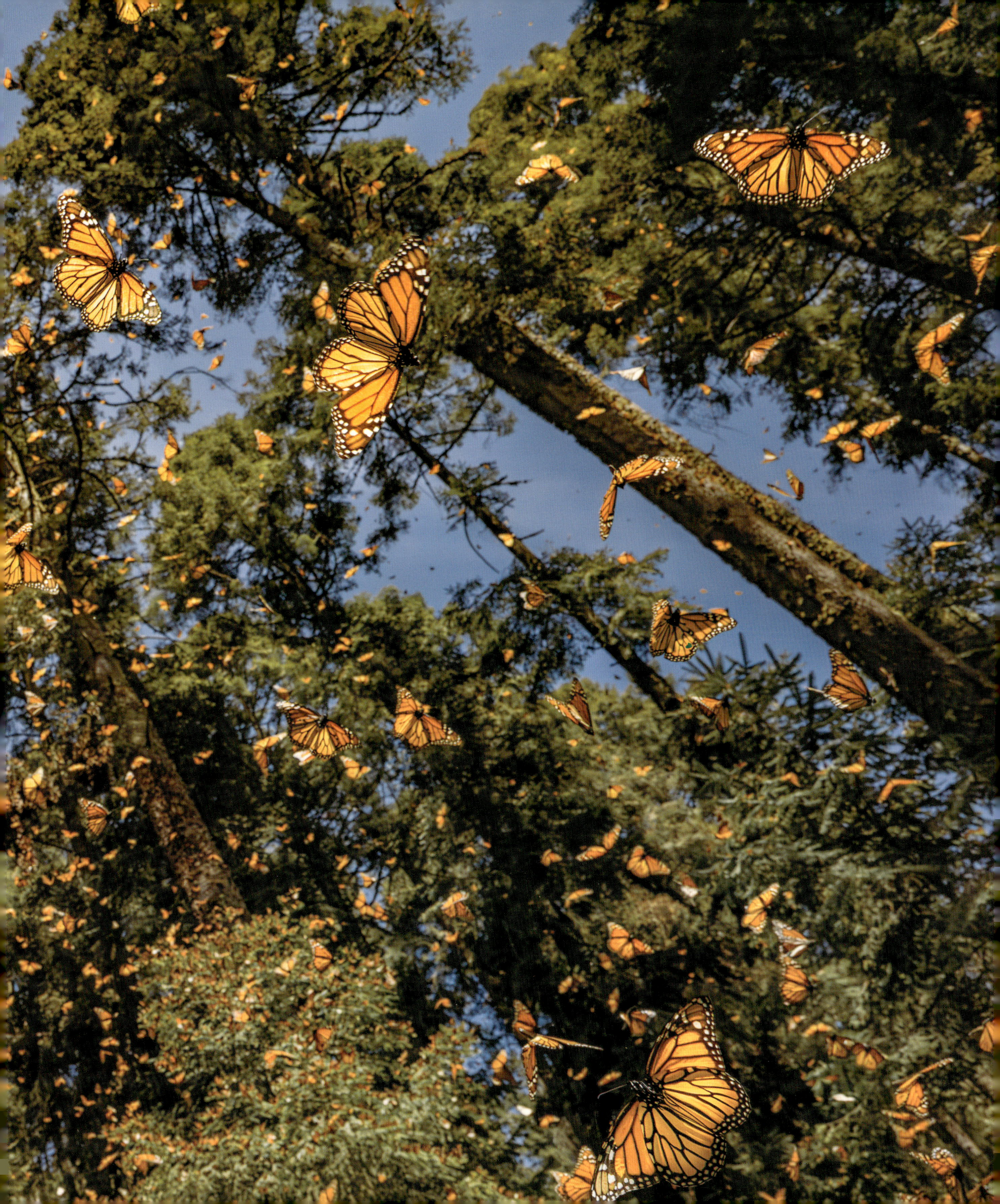

CLOSING THOUGHTS

TWO DECADES WITH THE MONARCHS

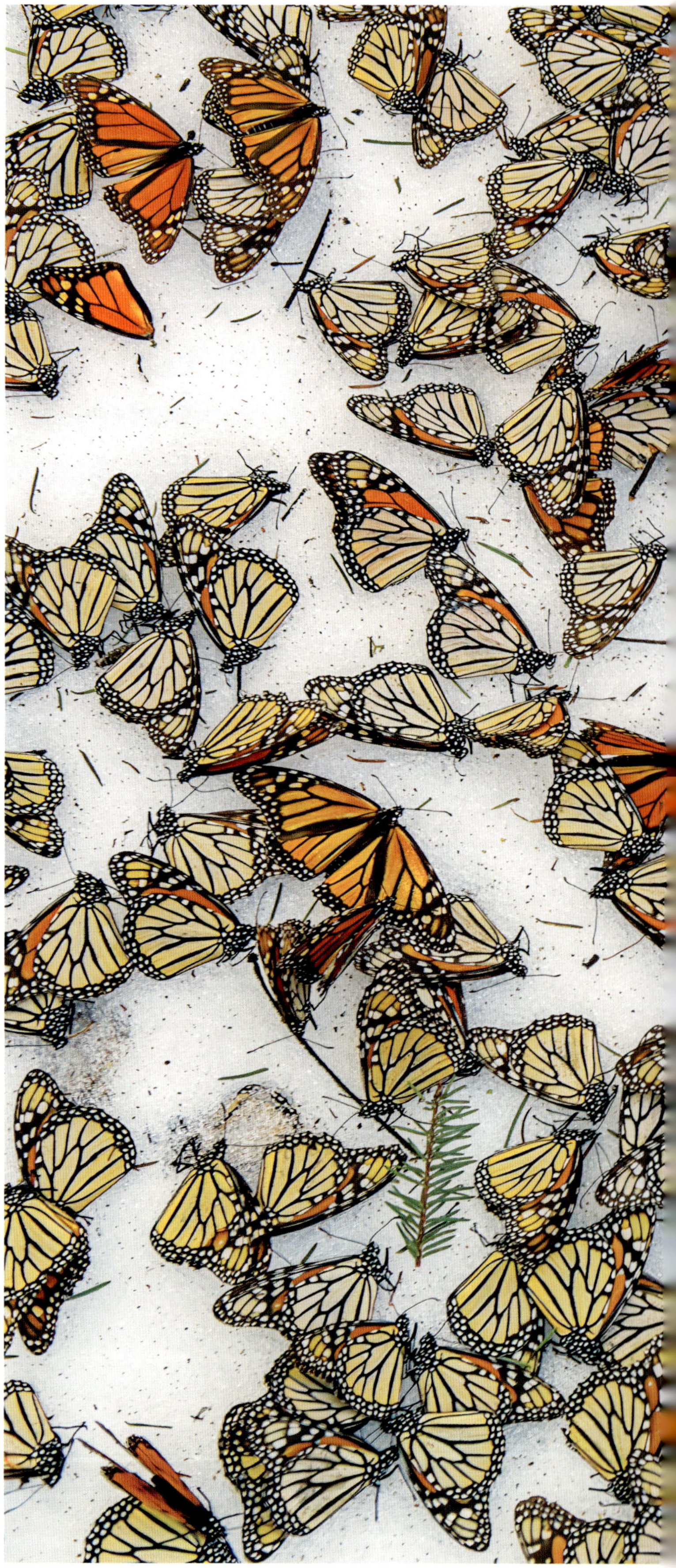

I'VE BEEN DRAWN FROM AN EARLY AGE to the wonders of migration and the wildlife concentrations of our planet. During my childhood in Spain, I watched in awe as cranes and geese arrived in the winters. My first awareness of the monarchs, and my fascination with the biodiversity of Mexico, came from the wonderful book *Diversidad de Fauna Mexicana* edited by Mexican photographer and conservationist Patricio Robles Gil, who was one of my earliest inspirations. I was about 12, and I devoured those pages filled with beautiful images of strange animals with even stranger names. Soon after, rummaging through an old closet inside my grandfather's studio, a room full of mysteries, I found the iconic August 1976 issue of *National Geographic,* which told the story of Fred and Norah Urquhart and their work with citizen scientists to track the monarchs to their wintering grounds in Mexico—places long known to local communities but a much awaited discovery for outside scientists. Those pages, combining beautiful pictures by Bianca Lavies and the gripping story of the Urquharts, were a mind-changing experience. I didn't know back then that I was setting off a chain reaction, the first step in a 20-year-long path that brought me to where I am today.

Twenty years is nothing in geological terms, but it is a quarter of an average human lifespan, and in the turmoil of what some are calling the Anthropocene, it is long enough to notice dramatic changes. The first time I visited the sanctuaries, in 2004, there was a two-foot-thick carpet of monarchs blanketing the forest floor. Some were dead; some had fallen and were too cold to fly. But that sort of mortality wasn't rare—it happened often, and in those years, it was a common sight. There were simply so many butterflies that those that had died of natural causes were piled on the ground. The abundance was astonishing. I am afraid I will never see that again.

That encounter is vividly described in my journals, which I had kept ever since I was a kid inspired by naturalists such as Gerald Durrell and Félix Rodríguez de la Fuente. The pages of those years were filled with enthusiastic pencil sketches, but I was never a gifted artist, or perhaps I didn't try hard enough, and my drawings were always a source of frustration. Photography came to the rescue as a way to share my observations with friends and relatives. Although I started very early, it took me a long time to become a professional. I didn't have much of a budget, and for years I was shooting only one or two rolls of film in an entire year. Unsure that I could make a living with photography, I decided to study environmental sciences and pursue a career in conservation, my other passion.

FROM WONDER TO WITNESS

I moved to Mexico to work in conservation, but that decision set my career as a photographer in motion, too. I began working with Patricio Robles Gil, whose work I had admired since childhood. The learning curve was steep but incredibly rewarding, and soon enough I was flying solo, caught in the balancing act of being a conservation photographer.

A carpet of monarch butterflies covers the forest floor of El Rosario Sanctuary after an unusually intense snowstorm that hit Mexico's state of Michoacán in March 2016.

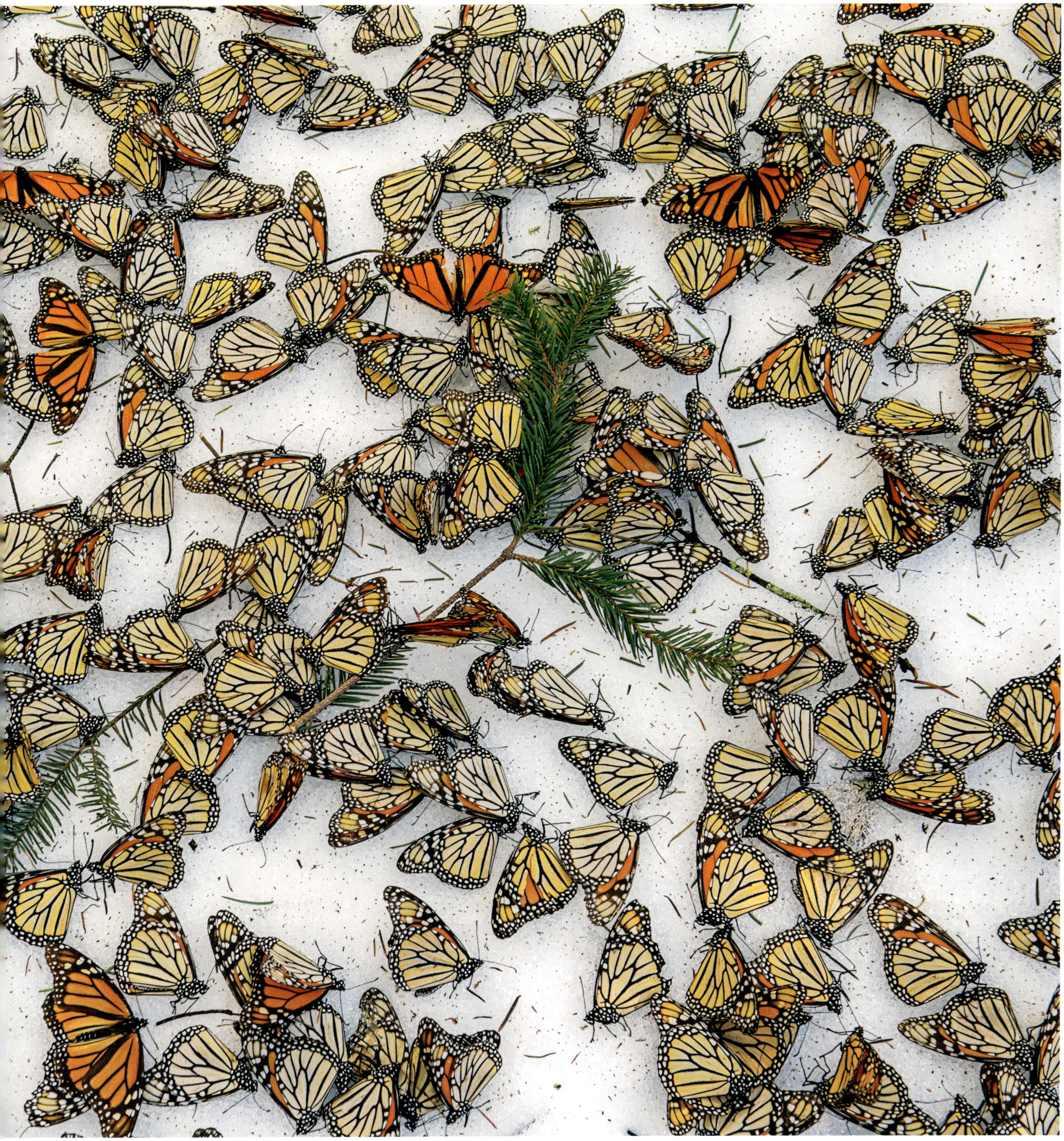

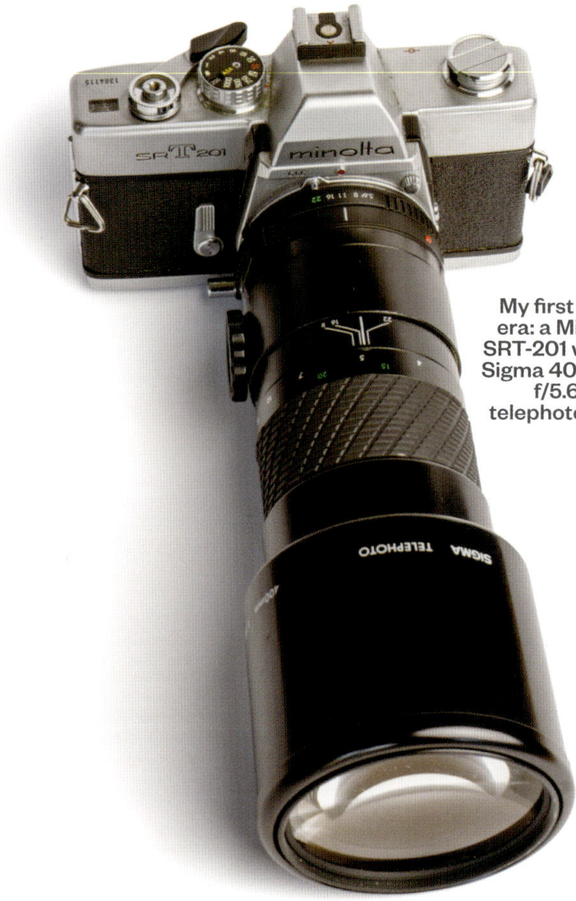

My first camera: a Minolta SRT-201 with a Sigma 400mm f/5.6 APO telephoto lens

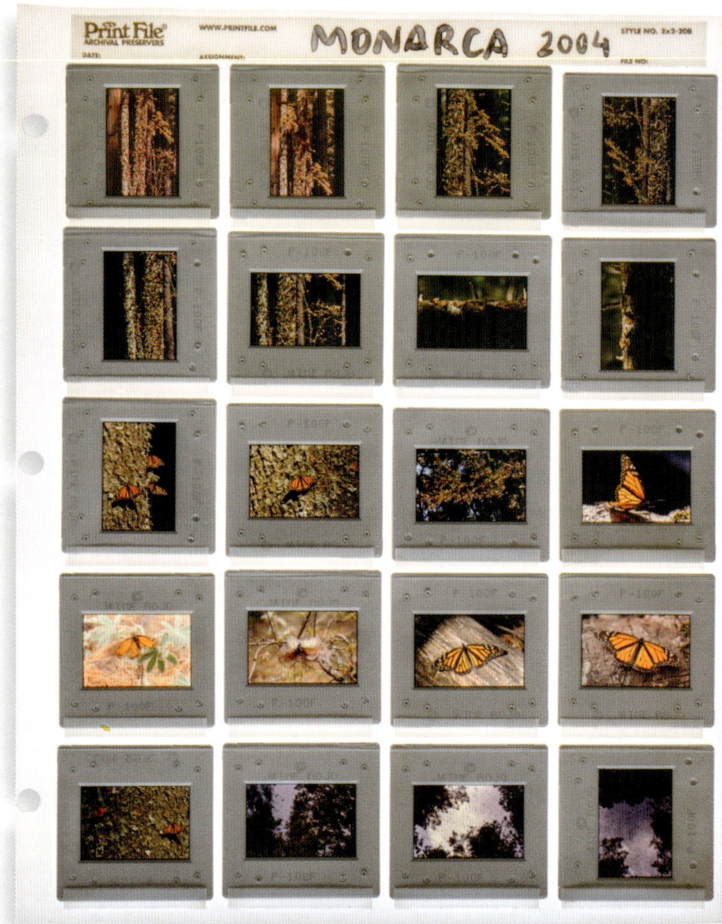

My first slides of monarchs, 2004

My first monarch images were rough attempts at finding order in the chaos, and I wasn't yet skilled enough in some basic techniques. Later, when technique was no longer a problem, I spent years focusing on the raw beauty of the natural world. Then came a period of creative exploration, when I tried many unusual ideas. In 2008, I met for the first time with Kathy Moran, the longtime natural history editor of *National Geographic* magazine. It was a 10-minute meeting, and I was so nervous that I talked too much. By the end she pointed out, "Are you meeting with a Nat Geo editor and not showing any images?" I had completely forgotten! I shared my monarch portfolio with her, and she had some kind words but also an intriguing question: "What is the story?" It didn't mean much to me back then, when I was just focused on capturing the beauty of this planet. But the question lingered, and I would revisit it years later, when I first got interested in visual storytelling.

I have pinpointed that moment to the 2013–2014 season, when the monarch population reached its lowest historical minimum. I had visited the sanctuaries many times before, but that season, I noticed for the first time visible signs of the decline—just a few trees had clusters, when in previous years the whole forest seemed to be alive with butterflies. That got me interested in what was happening with the monarchs and why the numbers were dropping so quickly. On that visit I made one of the most deceptive photos of my career (pages 8–9). It is an image of great beauty, in which you see a perfect oyamel forest at sunset, with the iconic shapes of fir trees, all covered in a dense, thick, healthy layer of wintering monarchs.

But the reality was quite different. Not many trees had butterflies, and I had to walk all over the forest to find these few meager clusters. The photo received a recognition at the Wildlife Photographer of the Year competition, and for the first time, I had to face the incongruence of my craft. I was able to create beautiful images that made the audience fall in love with the monarchs, but the real story was going in a very different direction. That is how I became interested in photojournalism, in telling stories with my pictures and bearing witness. My career took a new direction.

The decisive moment came in 2016, after a historic snowstorm in the forests of Michoacán. One of the photos I made that day won a World Press Photo, an influential award in photojournalism. It was a beautiful photo that also told a powerful story—the best of both worlds (pages 246–247). It was published in *National Geographic,* and soon after, I was invited to a meeting at the magazine's headquarters in Washington, D.C. I showed the editors my best work and pitched a story about monarch migrations. The idea was rejected because they thought I wasn't ready. My people photography felt too stiff, too forced, and they suggested I needed to improve that. They were right. I left the building feeling totally defeated, but I kept working and honing my storytelling skills.

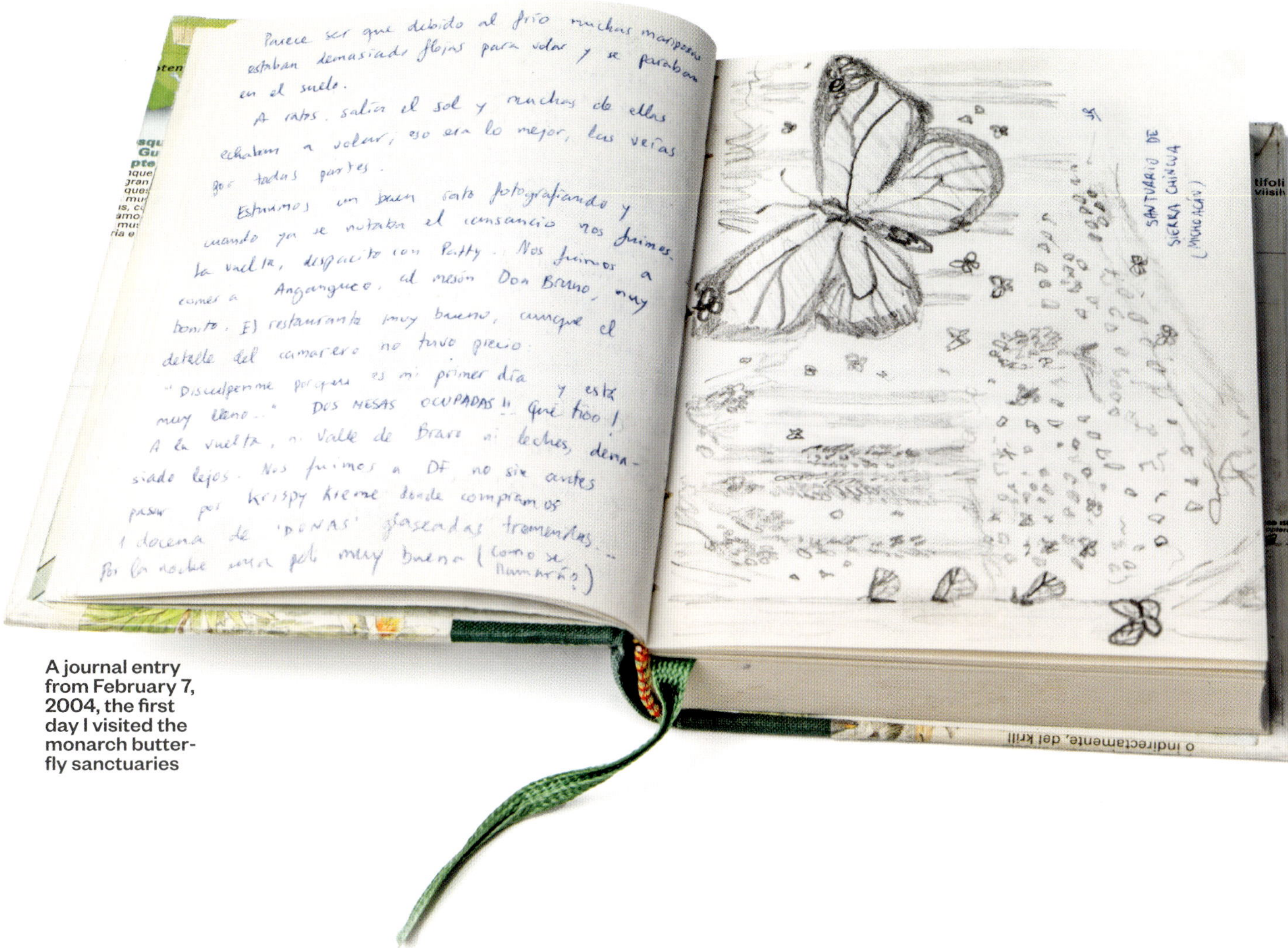

A journal entry from February 7, 2004, the first day I visited the monarch butterfly sanctuaries

THE TURNING POINT

The following years were a turbulent period. My wife and I had just become parents of our first daughter when a major earthquake hit Mexico City, and our neighborhood was severely affected. The experience was traumatic for my newly created family, and we moved back to Spain. But after more than 15 years living abroad in Mexico, I was having trouble finding myself at home there. I found solace in my long-term work with the monarchs. I was convinced that the story was worth telling, so I kept working on it by myself, developing a stronger storytelling focus. Then things happened so quickly. I got a grant from the National Geographic Society to expand my visual coverage of the monarchs, and almost immediately after, I got an assignment from *National Geographic* magazine.

Much of what happened in the two years I worked on that assignment, 2022 and 2023, fills this book. My story of monarch migration and its challenges appeared as a cover feature in January 2024. It was warmly received and triggered a cascade of attention: international press coverage, photo awards, invitations to speak at the National Geographic Storytellers Summit, then on the TED stage in Vancouver, and eventually speaking tours across North America as part of National Geographic Live. It also set in motion the book you now hold, something I have dreamed of for years. The ride has been dizzying, and I still struggle to believe how far these orange wings have carried me.

LESSONS LEARNED

One of the takeaway messages from the year I participated in TED came from social psychologist Brian Lowery's talk, suggestively titled "How to Live a Meaningful Life." He outlined three pillars: coherence (envisioning a world that makes sense), significance (mattering beyond oneself), and purpose (establishing clear goals). This idea made me think a lot, and it continues to do so.

Coherence is a challenge; my world *does not* make sense. For someone who for years seriously studied social ecology and the degrowth movement, which questions our obsession with endless growth on a planet with finite resources, my current lifestyle of international travel and a massive gear-related ecological footprint is and will always be a burden. I have been working on it for years, and I go through cycles of expansion and compression, but there is still much to do on that front.

Significance is indeed important to me. I have always believed in the role of mentorship—it made a huge difference in my early years, and that inspired me to give back by mentoring others later in my career—but it jumped to a whole new level when I became a father. I am learning along the way that contributing to others' life journeys creates a deeper sense of meaning, even if it doesn't always bring happiness. We are a tightly knit family of four, we get along very well, and we live an unusual life in many aspects, partly

(Continued on page 252)

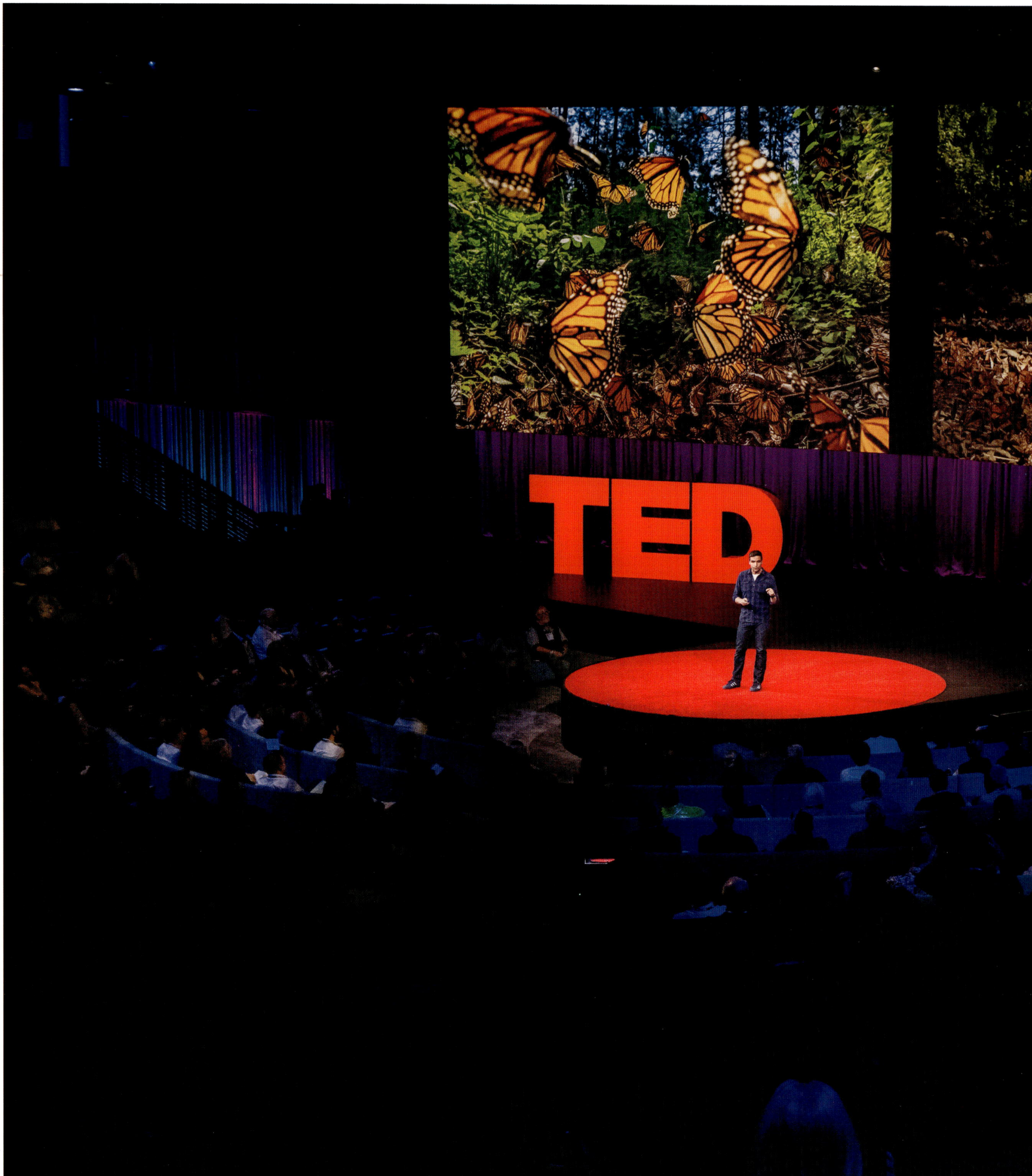
TED

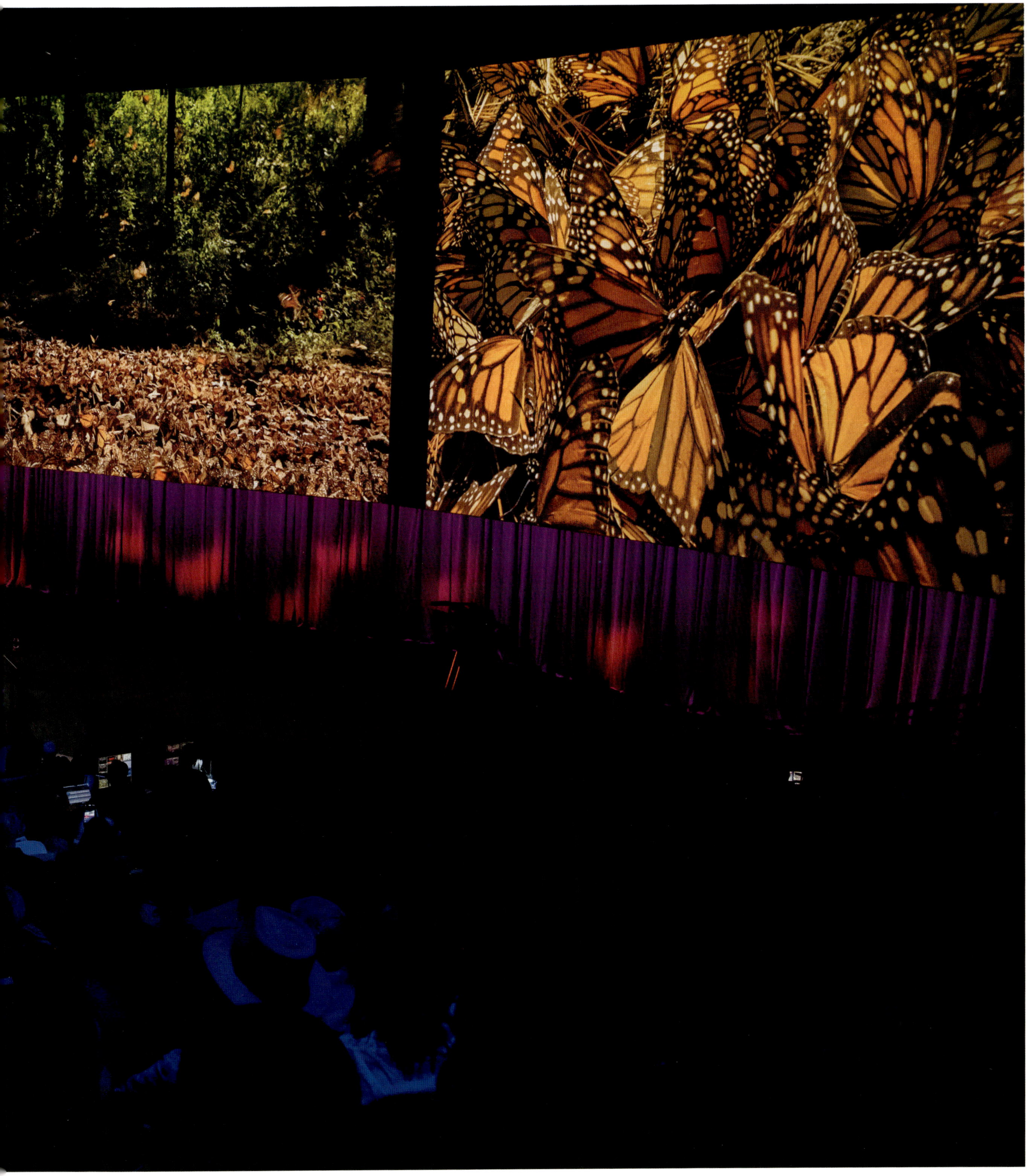

I presented the story of the monarchs to a global audience at TED2024 in Vancouver, British Columbia.

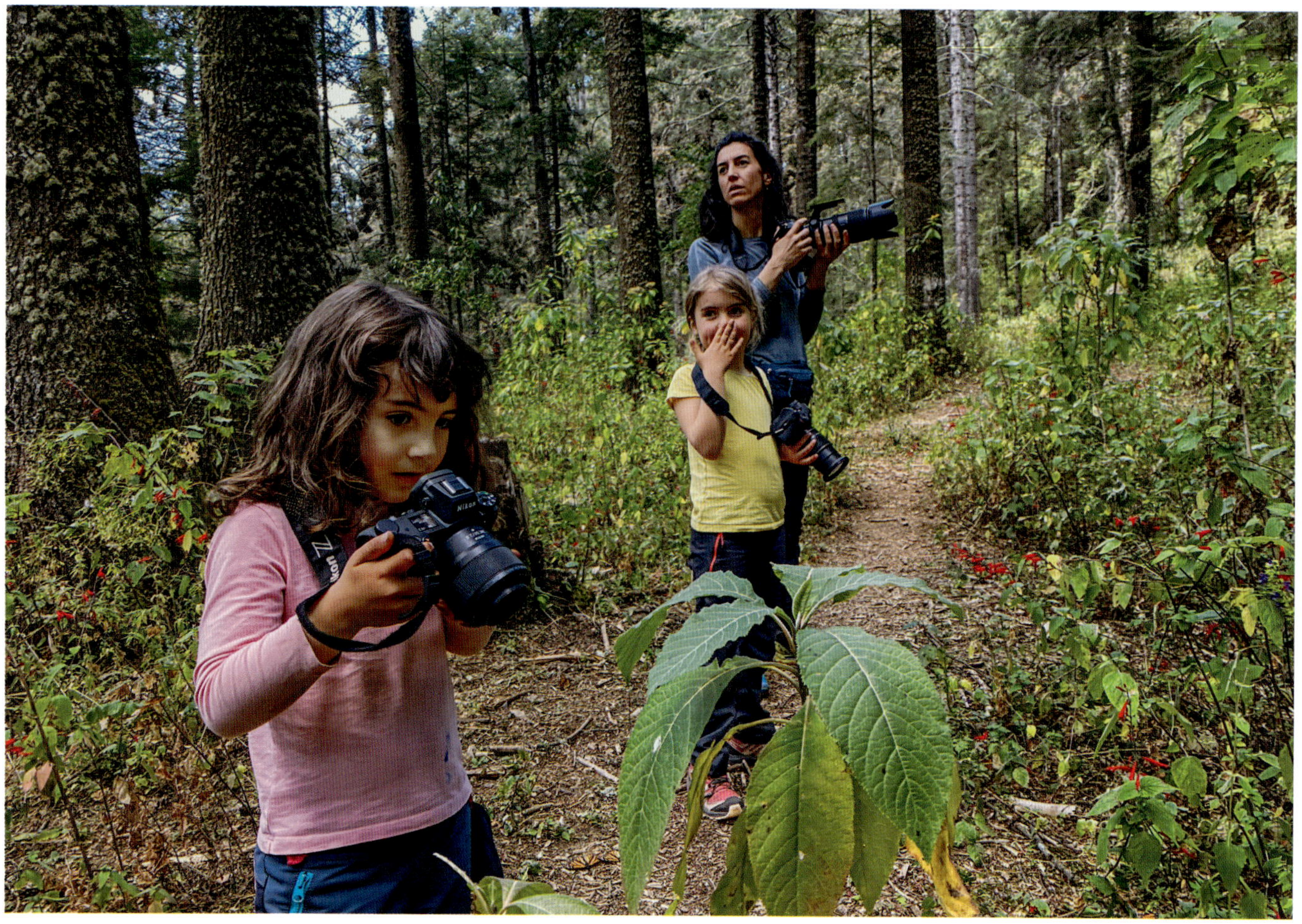

My daughters, Lola and Candela, and my wife, Fátima, at the monarch sanctuaries in 2023. Traveling together has helped us strengthen our family bonds.

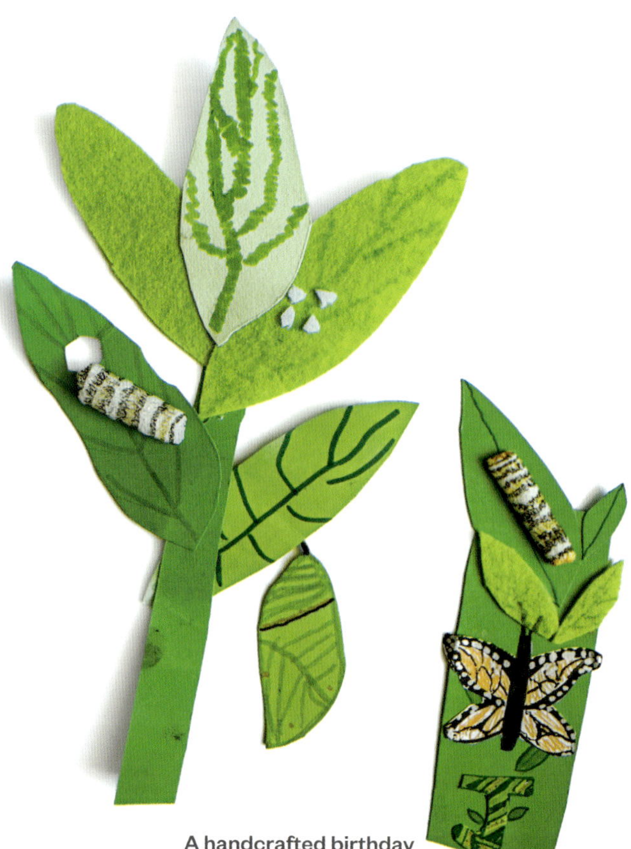

A handcrafted birthday present from my daughters representing milkweed and monarch eggs, caterpillars, and adults

(Continued from page 249)
due to the massive sacrifices my wife and I have made so that our daughters can be part of our projects. As you can imagine, monarchs have been a big part of our recent life story. From art projects to handmade birthday presents, from school presentations to tagging events, from being apart for weeks to intense trips together, monarchs have helped us grow together, helped us understand each other, and reinforced our decisions as a family.

I have always felt a strong sense of purpose. My transition from conservation to full-time photography was driven by the conviction that I could contribute more to protecting this planet if I invested all my energy and passion in what I truly wanted to do: visual storytelling. The kind of projects I work on—long-term, often transboundary, complex conservation stories—are rarely finished, and they all lead naturally into what is starting to feel like the next chapter of my life: ensuring that my images do more than inspire—that they help drive real change for our planet, whether by encouraging others to join the movement to protect the monarchs, or campaigning to mitigate the impact of the border wall between the U.S. and Mexico on wildlife corridors, or supporting the creation of new protected areas around the world.

THE JOURNEY CONTINUES

My journey with these extraordinary butterflies started in Mexico

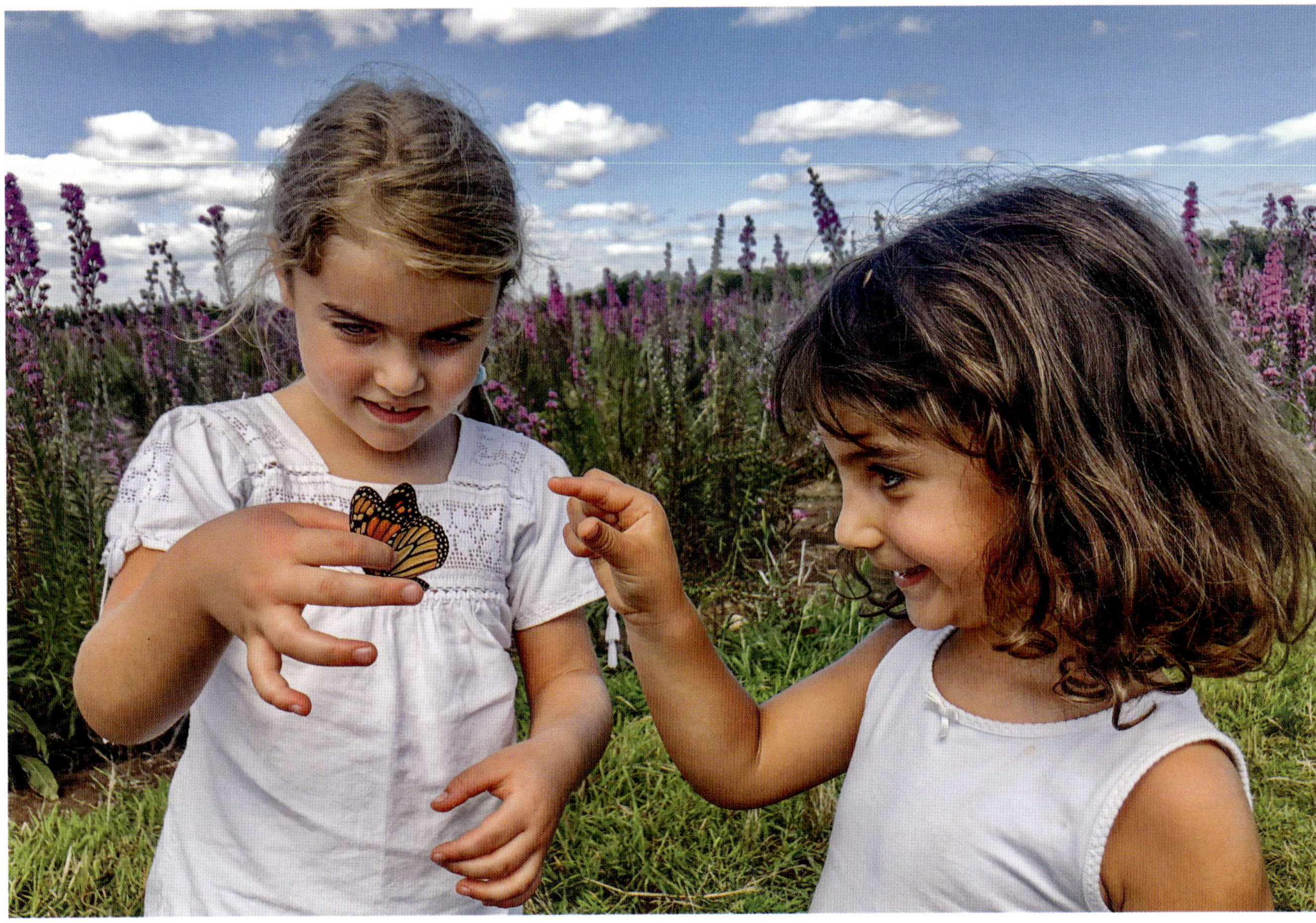

My daughters delight in a close butterfly encounter during a monarch-tagging event in Foley, Minnesota, in August 2022.

but has taken me to the United States and Canada, and even back to Spain, as I immersed myself further into the complexities of the story. Along the way, I have met wonderful people: brilliant researchers using advanced science to solve the mystery of the monarchs' migration; passionate volunteers with decades-long commitments to citizen science; community leaders who have risked their personal safety to protect the oyamel forests in Mexico; and farmers leading the way toward a more sustainable food industry. Meeting them has reshaped how I understand both conservation and storytelling.

I am humbled to share these pages with some of the most prominent monarch experts in the world, whose knowledge and experience enrich this work. Karen Oberhauser, Chip Taylor, and Eduardo Rendón Salinas are not only extraordinary scientists; they have dedicated their lives to protecting the monarchs and generously provided guidance for my own journey.

I don't know how much my own work can help protect the monarchs, and perhaps I never will. Open endings have their beauty. Our story is far from over. Like the monarchs themselves, I will keep moving, adapting, and returning. They are a symbol of North America, and their story can bring unity across borders, inspire hope, and reconnect us to the natural world. My hope is that these pages carry forward the same wonder I felt that cold morning in Sierra Chincua, and that they help keep alive the growing movement to protect this iconic migration.

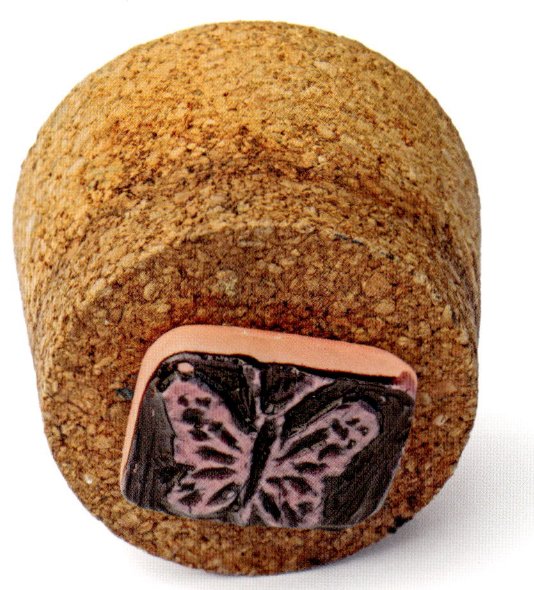

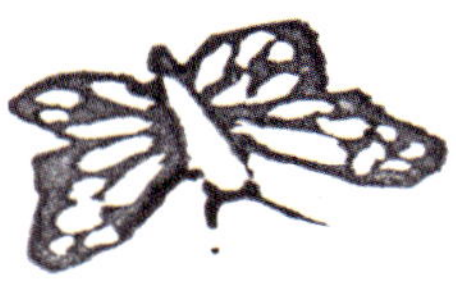

A handmade ink stamp featuring a monarch butterfly motif, made by my daughters for Christmas

ACKNOWLEDGMENTS

A VERY SPECIAL THANKS to Fátima Andrade for her enthusiasm and creativity in the amazing journey we are sharing; to Candela and Lola, who inspire me every day with the extraordinary way they see the world; to my parents, Roberto Rojo and Ana López, who always pointed me in the right direction; and to Patricia Rojo and Patricio Robles Gil, who, in the most unusual way, guided me in my first steps as a photographer.

Over the past 20 years, many people have shared their time, knowledge, and trust. Given the limited space on this page, I apologize in advance for any names that may be missing: Moisés Acosta, Anurag Agrawal, Gael Almeida, Carmelo Álvarez, Clemente Álvarez, Cosima Amelang, the Andrade Macho family, Ezequiel Andreu, Andrea Aneas, Marino Argueta, John Hart Asher II, Martha Askins, Zack Ater, Danae Azuara, Kristen Baum, Peter Berthelsen, Isa Betancourt, Leone J. Bicchieri, Joe Billings, David Blaauw, David Bohlken, Steve Bradbury, Elaine Bradley, Briley Bray, Cassandra Bray, Dustin Bray, Jane Breckinridge, the late Lincoln Brower, Skye Bruce, Andrew Buchanan, the late Darlene Burgess, Kianna Burtle, Wendy Caldwell, Reynaldo Camas, Marco Castro, Aldo Ceballos, Sonya Charest, Christine Chase, Jerónimo Chávez, Mary Paz Chávez, Matt Christiansen, Leah Churner, Nick Clark, Adrian Coakley, June Cohen, Héctor Colín, Juan Cornejo, Abel Cruz, Horacio Cruz, Martín Cruz, Donald Davis, Lizbeth Delgado, Roberto Díaz, Nina Dueñas, Carol Dunn, Keith Dunn, Fulvio Eccardi, Chris Eckstrom, Mike Edsey, Ernesto Enkerlin, Martín Esquivel, Cayetana Estrada, Anne Farrar, Amado Fernández, Juan Fernández Haeger, Jackelyn Ferrer, Linda Fink, Ashley Fisher, Kelsey Fisher, Linabeth Flores, Eric Flynn, Tom Frantzen, Timothy Fredricks, Keizo Fujii, Claudia Galeno, Carlos Galindo, Alexandra García, Eligio García, José Humberto García, Brandon Gibson, Elisa Gibson, Helen Gilks, the late Homero Gómez, Mauricio Gómez, Paula Gómez, Israel González, Ruby Graham, Rusty Graham, D. André Green II, Ramón Guadarrama, Omar Guzmán, Melissa Gwelch, Calvin Hall, Katherine Haughton, Tara Haughton, Tasha Haughton, Floriberto Hernández, Mirna Hernández, Emil Herrera, Kandice Hines, Susan Tyler Hitchcock, Eva Horne, Elizabeth Howard, Megan Huck, Karen Ibarra, Jasmine Ingram, Adrienne Izaguirre, Laura Jackson, Pablo Jaramillo, Diego Jordano, Pam Karlson, Rosamund Kidman Cox, Hun-Seok Kim, Jade Kochanski, Mara Koenig, Elizabeth Kolbert, Casey Kong, Esther Kong, Julie Kong, Joe Kooiker, Audrey Kouyoumdjian, Isabelle Labus, David La Puma, Frans Lanting, Javier Lascurain, Allen Lawrance, David Lee, Inhee Lee, Sarah Leen, Linda Lehmusvirta, Jim Lovett, Nathan Lump, William Maidment, Raed Mansour, Ganesh Marín, Sabino Marín, Felipe Martínez, J. Carmen Martínez, Markus Martínez, Omar Martínez, Betty McCulloch, Stacey McLain, Claire McNulty, Justin Meissen, Owen Meissen, Roberto Méndez, Leticia Mendoza, Brisa Mercado, Christine Merlin, Laurel Merriam, Jennifer Meyer, Eileen Mignoni, Marc Miller, Mónica Missrie, Genaro Mondragón, Thor Morales, Kathy Moran, Luis Morgado, Virginia Nelson, William Nelson, Kristine Nemec, Kyle Nessen, Amanda Newton, Timothy Newton, Michelle Nijhuis, Susan Norton, Karen Oberhauser, Tom O'Brien, Xavier Olazábal, Andy Olson, Isabel Ortiz Soto, Clint Otto, Carol Pasternak, Chee Pearlman, Emma Pelton, Legna Pérez, Luz Pérez, Ella Phillips, Charlie Plimpton, Chelsea Plimpton, Dena Podrebarac, Rémi Pognonec, Alex Pollack, Jean Ponzi, Sadie Quarrier, Michelle Rabkin, Isabel Ramírez, Juan Ramírez, Buffy Redsecker, Hugh Reed, Susan Rendall, Eduardo Rendón Salinas, Angélica Reyes, Óscar Reyes, Michael Rizo, Bradley Roback, Gerardo Robles Gil, Alejandra Rodríguez, April Rohlich, Luis Antonio Rojas, the Rojo López family, Estela Romero, Guadalupe Romero, Lorenzo Rosenzweig, Susan Ross, Ann Ryan, Cuauhtémoc Sáenz, Jeanne Salonek, Hilaria Sánchez, Jesús Sánchez, Peter Sanderson, Pedro Sanvicente, Florian Schulz, Michael Simanonok, Cecilia Simón, Margaret Sisk, Richard Sloan, Drew Smith, Alison Snell, Dena Spatz, Collin Spriggs, Bruce Stacken, Rachael Stretcher, the late José Luis Suero, Brenda Taggart, Levi Taggart, Douglas Tallamy, Doug Taron, Gloria Tavera, Chip Taylor, Sergio Téllez, Dan Terpstra, Margy Terpstra, Lisa Thomas, Jill Tiefenthaler, Rocío Treviño, Ignacio Úbeda, Sana Ullah, Katherine Underwood, Anand Varma, Carlos Velazco, Omar Vidal, Alexa Villalobos, Chris Waas, Craig Walker, Erik Wallsten, Gemma Ward, Beau Whetstine, Bryce Whetstine, Beth Whitaker, the late Terry Whitman, Richard Williams, Jamante P. Winston-Smith, Dawn Wormington, Kaitlin Yarnall, Fernando Zamácola, and Andrés Zárate.

Thanks also to the many organizations that supported this project: Atlautla, Alternare, Bosque Otomí-Mexica, California State Parks, Center for Research in Environmental Geography (National Autonomous University of Mexico), Cerro Altamirano, Cerro del Amparo, Cerro Pelón, Cerros Chivatí-Huacal, City of Chicago, Correo Real, El Valor, ENGIE, Euchee Butterfly Farm, Fort Hood, Gibraltar Botanic Gardens, Hotel Don Bruno, Instituto Americano Leonardo da Vinci, International League of Conservation Photographers, Journey North, Lighthouse Field State Beach, Lurie Garden, Mexican Fund for the Conservation of Nature, MNL (Minnesota Native Landscapes), Monarch Butterfly Biosphere Reserve, Monarch Joint Venture, Monarch Watch, Muscogee (Creek) Nation, National Commission of Protected Natural Areas (CONANP), National Geographic Partners, National Geographic Society, Neovolcanic Axis Conservation Fund, Ney Nature Center, Nikon Mexico, Nikon Spain, Palomas, Papalotzin Environmental Education Center, Parks Canada, Peggy Notebaert Nature Museum, Piedra Herrada, Point Pelee National Park, Presqu'ile Provincial Park, Profauna México, Resource Environmental Solutions, Rosetta McClain Gardens, Sierra Chincua, Sierra el Campanario, Swarovski Optik, Tallgrass Prairie Center, TED, Think Tank Photo, Tommy Thompson Park, Tribal Alliance for Pollinators, University of Northern Iowa, University of Wisconsin–Madison Arboretum, WILD Foundation, Wildlife Photographer of the Year, World Press Photo, World Wildlife Fund in Mexico, World Wildlife Fund (U.S.), Xerces Society, and Zapalinamé State Park.

ABOUT THE CONTRIBUTORS

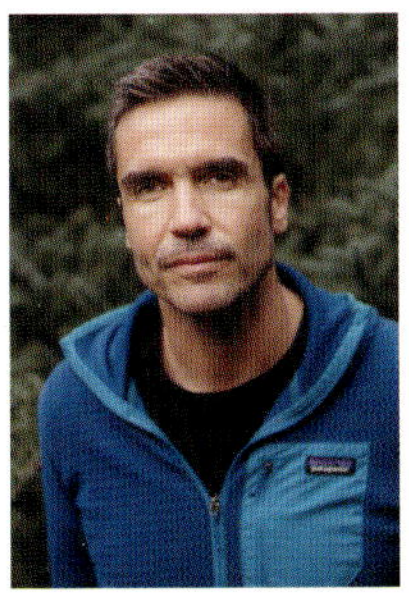

JAIME ROJO is a photographer and National Geographic Explorer with a background in environmental sciences who focuses on conservation stories about wilderness and wildlife. A senior fellow of the International League of Conservation Photographers and trustee of the WILD Foundation, he has earned honors from World Press Photo and Wildlife Photographer of the Year. He also works as a communications adviser to environmental, educational, and government institutions and is a frequent speaker on visual storytelling for conservation.

ELIZABETH KOLBERT is the best-selling author of *The Sixth Extinction,* for which she won a Pulitzer Prize; *Under a White Sky,* which was named one of the top 10 books of 2021 by the *Washington Post;* and, most recently, *Life on a Little-Known Planet.* For her work at the *New Yorker,* where she's a staff writer, she has received two National Magazine Awards and the Blake-Dodd Prize from the American Academy of Arts and Letters. She lives in Williamstown, Massachusetts, with her husband and children.

Monarch migration map by *National Geographic* magazine staff. Sources: Chip Taylor, University of Kansas; Karen Oberhauser, University of Wisconsin–Madison Arboretum; Christine Merlin, Texas A&M University; Wayne E. Thogmartin, USGS; USDA; Journey North; Monarch Joint Venture; Western Monarch Count; Biota of North America Program; Nathan P. Lemoine. Monarch butterfly art by Lucy Newton.

All photos by Jaime Rojo except the following: 250–251, Jasmina Tomic; 253 (UP) and 255, Fátima Andrade.

Since 1888, the National Geographic Society has funded more than 15,000 research, conservation, education, technology, and storytelling projects around the world. National Geographic Partners distributes a portion of the funds it receives from your purchase to National Geographic Society to support their mission to illuminate and protect the wonder of our world.

National Geographic Partners, LLC
1145 17th Street NW
Washington, DC 20036-4688 USA

Get closer to National Geographic Explorers and photographers, and connect with our global community. Join us today at nationalgeographic.org/joinus

For rights or permissions inquiries, please contact National Geographic Books Subsidiary Rights: bookrights@natgeo.com

ISBN: 978-1-4262-2539-0

The authorized representative in the EU for product safety and compliance is Disney Trading B.V., Asterweg 15S, 1031 HL, Amsterdam, The Netherlands
email: DCP.DL-EU.bookscontact@disney.com

Printed in China

26/RRDH/1

ABOUT THE NATIONAL GEOGRAPHIC SOCIETY

NATIONAL GEOGRAPHIC

THE NATIONAL GEOGRAPHIC SOCIETY is a global nonprofit organization that uses the power of science, exploration, education, and storytelling to illuminate and protect the wonder of our world.

Since 1888, the Society has pushed the boundaries of exploration to better understand our world. We have awarded more than 15,000 grants to National Geographic Explorers—scientists, conservationists, innovators, educators, and storytellers—for work across all seven continents. Today, Explorers are advancing new knowledge and leading conservation programs with outsize impact to protect nature, wildlife, historical places, and communities. They're documenting the wonder of our world—including its beauty, its mystery, and the threats it faces—and inspiring people to care and act on behalf of our planet and its people.

Over the years, monarch butterflies have captivated *National Geographic* magazine readers and Explorers. A 1937 article in *National Geographic* described the butterflies' epic fall migration as observed in the field: "The southward-flying swarms are often very conspicuous, as they may consist of tens of thousands of butterflies flying up to three hundred feet or more in the air ..." In 1963, monarchs were the feature of an article that shed light on the beginnings of the practice of delicately tagging butterflies with paper to learn their migration stories.

As methods evolved, the National Geographic Society supported Fred and Norah Urquhart for further monarch butterfly research, which led to years of fieldwork. In 1975, their team members Catalina Aguado Trail and Kenneth Brugger located the monarchs' winter habitat in the mountains of central Mexico. Trail appeared on the cover of *National Geographic* magazine in August the following year, surrounded by monarch butterflies. In 2008, UNESCO named the Monarch Butterfly Biosphere Reserve a World Heritage site.

To expand on the 1975 findings on the species' winter habitat, in 2019 National Geographic Explorer D. André Green II received funding to develop microcomputing sensors to track the entire migratory path of individual monarch butterflies, seeking to further understand the delicate creatures. Today, the National Geographic Society continues to support monarch conservation and habitat protection through dedicated Explorers' use of emerging technologies like microtracking, as well as radio telemetry and genetic analysis.

Explorer Jaime Rojo has photographed monarch butterflies for nearly two decades. After becoming a National Geographic Explorer in 2020, he continued to build on the legacy of the organization's monarch research, now expanding his work to understanding the benefits of wildlife corridors for a variety of species, with the National Geographic Society supporting his work to document the monarch butterfly migration. Nearly 50 years after the announcement that monarchs were overwintering in the mountains of Mexico, the January 2024 cover story of the magazine featured Rojo's photographs accompanied by highlights of the 21st century's advances in monarch conservation, even in the midst of their population decline.

Celebrating excellence and fostering innovation are at the heart of the National Geographic Society's mission. The Society supports remarkable individuals whose groundbreaking work is shaping the future of exploration, science, education, and storytelling.

Explorers like Rojo are not only illuminating and protecting the wonder of our world through their work, but they are also inspiring the next generation of Explorers through efforts like the National Geographic Society's Slingshot Challenge, a global call for youth to describe their solutions for environmental issues. Recent award recipients have included youth-led monarch butterfly conservation efforts.

The more opportunities people have to understand the nature and cultures of the world, the more inspired they are to protect them. Today, National Geographic Society reaches millions of students each year through educational offerings and engages audiences around the globe through signature experiences, stories, and content.

To learn more about our work and how you can join us, visit *natgeo.com/impact*.